New Genetics, New Social Formations

The genomic era requires more than just a technical understanding of gene structure and function. New technological options cannot survive without being entrenched in networks of producers, users and various services. New genetic technologies cut across a range of public domains and private lifeworlds, often appearing to generate an institutional void in response to the complex challenges they pose. Chapters in this volume discuss a variety of these novel manifestations across both health and agriculture, including:

- gene banks
- intellectual property rights
- committees of inquiry
- non-governmental organisations (NGOs)
- national research laboratories

These are explored in such diverse locations as Amazonia, China, Finland, Israel, the UK and the USA. This volume reflects the rapidly changing scientific, clinical and social environment within which new social formations are being constructed and reconstructed. It brings together a range of empirical and theoretical insights that serve to help better understand complex, and often contentious, innovative processes in the new genetic technologies.

Peter Glasner is Professorial Fellow in the Economic and Social Research Council's Centre for Economic and Social Aspects of Genomics at Cardiff University. He is Co-editor of the journals *New Genetics and Society* and *21st Century Society*. He has a longstanding research interest in genetics, innovation and science policy. He is an Academician of the Academy of Learned Societies in the Social Sciences.

Paul Atkinson is Distinguished Research Professor in Sociology at Cardiff University, where he is Associate Director of the ESRC Centre for Economic and Social Aspects of Genomics. He has published extensively on the sociology of medical knowledge and qualitative research methods. He is Co-editor of the journal *Qualitative Research*. He is an Academician of the Academy of the Social Sciences.

Helen Greenslade is Editorial Assistant for CESAGen's Genetics and Society Book Series. She graduated from Cardiff University with a degree in Italian and Spanish, and holds an MA in European–Latin American Relations from the University of Bradford.

Genetics and Society

Series editors:
Paul Atkinson, *Associate Director of CESAGen, Cardiff University*;
Ruth Chadwick, *Director of CESAGen, Cardiff University*;
Peter Glasner, *Professorial Research Fellow for CESAGen, Cardiff University*;
Brian Wynne, *member of the management team at CESAGen, Lancaster University*.

The books in this series, all based on original research, explore the social, economic and ethical consequences of the new genetic sciences. The series is based in the ESRC's Centre for Economic and Social Aspects of Genomics, the largest UK investment in social-science research on the implications of these innovations. With a mix of research monographs, edited collections, textbooks and a major new handbook, the series will be a major contribution to the social analysis of new agricultural and biomedical technologies.

New Genetics,
New Social Formations

Edited by Peter Glasner, Paul Atkinson
and Helen Greenslade

Routledge
Taylor & Francis Group

LONDON AND NEW YORK

First published 2007
by Routledge
2 Park Square, Milton Park, Abingdon, Oxon OX14 4RN

Simultaneously published in the USA and Canada
by Routledge
270 Madison Ave, New York, NY 10016

Routledge is an imprint of the Taylor & Francis Group, an informa business

© 2007 Peter Glasner and Paul Atkinson, editorial content and selection;
individual chapters, their contributors

Typeset in Sabon
by Taylor & Francis Books
Printed and bound in Great Britain
by MPG Books Ltd, Bodmin

British Library Cataloguing in Publication Data
A catalogue record for this book is available from the British Library

Library of Congress Cataloging in Publication Data
A catalog record for this book has been requested

ISBN10: 0–415–39323–X   ISBN13: 978–0–415–39323–2 (hbk)
ISBN10: 0–203–96289–3   ISBN13: 978–0–203–96289–3 (ebk)

Contents

Illustrations

Contributors

Kean Birch is a research fellow in the Centre for Public Policy for Regions (CPPR) at the University of Glasgow, and a doctoral student in the Department of Planning at Oxford Brookes University. He has a background in sociology and technology studies, and more recent interests in bioethics, economic geography and economic sociology. Currently his research, including his thesis, focuses on the biotechnology industry, although he has a more general interest in issues around the development of less-favoured regions in Europe, and the expansion of the social economy in the UK.

Henrik Bruun, Docent, PhD, is a Senior Researcher at the Helsinki Institute of Science and Technology Studies (HIST). His research focuses on knowledge networking in science and innovation. Recent projects include a study of distributed problem-solving in a Finnish engineering project, an assessment of the Academy of Finland's practices for supporting interdisciplinary research, and a comparative case study of knowledge networking in biotechnology companies. Henrik is the Chief Editor of *Science Studies*, an interdisciplinary journal for science and technology studies.

Lesley Henderson is Lecturer in Sociology and Communications in the School of Social Sciences and Law at Brunel University, West London. Prior to this she was Senior Research Fellow in the Department of Sociology at the University of Glasgow and a member of the Glasgow Media Group. She has published several articles and papers on the content, transmission and reception of media messages mainly concerning science, health and social problems (for example genetics, breast cancer, the mumps, measles and rubella vaccine, mental health). She is currently developing research on young people, citizenship and television news for a large study funded by the Arts and Humanities Research Council, and continuing her work in the area of international media and mental distress with the Swedish Association for Mental Health. She regularly acts as consultant to non-academic organisations such as

Department of Health, and is involved with developing challenging representations of science in media (PAWS). She has three sole-authored books forthcoming: *Social Issues in Television Fiction* (Edinburgh University Press); *Researching the Media: Issues, Ethics, Methods and Processes* (Open University Press); and *QualitativeResearch Design* (Sage).

Peter Hopkinson is senior lecturer in environmental management. As an environmental scientist his research and teaching interests span the physical, biological and social sciences. His main focus of research is on the environmental impacts of technological and organisational change. He is working with Liz Sharp to enable the University of Bradford to achieve a Sustainable Campus programme. The son of a beekeeper, he spent his formative years surrounded by honey, bee stings and conversations about waggle dances.

Mavis Jones's scholarly focus could be described as STS-informed policy research. She is particularly interested in the definition of 'relevant' expertise in technology policy development, and the circulation and mobilisation of discourses by engaged actors. Her past work has dealt with deliberative democratic mechanisms (specifically, a consensus conference on GM crops and food) and public engagement in reproductive and genetic technologies policy. In her most recent role as a Senior Research Associate at the Centre for Environmental Risk (School of Environmental Sciences, University of East Anglia), she has been involved in comparative research on risk governance across issues and political domains. She is currently conducting a study of policy learning between the UK and Canada in the governance of assisted human reproduction.

Hyo Yoon Kang is currently a researcher in law at the European University Institute in Florence, and a teaching fellow at the London School of Economics where she teaches property law. Her PhD was entitled 'Processes of Individuation and Multiplicity: the Human Person in Patent Law relating to Human Genetic Material and Information'. She has been awarded a postdoctoral fellowship at the Max Planck Institute for the History of Science in Berlin for the study of the relationship between patent law and biological taxonomies, starting in October 2006. She was a fellow of the German Academic Exchange Service and visiting scholar at University of California at Berkeley.

Loes Kater is involved with quality assurance projects in academic education and scientific research at QANU (Quality Assurance Netherlands Universities) in the Netherlands. Until recently she was employed as postdoctoral researcher at the University of Twente. She carried out a comparative research project between the United Kingdom and the Netherlands on developments in stem cell research, financed by the

Netherlands Organisation for Scientific Research (NWO). In 2002 she completed her PhD thesis on end-of-life decisions at the University of Maastricht. She analysed the role of bio-ethicists and medical jurists in the Dutch debate on euthanasia. She holds a Master's degree in health sciences (University of Maastricht) and a Master's degree in philosophy (University of Amsterdam).

Jenny Kitzinger is Professor of Media and Communication Research at Cardiff University. She specialises in research into the media coverage and audience reception of social, health and scientific issues. She has also written extensively about sexual violence. Her most recent book, *Framing Abuse: media influence and public understanding of sexual violence against children*, was published by Pluto Press (2004). Jenny is also co-editor of *Developing Focus Group Research: politics, theory and practice* (Sage, 1999), and co-author of *The Mass Media and Power in Modern Britain* (Oxford University Press, 1997), *Great Expectations* (Hochland & Hochland, 1998) and *The Circuit of Mass Communication in the AIDS Crisis* (Sage, 1999).

Ruth McNally has a BSc inGenetics, an MA in Socio-Legal Studies and a PhD in Science and Technology Studies. She is a Senior Research Associate at the ESRC Centre for Economic and Social Aspects of Genomics (CESAGen), Cardiff University. Her current research projects are the CESAGen Flagship Project on proteomics and an ESRC project on sustainable technology transitions. She is experimenting with the use of the IssueCrawler and 'PROTEE' as methods for reassembling the social.

Paul Oldham is a social anthropologist specialising in issues surrounding the human rights of indigenous peoples and biodiversity. Paul trained at Lancaster University (BA Hons) and Cambridge University (MPhil), and carried out doctoral research at the London School of Economics and Political Science. Paul has carried out extensive fieldwork with the Piaroa (Wothïha) in the Venezuelan Amazon and in 1993 worked to establish the Regional Organisation of Indigenous Peoples of Amazonas (ORPIA) for whom he continues to serve as an independent adviser. In 2002 Paul returned to academia as a member of the Anthropology Department at the University of Durham before joining CESAGen in 2003 to work on the flagship project Indigenous Peoples and Globalisation of Genomics in Amazonia. His research interests principally focus on the rights of indigenous peoples and the United Nations Convention on Biological Diversity.

Nick Pidgeon is Professor of Applied Psychology at Cardiff University. His research looks at how public attitudes and institutional responses form a part of the dynamics of a range of risk controversies, including those of radioactive waste, climate change, GM agriculture and nanotechnology. In his research he has argued that public policy decisions about controversial technologies need to be sensitive to public values if fair and equitable outcomes are to be found. However, this does require robust methods for eliciting such values and for promoting a genuine dialogue between scientists, policy-makers and civil society about emerging science and technology issues. He was first author of the chapter on risk perception and communication in the Royal Society's 1992 *Report on Risk*, co-author (with B. Turner) of the book *Man-Made Disasters*, Butterworth-Heinemann, 1997 (2nd edn), and (with R. Kasperson and P. Slovic) of *The Social Amplification of Risk*, Cambridge, 2003.

Wouter Poortinga is an Academic Fellow at the Welsh School of Architecture and the School of Psychology, Cardiff University. Wouter's research focused on people's responses to various environmental and technological risks. One of his recent topics was on the role of trust in the perception and acceptability of risks. His wider research interests are in studying the interaction between the psychological and social/environmental basis of people's health, well-being and quality of life.

Barbara Prainsack is a lecturer in comparative politics, and a postdoctoral researcher at the Department of Political Science, University of Vienna, Austria. Her work focuses on biotechnology regulation, bioethics, and the relationship between religion and politics. Her recent project deals with biobanks and the transformation of health governance, and is funded by the GEN-AU programme of the Federal Austrian Ministry of Science, Education and Culture.

Angela Procoli is a Researcher in Social Anthropology at the Laboratoire d'Anthropologie Sociale, Collège de France. She is the author of *Anthropologie d'une Formation au Conservatoire National des Arts et Métiers* (Education Pedagogie, 2001) and editor of *Workers and Narratives of Survival in Europe* (SUNY Press, 2004).

Tee Rogers-Hayden, of Cardiff University, is an affiliated Research Fellow to the Center for Nanotechnology in Society at the University of California, Santa Barbara. She is interested in relationships between science and society, especially regarding the introduction of new technologies. She has been researching public debates on GM, in New Zealand and in the UK, analysing New Zealand's Royal Commission into Genetic Modification, and furthering research from the official evaluation of the UK's GM Nation? Currently she is undertaking research into public participation and stakeholder deliberation regarding nanotechnology in the UK, Canada and the US.

Liz Sharp is a senior lecturer in environmental policy at the University of Bradford's Department of Geography and Environmental Science. Having trained as a geographer and planner, her research has concentrated upon the way the public interacts with institutions in the processes of developing and implementing environmental policy in the UK, most recently in relation to waste and water management. She lives near Bradford and enjoys walking in the hills.

Margaret Sleeboom-Faulkner is Research Fellow at the International Institute for Asian Studies (IIAS, Leiden) and at the Amsterdam School for Social Science Research (ASSR), and a lecturer in the fields of Asian studies and cultural anthropology at the University of Amsterdam. Her work focuses on the two areas of nationalism and processes of nation-state building in China and Japan, and on biotechnology and society in Asia. She has set up and directs the Socio-genetic Marginalization in Asia Programme (SMAP) in cooperation with the Netherlands Science Organization (NWO), the IIAS and the ASSR (2004–9). She publishes widely on the social-science aspects of new genetic and medical technologies, on issues of ethnic and national identity in East Asia, and on the history of science education in China, such as *Academic Nations in China and Japan* (Nissan/RoutledgeCurzon, 2004) and *Genomics in Asia: Cultural Values and Bioethical Practices* (Kegan Paul, 2004).

Chie Ujita is a research student at the University of Bradford. Her main area of research is public awareness of risk and the implementation of the precautionary principle in UK society. She is interested in the UK court system, which is very different from that in Japan, her country of origin.

Clare Williams is Reader in the Social Science of Biomedicine in the School of Nursing and Midwifery, King's College London. Her key research interests include the clinical, ethical and social implications of innovative health technologies, particularly from the perspective of healthcare practitioners and scientists. Her two current research projects focus on pre-implantation genetic diagnosis, and stem cell treatment for liver disease and diabetes. She has written numerous publications for social science and practitioner journals, mainly in the area of prenatal screening and foetal medicine. Clare is author of *Mothers, Young People and Chronic Illness* (Ashgate Press, 2002); and (with S. P. Wainwright) co-author of a forthcoming book, *The Body, Biomedicine & Society: Reflections on High-Tech Medicine* (Palgrave Innovative Health Technology Book Series).

Acknowledgements

The support of the Economic and Social Research Council (ESRC) is gratefully acknowledged. The work was part of the programme of the ESRC Research Centre for Economic and Social Aspects of Genomics.

CESAGen

The ESRC Centre for Economic and Social Aspects of Genomics (CESAGen) was established in October 2002 as a collaboration between the Universities of Lancaster and Cardiff. CESAGen's main objective is to work with genomic science while investigating the economic and social factors that shape natural knowledge.

The CESAGen Book Series

The General Editors of the series are Paul Atkinson (Cardiff), Ruth Chadwick (Cardiff), Peter Glasner (Cardiff), and Brian Wynne (Lancaster). Between them, the editors enjoy international reputations. Their expertise covers the entire spectrum of relevant research fields – from bioethics and research regulation to environmental politics and risk, to science and technology studies, and to innovative health technologies.

Artist in Residence

Paul Harrison is an artist and researcher with a background and prevailing interest in print, printmaking and publishing. His practice inherently combines the use of traditional print methods and materials with new and developing technologies. The focus of this practice is a developing dialogue and collaboration with laboratory and social scientists as an integral part of a visual investigation into the production of images emerging from new developments in genetics and cell research. He is interested in how this new information is processed and visualised in both a specialist and a public context. He is presently engaged in projects with scientists at the University of Dundee Biocentre, the Human Genetics Unit, MRC, Edinburgh and Cold Spring Harbor Laboratory (CSHL) New York. He is also artist in residence at the Human Genome Organisation (HUGO) and visiting fellow/artist in residence at the Centre for Economic and Social Aspects of Genomics (CESAGen) at the University of Cardiff. His work can be seen online at http://www.personal.dundee.ac.uk/~plharris (accessed 24 May 2006).

1 Introduction

New genetics, new social formations

Peter Glasner and Paul Atkinson

The mapping and sequencing of the genome of human beings and other forms of life in the first decade of the new millennium is leading to a reassessment of genomics as systems biology, with a new emphasis on function rather than structure. The result of this development has been the formation of new kinds of knowledge about living things (sometimes described as the 'omic' revolution) with the establishment of new disciplinary boundaries (as, for example, in proteomics), and, significantly, with the development of new intellectual and social spaces within which these events occur. The post-genomic era requires more than just a technical understanding of gene structure and function. New technological options cannot survive without being entrenched in networks of producers, users and various services. A new research system is coming into being centred on the production, use and commodification of genetic knowledge, based on new sets of knowledge, technologies and commodities, and embodying a new set of socio-technical relations involving new groups of actors (Glasner 2002). Innovation, as Brown and Webster (2004: 162) describe it, is a 'melange of knowledge, technology, organisation and wider socio-political activity' operating in the complex networks of this new research system. The governance of genomics is being reshaped by a new culture which reflects the changing relationships between government, industry and techno-scientific development (Gottweis 2005). This volume attempts to explore the contours of the new social formations that are co-constructed in, and embodied by, this post-genomic enterprise.

The new biotechnologies are fundamentally constitutive of the biological in that they are both a *tool* and a part of the *process* of biological development, resulting in unique configurations of (to follow Latour 1993) hybrid formations. The accepted view of nature–society relations, that society is inherently plastic and pliable while nature is remote and autonomous, is turned on its head. The new biotechnologies have made nature pliable and society remote and difficult to change (Brown and Michael 2004: 15). They are neither simply opposed to nature, nor even external to it. They are clearly still tools that are objects to regulate, produce or regenerate nature. But they are also constitutive of defining nature itself, framing it through active participation (Thacker 2005). In this sense they are part of the process that Jasanoff (2004: 2) describes as the co-production of nature and society. The ways in which the world is apprehended and represented by individuals is inseparable from the ways in which they inhabit it. Socio-technical knowledge thus both 'embeds and is embedded in social practices, identities, norms, conventions, discourses, instruments and institutions'. Such an approach, while not a fully fledged theory, provides a useful way of organising and interpreting complex phenomena in an area such as biotechnology which is rapidly changing, and has deep moral and ethical, as well as practical implications, for society as a whole.

In their discussion of the 'risky creatures' created through, and governing the development of, xenotransplantation technologies, Brown and Michael (2004: 208) highlight the need for new regulatory bodies to reflect, at least in part, some of the crucial features of the new objects of regulation. They suggest that related genetic technologies such as pharmacogenomics, tissue engineering and stem cells also challenge the boundaries of existing institutional corporealities and identities. Tissues and genes are potentially fragmented from conventionally understood species boundaries by new innovations in genomic technologies (Waldby 2002). The products of the innovation process then combine human actors, natural phenomena and socio-technical production in a variety of relatively unstable (in the sense of being continually co-constructed) hybrid social formations (Brown and Webster 2004). Such co-constructions need to be stabilised (albeit only for a short time) if they are to effectively mobilise actors to create novel institutions in the process of innovation. Some of this occurs through the defining 'intermediaries' that pass between actors, such as texts (for example scientific papers), things (for example computer software), or skills (for example clinical knowledge). These appear in a variety of contexts, including public engagement, techno-scientific economies, socio-technical platforms and social representations.

The shift from reliable to socially robust knowledge, recognised by Nowotny, Scott and Gibbons (2001), provides examples of both new social formations and the intermediaries that contribute to their stabilisation. Recognising that scientific knowledge is always incomplete knowledge, they argue that it is also always therefore contested knowledge. Technosciences, such as the new biotechnologies, illustrate the extent to which the conceptual boundaries within which scientific disputes have conventionally been resolved are now obsolete. While objectivity, proof and verification continue to be valued, these are inextricably entwined with shifting local practices which value them differentially, and are more or less robust and reliable depending on context. Progress is no longer viewed as a linear phenomenon, but messily contingent, co-constructed and contextualised. They describe the new social formations within which science now operates as part of the *agora*, which facilitates the greater involvement of non-experts in decision-making as an integral part of producing socially robust science.

A variety of new decision-making structures has been harnessed by governments and NGOs to involve the 'public' over the last fifty years. These include, among others, deliberative polls, citizens' juries, consensus conferences, science courts, and focus groups. However, society has developed a number of ways of translating the goals of those in authority into the choices of individuals, effectively acting as a form of social control and a legitimation for commercial interests. In the case of citizens' juries, for example, a combination of transposing the symbolic baggage associated with the legal framing of the judge and jury system to debates about innovative technologies, and introducing rituals of precision which inadequately mimic those in Courts of Law, serve only to give the appearance of accommodating non-experts into the decision-making process. In effect, these trappings sit comfortably within the existing relations of production and so do little to co-construct a socially robust techno-science (Glasner and Rothman 2004: 111 *et seq.*).

In part this is due to the creation of an 'imagined' public, rooted in persistent concerns about public ignorance, and the public mistrust of science. The continuing reinvention of the 'deficit model' depends on the myth that, because publics mistakenly expect certainty and risk from science, science is obliged to delete reference to these in policy debates and focus instead on imagined and unacknowledged audiences (Wynne 2001). It is also due to agencies within science where, following the mapping of the human genome, turf wars continue. These actors are competing not only for positions on the reductionist/determinist and holistic/functionalist spectra, but also for regulatory freedom and financial support. Together, it can be argued, the real character of innovative trajectories has been masked through intermediaries embedded in the public involvement over the introduction of GM food or crops, or the development of downstream therapeutic applications from pharmacogenetic or stem cell technologies (Wynne 2005).

In the United Kingdom, one major attempt to engage a wide variety of people on a complex issue of biotechnology is evaluated by Pidgeon and Poortinga in their chapter on the 'GM Nation?' public debate. This was one of a variety of different ways which the UK government used to consider a decision on approving the commercialisation of genetically modified crops. It was unusual in that the results of a public dialogue would be placed alongside more traditional sources of evidence and expertise. The authors compare its findings with those of an extensive survey of a representative sample of British public opinion obtained shortly after the debate concluded. The results of 'GM Nation?' are seen as borne out to a greater degree than its critics have suggested, implying that it could form an important intermediary in future large-scale public engagement exercises.

In her chapter, Loes Kater discusses the nature of public participation during the establishment of the UK Stem Cell Bank. She recognises that the precise role played by the 'public' remains a contested issue, especially when, as in this case with its associated complex ethical and legal issues, science–society relationships are under strain. She suggests two roles became clearer as the Stem Cell Bank progressed. In one, the 'public in general' was enrolled in the Bank's emerging network of allies. In the second, a process of consultation took place, specifically through the intervention of the British Medical Research Council, aimed at 'specific publics'. Kater concludes, however, that with regard to the 'public in general' it was only *representations* of public expectations that were enrolled – effectively creating just the sort of intermediary discussed by Wynne (2005). The consultation with more specific public groups on the other hand at least allowed for a more interactive dialogue where real differences could be allowed to surface.

Rogers-Hayden and Jones also argue for a more reflexive approach to public engagement in their comparison between Canada and New Zealand of their Commissions on reproductive technologies and genetically modified organisms respectively. The authors chart the many similarities between the two exercises while recognising the inherent limitations of Commissions as a means of engaging the public in decision-making. They conclude that overcoming the modernist premise on which Commissions are based by embedding 'reflexive, human-based values' early into their design may serve to produce the necessary new social formations required by contemporary liberal democracies.

One of the premises highlighted during the Commission debates in both countries was that of the precautionary principle. In their fine-grained analysis of its role in the trial of anti-GM activists following their destruction of GM crops, Ujita and her colleagues show how the institutional setting of the court modified the way in which actors shaped their views and arguments. They suggest that such trials serve an important symbolic function in policy-making, particularly for the activists, who can use them as platforms for disseminating their views. Hence, while not necessarily an ideal forum for public debate, such trials, and the way in which the precautionary principle is played out in them, provide interesting examples of one *agora* as identified by Nowotny *et al.* (2001).

Globalisation has highlighted the need to recognise that markets in scientific and technological downstream applications from genomic innovations exhibit similar characteristics to existing markets in the wider knowledge economy (Glasner and Rothman 2004). There has been an enormous increase in the codification of knowledge, which, together with networks and the digitalisation of information, is leading to its increasing commodification. There is increasing interdependence of international flows of goods and services, direct investment, and technology and capital transfers associated with increasing specialisation, and chains of production crossing international boundaries (Appadurai 1996). There is a substantial national and regional structural adjustment, with an emphasis on flexibility and networking built through. Time has now become, alongside knowledge, a new factor of production, essentially compressing and reordering existing conceptions of what is understood by the production process as shown in the freezing or banking of 'immortal' stem cell lines (Glasner 2005). Together, these elements suggest that the transition to a knowledge-based economy is so fundamentally different from the resource-based system of the last century that conventional economic understanding must be re-examined (Barry and Slater 2005).

Birch uses the ongoing construction of a multi-national biotech industry to illustrate these developments in the knowledge-based economy. He argues that the USA constructed a global market through deliberate changes to law and industrial policy designed to benefit its local industry and national interest. The resultant changes, first to national, and subsequently to international definitions of patentability, have resulted in financial accumulation and economic growth at a disproportionate level suggesting that any study of the knowledge-based economy must consider its wider social and cultural relations. Today, the gap between the richer nations of the global North and the poorer ones in the South is increasing just as rapidly in biotechnology and its applications as elsewhere. These global divisions have been exacerbated as huge transnational biotechnology corporations have developed from mergers and acquisitions, located mainly in Europe and North America. The top five biotech companies now own about 95 per cent of all gene transfer patents. Global governance agreements such as TRIPS form intermediaries which co-construct both innovative scientific developments and their applications.

This suggests a need to focus on the issues of control, access and influence over agendas for the future innovation and exploitation of the new genetic technologies that such a polarisation implies. Such issues are firmly embedded within a variety of commercial, regulatory and governmental institutions. Oldham uses the debates surrounding biodiversity and intellectual property to discuss biopiracy and the bioeconomy. He identifies the emergence of the 'bioeconomy' as a result of the growing convergence between the biosciences (widely drawn) and broader local and global regulatory processes. Focusing on the Convention on Biological Diversity as a new social formation configured in a variety of ways, he suggests that the contested area of intellectual property rights has served to encourage biopiracy rather than the benefit sharing stated in its aims.

Kang, in her chapter on patenting human genetic material and information, questions the appropriateness of using a utilitarian economic justification in the equation that trades temporary monopoly rights to the patentee with free use of a patented invention by the public. She suggests, using the example of *Moore v the Regents of the University of California*, that current patent law constructs the human subject as a thing rather than a person in order for it to be an object of property relations. This process produces new legal objects that reside in the space between the legal and techno-scientific practices that take the human body as their object of knowledge. As a result human agents cease to be the primary field for study, to be replaced by a focus on the mediated relationships between self and the social networks within which the self is situated.

Sleeboom-Faulkner also discusses identity formation in the context of the new genetics, in her analysis of genetic population mapping in the People's Republic of China and Taiwan. However, the findings from such studies are not only contentious in themselves (as, for example, in the Human Genome Diversity Project), but generate disputes on aboriginal rights to territory, resources and self-determination. She suggests that, for some, the Taiwanese have become perceived as indigenous populations (isolates of historic interest) in the political sense. Clearly this perception of a Taiwanese identity is constructed both internally, and outwith the island using socio-cultural as well as genetic resources not limited just to the Chinese peoples. Its changing role as a hybrid formation is closely related to similar discussions about national identity on mainland China.

Styles of practice in the investigation of research problems, the results that are generated, and the ways in which the process is regulated, produce distinctive and novel institutional, epistemic and material configurations. In the case of biomedical advance, Keating and Cambrosio (2003) have suggested the term 'biomedical platform' to cover the range of activities in contemporary biomedicine, ranging from laboratory research to clinical trials and routine diagnosis. In a study of changing diagnoses of lymphoid tumours, for example, they show how three distinctive approaches (the morphological, immunophenotypic and molecular genetic) emerge chronologically, but do not result in the replacement of an earlier by a later platform. New platforms are integrated into an expanding set of clinical-biological strategies through complex realignments and articulations with earlier ones. However the key to how a new diagnostic platform is held together in practice is to be found in the protocols that specify *inter alia* sufficient sensitivity, specificity, reproducibility, robustness, reliability, accuracy, precision and clinical relevance – or, more broadly, regulation (Keating and Cambrosio 2004: 39). The development and approval of a regulatory protocol is itself a highly political process almost as complex as the application of the protocol itself.

Angela Procoli focuses on the new forms of knowledge production discussed by Nowotny and her colleagues. She uses her anthropologically informed methodology to analyse three case studies within the French scientific community: surface scientists, geneticists and breeders, and quantitative and molecular scientists. She suggests that the boundaries of these groups are as much produced by economic, social and cultural forces from outside in the wider society, as they are from within. This occurs as a result of such communities becoming 'hybrid fora' made up from local and professional groups as well as the scientists themselves. These groups bear many similarities to the bio-medical platforms suggested by Keating and Cambrosio.

In his chapter on bioinformatics challenge, Henrik Bruun identifies bioinformatics tools for the storage, manipulation and analysis of data as the basis for functional genomics, proteomics, and many other new research platforms in the bio-sciences. He discusses a particular research platform based on the use of micro-arrays, as a case study of the challenges faced by the 'new biology'. His approach illustrates the need to focus on transformations with educational, cognitive, epistemological and practical research implications for organisations, institutions, and for the scientists themselves. Bruun concludes that genomics is indeed undergoing a transformation in knowledge production, even though practices integrating bio-informatics into the laboratory vary greatly between laboratories.

Studies drawn from such areas as literary theory, deconstructionist methodologies and the sociology of scientific knowledge have shown how scientific rhetoric and forms of discourse contribute to legitimating scientific authority in complex social contexts (Lewenstein 1995; Miller *et al.* 1998). In tracing the patterns of representation of genomics across a range of media forms, and reviewing and reassessing broad conceptualisations concerning science on and in the media, it is possible to identify specific patterns with regard to genomics. This involves consolidating insights about key nodes of media theory, pertaining to textual analysis, including: genre, narrative continuities and transformations across media forms, iconic imagery, spectacle, and national and international specificities. It also involves explorations of the production and consumption of representations, where theories of media production, identification, audience and risk come into focus. Theories of globalisation and the internationalisation of media forms therefore also come under scrutiny, with explication of some national issues that are crucial in understanding contemporary public science.

The chapter by Kitzinger and her colleagues suggests, through a broad investigation of genomics, a theoretical framework for understanding genomics as a public and mediated science. This involves new insights about the relation between science, media and 'the public', serving to reconfigure understandings of public engagements with genomics and other sciences. As a case study, they examine the ways in which the embryo is imagined, visualised and represented in controversies over stem cell research, particularly in relation to a series of 'breakthroughs' between 2000 and 2005. What counts as an embryo has become a contested intermediary, with both proponents and critics of stem cell research mobilising metaphors and personifications through visual representations of its origins, destiny and death to attempt stabilisation. This occurs within a 'balanced' media coverage that uses 'breakthrough' science to systematically disregard wider scientific and socio-political issues and challenges.

Prainsack takes this further in her discussion of the regulation of genomic technologies in Israel. Israel has a relatively permissive approach to the new genetic technologies, including research in human embryonic cloning.

Looking at bioethics discourse as a hybrid form, she suggests that the moral debate in Israel uses similar terms to those used elsewhere, such as the 'sanctity of life', but often arrives at different conclusions. This does not mean that the Jewish religion can be conceived as immoral compared with, for example, Christianity. Rather, an analysis of the discourse of risk and nationhood suggests that sustaining life is a collective as well as an individual imperative. This can best be understood in the context of Israel's perception of a state under constant hostile political pressure through violent conflict from outside. Effectively, bioethics discourse is at the same time a discourse of the politics of survival in which the application and use of advanced genomic technologies to developing biomedical technologies plays a key part. This makes public objections to, for example, stem cell technology effectively 'inconceivable'.

McNally and Glasner look at a different form of discourse, as a defining intermediary concentrating on the concept of gene. They discuss the extent to which developments in the new biology either make its use redundant or an essential building block for the future. Using the framework of Evelyn Fox Keller's *The Century of the Gene* (Keller 2000), they analyse three 'visions' of how the new biology is to develop following the completion of the Human Genome Project. They discover that, just as gene talk is still ubiquitous in spite of the gene losing its pre-eminence as an explanatory tool, paradigm talk has now also become prevalent as a new intermediary alongside it, used, as gene talk once was, to mobilise resources and enrol new actors.

These chapters together reflect the rapidly changing scientific, clinical and social environment within which new social formations are being constructed and reconstructed. They bring together a range of empirical and theoretical insights that, like the chapters in the companion volume to this one, *New Genetics, New Identities*, serve to help better understand complex, and often contentious, innovative processes. They also provide insights into the relative instability of new hybrid forms and their defining intermediaries. In reporting on their work, the contributors have marked an important stage in the conduct of social science research and its relationship to techno-scientific and clinical practice.

References

Appadurai, A. (1996) *Modernity at Large: cultural dimensions of globalisation.* Minneapolis, MN: University of Minnesota Press.

Barry, A. and Slater, D. (eds) (2005) *The Technological Economy.* London: Routledge.

Brown, N. and Michael, M. (2004) 'Risky creatures: institutional species boundary change in biotechnology regulation', *Health, Risk and Society*, 6(3): 207–22.

Brown, N. and Webster, A. (2004) *New Medical Technologies and Society: reordering life.* Cambridge: Polity.

Glasner, P. (2002) 'Beyond the genome: reconstituting the new genetics', *New Genetics and Society*, 21(3): 267–77.

—— (2005) 'Banking on immortality? Exploring the stem cell supply chain from embryo to therapeutic application', *Current Sociology*, 53(2): 355–66.

Glasner, P. and Rothman, H. (2004) *Splicing Life? The New Genetics and Society*. Aldershot: Ashgate.

Gottweis, H. (2005) 'Governing genomics in the 21st century: between risk and uncertainty', *New Genetics and Society*, 24(2): 175–93.

Jasanoff, S. (ed.) (2004) *States of Knowledge: the co-production of science and social order*. London: Routledge.

Keating, P. and Cambrosio, A. (2003) *Biomedical Platforms. Realigning the Normal and the Pathological in Late-Twentieth Century Medicine*. Cambridge, MA: MIT Press.

—— (2004) 'Signs, markers, profiles, and signatures: clinical haematology meets the new genetics (1980–2000)', *New Genetics and Society*, 23(1): 15–45.

Keller, E. F. (2000) *The Century of the Gene*. Cambridge, MA: Harvard University Press.

Latour, B. (1993) *We Have Never Been Modern*. Cambridge, MA: Harvard University Press.

Lewenstein, B. (1995) 'Science and the Media', in S. Jasanoff, J. C. Petersen, T. Pinch and G. E. Markle (eds), *Handbook of Science and Technology Studies*. Thousand Oaks, CA: Sage.

Miller, D, Kitzinger, J. and Beharrell, P. (1998) *The Circuit of Mass Communications*. Thousand Oaks, CA: Sage.

Nowotny, H., Scott, P. and Gibbons, M. (2001) *Rethinking Science: knowledge and the public in an age of uncertainty*. Cambridge: Polity.

Thacker, E. (2005) *The Global Genome: biotechnology, politics and culture*. Cambridge, MA: MIT Press.

Waldby, C. (2002) 'Stem cells, tissue cultures and the production of bio value', *Health: An Interdisciplinary Journal for the Social Study of Health, Illness and Medicine*, 6(3): 305–23.

Wynne, B. (2001) 'Creating public alienation: expert cultures of risk and ethics' in GMOs', *Science as Culture*, 10: 445–81.

—— (2005) 'Reflexing complexity: post-genomic knowledge and the reductionist returns in public science', *Theory, Culture and Society*, 22(5): 67–94.

2 British public attitudes to agricultural biotechnology and the 2003 GM Nation? public debate
Distrust, ambivalence and risk

Nick Pidgeon and Wouter Poortinga

Introduction

There is considerable emphasis being placed in the UK today upon stake-holder and public dialogue in relation to science and technology issues. Such processes have in many respects overtaken attempts to promote increased 'public understanding of science' and greater science literacy through more traditional science communication methods. When discussing the move to greater public engagement in the UK at this moment in time, it is important to take account of the ways in which the UK had been impacted by a recent history of controversy concerning science, technology and risk issues. Above all, two issues have dominated the public policy agenda and thinking. The Bovine Spongiform Encephalopathy (BSE or 'mad cow') crisis, and the initial controversy over genetically modified (GM) crops, both occurring in the mid-to-late 1990s, marked a turning point in the way UK science policy was viewed. Both the independent inquiry into the causes of BSE (Phillips, Bridgeman and Ferguson-Smith 2000) and a wider House of Lords Select Committee on Science and Technology (2000) report on *Science and Society*, argued that there existed a crisis of trust in UK science policy-making. The question of whether a crisis of trust in *science itself* actually existed at this point in time is debatable. At the level of general beliefs about the contribution of science and technology to society, public attitudes have remained stable and highly favourable in the UK in recent years (see OST/Wellcome 2000; Poortinga and Pidgeon 2003a; DTI/MORI 2005). Where strong concerns are expressed by people, it is typically with respect to operation of the science policy process in relation to much more *specific* and controversial issues (radioactive waste and nuclear power, GM food, the mumps, measles and rubella vaccine). A contributory factor in the presumed crisis of legitimacy surrounding science policy was assumed to be a failure of the traditional one-way 'deficit model' of science risk communication, as advocated in the earlier Royal Society (1985) report on public understanding of science. In response, both the BSE and Lords reports stressed the importance of *openness* in government and the science community as a precondition to re-establishing credibility and trust in risk management and policy. The Lords report also highlighted a need to broaden the base of public consultation and dialogue on controversial science policy issues (see also POST 2001). The implication was that we need to move beyond traditional public understanding of science efforts if we are to resolve some of the most contested issues of science policy.

Accordingly, the Lords report recommended:

That direct dialogue with the public should move from being an optional add-on to science-based policy-making and to the activities of research organisations and learned institutions, and should become a normal and integral part of the process.

(House of Lords Select Committee on Science and Technology, 2000, paragraph 5.48)

In parallel with these important national policy developments in the UK, dialogue-based approaches involving diverse groups of citizens and stakeholders have gained in popularity for environmental and risk decision-making and conflict resolution in both continental Europe (Renn *et al.* 1995; Joss 1998) and North America (Beierle and Cayford 2002). In an important conceptual contribution to this evolution of approach the 1996 US National Research Council report on *Understanding Risk* (Stern and Fineberg 1996) had also developed a detailed set of proposals which they termed the 'analytic-deliberative process'. This combines sound science and systematic uncertainty analysis with participatory deliberation by an appropriate representation of affected parties, policy-makers and specialists in risk analysis. According to the authors, dialogue, participation and deliberation should occur *throughout* the process of risk characterisation, from problem-framing through to detailed risk assessment and then on to risk management and decision implementation.

Despite recent case examples, much good theoretical work, and conceptual analysis of the normative and substantive reasons for engaging in participatory processes (e.g. Fiorino 1990; Pidgeon 1998; Stirling 2005), there remains far less agreement upon the precise methodologies for achieving these ends. Renn *et al.* (1995) distinguish between three broad classes of citizen participation: genuine *deliberative methods* which allow for fair and competent debate and discussion between all parties, such as consensus conferences, citizen juries and planning cells; traditional *consultation methods*, including public meetings, surveys, focus groups and mediation, where there is little or no extended debate; and finally *referenda* in which people do have democratic power but which are not generally deliberative in nature. All of these approaches have contrasting strengths and weaknesses, and one of the lessons to have emerged from initial experiments with participation is that for some complex or long-running issues hybrid or multi-stage methodologies may often be necessary (see, e.g., Renn 1999) in order to: (a) accommodate specific requirements (for example framing, options appraisal, choice) at different points in time within an extended deliberative process; (b) compensate for weaknesses in any singular approach to participation through 'triangulation' of outcomes and other data; or (c) accommodate different groups of stakeholders or citizens depending again upon specific requirements of sponsors or partici-pants at particular points in time. The development of analytic-deliberative methods will also require evaluation of both process and outcomes (see Chess and Purcell 1999; Rowe and Frewer 2000).

This chapter takes as its case-study the GM Nation? public debate, a major public participation exercise on the commercialisation of agri-cultural biotechnology which occurred in Britain during the summer of 2003. GM Nation? was important because it represented a genuine attempt to engage a wide range of people on the complex issue of biotechnology, and was relatively novel in the British context. It also formed one of four main evidence streams (the other three being a general science review, an economic analysis, and specific information from the outcomes of farm-scale trials of GM crops that had recently been completed) which the UK government was to consider when deciding upon the commercialisation of GM. In the event, that synthesis of evidence and decision came in early 2004 (DEFRA 2004). What was particularly unusual for the whole process was the explicit attempt to place public dialogue alongside the more tradi-tional evidence streams from the scientific and economic analyses.

As noted above, major engagement exercises are run for a variety of reasons – some quite complex – and any single approach is rarely capable of meeting all objectives fully. In the case of GM Nation?, a range of objectives was discernible, including to allow participants the opportunity for debate and deliberation, but crucially also to access what questions a 'grass roots' public might have about the commercial growing of GM crops in the UK. Of course, the idea of a single 'public' with readily accessible 'attitudes' to risk issues is something of a misnomer (Pidgeon *et al.* 1992): there are multiple groups in society, each with complex identities and commitments, and a series of overlapping framings and representations circulating at any one time which people might draw upon to debate any issue. And fundamentally, GM Nation? was *not* intended as an opinion poll, but as a set of interlinked dialogue activities. These issues notwith-standing, the debate organisers did set themselves the task of accessing and reporting a range of viewpoints, through the debate activities, which included a large number of open public meetings, an interactive website, and a small set of specially convened focus groups (so-called 'narrow-but-deep' groups) of pre-selected individuals.

A number of questions have been raised regarding the methodology adopted for obtaining and representing public views on GM food and crops during the GM Nation? process. One such has been whether questionnaire responses obtained from over 36,000 participants attracted to the open debate meetings and interactive website might have exhibited bias because the sample was self-selecting in nature – primarily in the direction of a highly anti-GM standpoint – a possibility raised by commentators shortly after the publication of the debate final report (see Campbell and Townsend 2003). There is, however, a wider and long-standing question as to whether open deliberative activities, such as public meetings (which in the event comprised a core component of the GM Nation? activity), do indeed generate outcomes which 'represent' wider public attitudes and opinion in some meaningful sense. On the one hand, open meetings and events are important as a means of maintaining transparency of process, and in theory at least allow anybody who wishes to contribute to a debate to do so. However, it has also been known for many years that open events tend to attract those who have more invested in the issue, are demographically or attitudinally different in some way, or hold more polarised views than would a representatively sampled group of the general or affected public (e.g. Jackson and Shade 1973; Heberlein 1976). As a result Glass (1979) has argued that open meetings are not useful tools for providing representative input to public decisions. However, empirical studies have also found that under some circumstances the views and demographic characteristics of those attending public meetings, even for hotly contested issues, can indeed accord with those of the general or otherwise affected public (Gundry and Heberlein 1984). Other studies present a more complex picture. For example, McComas reports both similarities and differences between public meeting participants and the locally affected (non-participant) citizens in her study of local waste management participation. She concludes, in relation to the wider literature on the subject, that 'lack of clear patterns with respect to representation may be a result of researchers' different conceptualisations of the public or varying contexts of the research' (McComas 2001: 137).

In the current chapter we discuss the value of the GM Nation? outcomes (the debate Steering Board concluded that in total seven broad findings, or 'messages', could be drawn) by comparing them with data obtained from a representative sample of British public opinion, obtained shortly after the debate was concluded. Although not designed as direct tests of the GM Nation? outcomes (which were formulated and published after our survey had completed fieldwork), the findings we report here nevertheless give some important additional evidence on the validity of the conclusions drawn during the GM Nation? debate process.

The GM Nation? public debate and findings

The GM Nation? public debate arose as a direct result of a recommendation to the British government by the Agriculture and Environment

Biotechnology Commission (AEBC), a multi-stakeholder consultative body, in its report of 2001 titled *Crops on Trial* (AEBC 2001). This report had considered the controversy generated by the government's farm-scale trials for GM (FSEs), concluding that public policy on GM crops should 'expose, respect and embrace the differences of view which exist, rather than bury them'. It went on to call for 'an open and inclusive process of decision-making around whether the GM crops being grown in the FSEs should be commercialized, within a framework which extends to broader questions'. The report also called for an improved understanding of the basis of public views on these matters, and for the future role of GM crops within UK agriculture to be considered in 'a wider public debate involving a series of regional discussion meetings'. In response the Government asked the AEBC to:

> Advise, by the end of April 2002, on how and when to promote an effective public debate on possible commercialization of the FSE crops and how to make the best use of the results of such a debate. The advice should also cover how to determine the public accept-ability of GM crops.

Subsequently, in her letter to the AEBC of 25 July 2002, the Secretary of State responsible for the debate, Margaret Beckett, confirmed the government's wish that a 'public dialogue on GM' should take place. The government, she stated, was committed to a 'genuine, balanced discussion, and also to listening to what people say'.

While extensively used in some other European nations such as Denmark, the Netherlands and Switzerland, major deliberative processes have been less common in the UK. Consensus conferences had occurred in 1994 for plant biotechnology (POST 1995) and in 1999 for radioactive waste management (UKCEED 1999). A criticism of many deliberative processes held at national level – and a source sometimes of considerable 'stakeholder fatigue' – is that outcomes may not have a direct bearing upon relevant decisions or policy. As noted above, this was clearly not intended to be the case for the GM Nation? debate, with a clear route for the outcomes to input to the policy process being envisaged by AEBC, and a commitment by government (at least as expressed) to take account of those findings.

In terms of scale, GM Nation? involved participants from every part of the UK, across a six week period from 3 June to 18 July 2003. It was overseen by an independent Steering Board comprised of the Chair of the AEBC together with deliberation specialists and stakeholders from across the spectrum of opinion on GM agriculture. Much of the day-to-day implementation of the debate was carried out by the Government's Central Office of Information (COI) as main contractor to the Debate Steering Board.

Figure 2.1 illustrates the main constituent elements of GM Nation?. In November 2002, and prior to the main debate process, a number of preliminary discussion groups (known as 'Foundation Discussion Workshops') had been convened. The primary aim here was to investigate how a cross-section of ordinary citizens would make sense of these issues. This was a practical attempt to generate resources such as stimulus materials for use in subsequent stages of the debate, so as to allow lay perspectives to shape the terms of the engagement in an innovative way.

The main 'debate' process in the summer of 2003 comprised three principal engagement mechanisms.

- The first was a series of open public meetings, which anybody could attend, organised into three levels or 'tiers'. Tier 1 meetings (three in England and one each in Wales, Scotland and Northern Ireland) were conceived of as 'national' high-profile events resourced by COI and professionally facilitated. These attracted approximately 1,000 participants in total. Tier 2 meetings, of which there were about forty, were typically hosted by a local authority or other major organisation, often with the assistance of COI. A large number, estimated to be in excess of 600, of Tier 3 events took place, although this estimate was based upon intentions expressed to COI to hold a meeting, rather than a confirmed tally of actual meetings held (PDSB 2003).
- The second way in which anybody could engage openly with GM Nation? was via a dedicated interactive debate website, which contained a range of debate materials and interactive resources. The hope was that people who could not attend the tier meetings would be able to register their views through this website.
- Finally, the third means of engagement comprised a series of ten closed-group discussions with ordinary members of the public known as 'narrow-but-deep' groups. 'Narrow' in this context refers to the scope of representation, since only 77 members of the public took part, albeit being recruited through standard market research criteria to reflect a broad demographic cross-section of the UK population. 'Deep' refers to the anticipated level of engagement and deliberation in these groups, in comparison to that available for the typical open-meeting participant. These groups met twice, with a gap of two weeks between the meetings, during which participants were invited to explore the GM issue individually, using official stimulus materials and any other information that they could access, and to keep diaries of their discoveries (including newspaper clippings, website downloads, etc.), thoughts, relevant conversations and so on. This process was also conceived of as a 'control' on the possibility that those attending the public meetings and other self-selecting participants in the debate, such as the website respondents, did not represent the views of a 'silent majority'. In the words of the Steering Board, the narrow-but-deep process was intended to give a

Foundation Discussion Workshops

November 2002

Total of 9 evening focus groups. 8 with ordinary citizens pre-selected to represent a spread of socio-demographic characteristics. 1 further meeting held with GM stakeholders.

Exploratory 'framing' of issues in preparation for the main debate process in the summer

Debate Open Meetings

Tier 1
Major 'kick-off' meetings organised by Steering Board (n=6 meetings)

Tier 2
Meetings organised by local councils or national organisations and supported by Steering Board (n= 40 meetings estimated)

Tier 3
Local meetings organised by interest groups, science and educational centres etc. (n=629 meetings estimated)

MAIN DEBATE July 2003

Interactive Debate Website

Including information on GM, and the opportunity to register views in qualitative and quantitative form.

Closed 'Narrow-but-Deep' Groups

10 groups held with 77 ordinary citizens pre-selected to represent a spread of socio-demographic characteristics

Each group met on two separate occasions to deliberate GM agriculture, with a period in between to gather information.

Transcripts and Rapporteurs' Reports

Questionnaire Responses

Emails/ Letters

Qualitative Analysis of Participants' Discourse

Steering Board Final Report (24 September 2003)

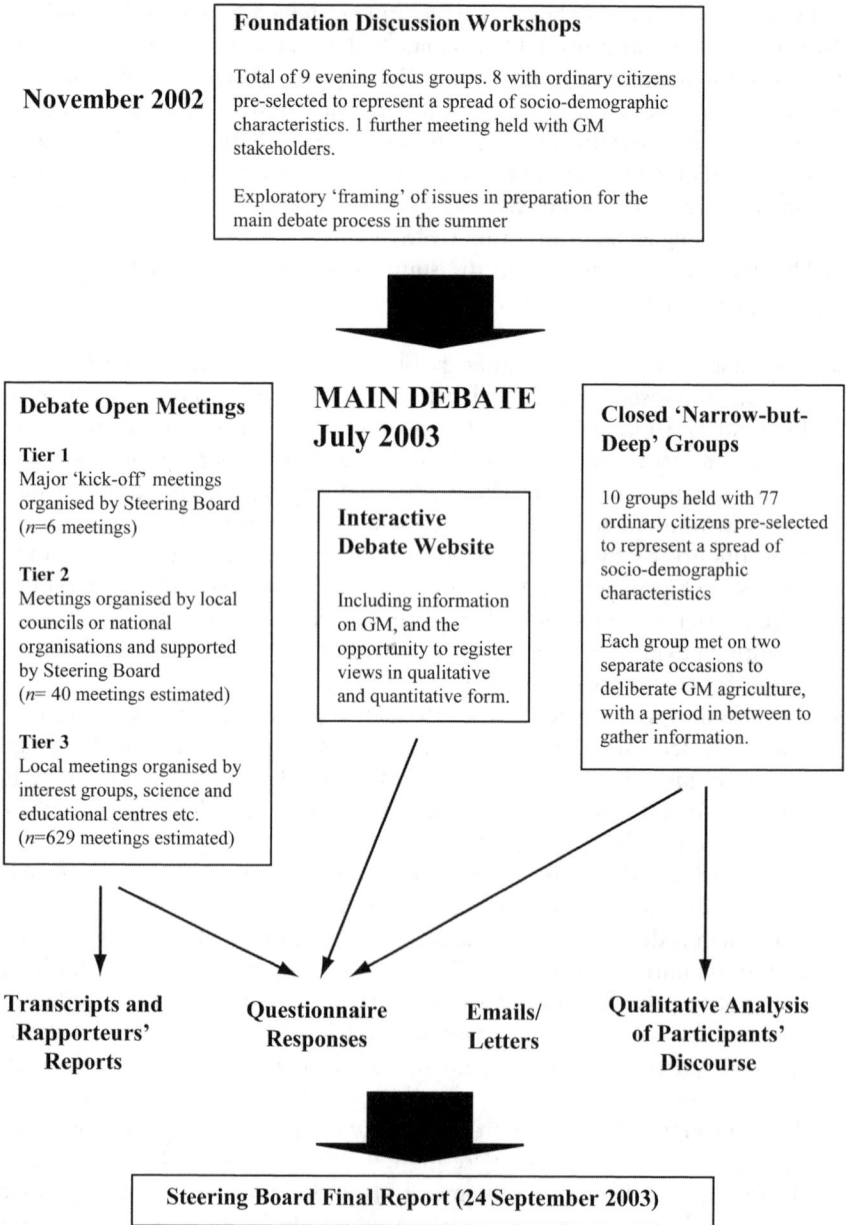

Figure 2.1 Main elements of GM Nation?

'*qualitative idea* of whether the general population might respond differently to GM issues to the self-selecting people who became active participants in the GM Nation? Programme' (PDSB 2003: para. 194, emphasis added).

The Steering Board's final conclusions and report were based upon a combination of several data streams: primarily qualitative, including rapporteurs' reports from meetings, qualitative analysis of the 'narrow-but-deep' discussions, and open-ended feedback responses (letters and emails received); but also quantitative, from thirteen attitude questions on a feedback questionnaire which was distributed in paper form but could also be completed on the debate website. The feedback questionnaire in particular proved popular, with a total of over 36,000 responses obtained in almost equal proportions from paper copies distributed to meeting organisers and their equivalent website versions (see Rowe *et al.* 2006). The 77 'narrow-but-deep' participants also completed this questionnaire, albeit twice: once at the commencement of their involvement and again at the beginning of their second meeting. In its final report the debate Steering Board (PDSB 2003: 51–3) summarised its seven main findings as the following 'key messages':

1 People were generally uneasy (regarding safety and risks to the environment, as well as wider social and political issues) about agricultural GM.
2 The more people engaged in agricultural GM issues, the harder their attitudes and more intense their concerns.
3 There was little support for early commercialisation.
4 There was widespread mistrust of government and multi-national companies.
5 There was a broad desire to know more and for further research to be done.
6 Developing countries have special interests (in using GM technology for food, medical and non-food applications).
7 The GM Nation? debate was welcomed and valued.

Despite this rather complex, and interrelated, set of findings, some sections of the national British media, although it should be said *not* the GM Nation? report authors themselves, focused upon the headline findings from the questionnaires completed by the 36,000 open debate participants by presenting these uncritically as if they were from a nationwide representative opinion poll. For example, the *Daily Mail* headline of 25 September 2003 read '9 out of 10 vote no to GM Crops'. However, the seven key findings go far beyond a simple 'anti- or pro-GM' position, reflecting a more complex set of concerns and discourses (and, as noted above, drawing upon a more extensive, and both qualitative and quantitative, data set).

In this chapter we triangulate these broader GM Nation? outcomes against responses to a major national opinion survey conducted immediately after the public debate had concluded. This survey formed part of an extensive independent evaluation of the public debate process, conducted by a consortium of academics, including the present authors (see Horlick-Jones *et al.* 2004, 2006; Rowe *et al.* 2005; Pidgeon *et al.* 2005).

The UEA/MORI survey methodology

The opinion survey was administered in England, Scotland and Wales by the market research company MORI for the 'Understanding Risk' programme at the University of East Anglia (UEA). Data were collected between 19 July and 12 September 2003, just after the GM Nation? public debate was concluded but before the debate Steering Board issued its final report at the end of September. A representative sample of 1,363 people aged 15 years and older was interviewed face-to-face in their own homes in Britain. The overall sample was made up of a core, nationally representative British (England, Scotland, Wales) sample of 1,017 interviews, with additional booster questionnaires then gathered in Scotland (151) and Wales (195) respectively. The booster surveys were added in order to have large enough samples in each of Scotland and Wales to conduct comparative analysis with the sub-set of responses drawn only from England. However, all frequency data reported here have been weighted back to the known profile of the British population in terms of age, gender, social class and region.

The survey questionnaire consisted of over 170 items. Alongside standard demographic questions, the main instrument consisted of three main sections. The first section examined public perceptions of GM food. More specifically, respondents were asked a set of questions similar to those asked in a survey conducted in summer 2002 (described in Poortinga and Pidgeon 2003a). In addition to public perceptions of GM food in general, this section was aimed at capturing possible shifts in public sensibilities, awareness and knowledge of risk issues in relation to GM food, as well as issues of trust in the governance of GM food. The second section contained questions that were adapted primarily from the GM Nation? public debate questionnaire (see PDSB 2003: 60–1). These questions were designed to measure specific risks and benefits associated with GM food and crops. The third section of the survey contained questions specifically developed to evaluate the GM Nation? debate itself. This section of the survey mainly focused on awareness of the debate as well as on people's views and understandings of the value and impacts of the debate process itself.

The full descriptive survey results are reported in Poortinga and Pidgeon (2004a). The survey sample did differ in two important ways from the process adopted in GM Nation? The GM Nation? data was collected across the UK, and hence included Northern Ireland. The UEA/MORI sampling was confined to Great Britain (England, Scotland, Wales) alone. The final sample characteristics for the survey are shown in Table 2.1. A second difference was that, for some of the items (particularly in the first section), where we wanted comparability with the earlier survey in 2002, we asked only about GM food. The GM Nation? debate itself was about the wider question of agricultural biotechnology (GM food *and* crops) more generally, although, as the Steering Board's final report (PDSB 2003) makes clear, GM food was a central part of these deliberations. Where this makes a material difference to the comparisons offered here, this is noted below. As discussed above, GM Nation? made seven key findings, of which two cannot be addressed directly with the survey data. These were: finding 2, that the more people engaged the harder their attitudes became; and finding 3, that there was little support for early commercialisation. The remaining five findings can be addressed, and these are discussed in the following section in turn.

Table 2.1 Characteristics of the 2003 UEA/MORI GM survey sample

Characteristic		%	Characteristic		%
Gender	Male	49	Level of	No formal	24
	Female	51	education	GCSE	25
				Vocational/ NVQ	8
Age	15–24	13		A-level	15
	25–34	21		Bachelor degree	16
	35–44	18		Postgraduate	4
	45–54	16		Other/ Don't know	12
	55–64	13			
	65 and older	19	Marital Status	Married	47
				Cohabiting	11
Class	AB	22		Single	23
	C1	30		Widowed	8
	C2	19		Divorced	6
	DE	28		Separated	2
Income[1]	Low	18	Employment	Full-time	45
	Average	27	Status	Part-time	11
	High	21		Unemployed	7
	Don't know/ Refused	36		Retired	23
				Student	5
Ethnic	White	93		Disabled	3
background	Black	2		Looking after	6
	Asian	3		children	
	Other	1			

Source: UEA/MORI GM Survey 2003 (weighted dataset, *n*=1,363).

Notes:
[1] Low: <£11,500 gross per annum; Average: £11,500 to £30,000; High: ≥£30,000.

Results

Concern: 'People are generally uneasy about GM' (Steering Board Finding 1)

This finding represented the first and most consistent message of the Steering Board's final report. The Board argued that most people were not only uneasy about such things as eating GM food, and the impact of GM crops on the environment, but were also concerned with wider social and political questions, with the mood ranging from 'caution and doubt, through to suspicion and scepticism, to hostility and rejection' (PDSB 2003: 6). The Steering Board also noted that such concern (especially for risks) was highest amongst open-debate participants, who also rejected any possible benefits from GM. By comparison, the 'narrow-but-deep' participants expressed similar anxieties about potential risks (albeit to a lesser degree) whilst being more willing to entertain the notion that GM might bring at least some benefits.

The UEA/MORI survey included items addressing general feelings towards GM food (specifically positive through to negative beliefs). There is growing evidence that people's initial 'affective' response is an important part of the way in which representations of risk issues are constructed (see, e.g., Loewenstein *et al.* 2001; Finucane *et al.* 2000; Langford 2002; Slovic *et al.* 2004). People's general orientation towards an issue – whether it is seen as 'good' or 'bad' – may function as a key filter influencing the way subsequent information is processed, such as perceptions of potential benefits, communications about the issue from others, or even trust in risk managers (see, e.g., Poortinga and Pidgeon 2004b, 2005). Our own results suggest that along such measures a proportion of the British public feel particularly uncomfortable about GM food. Accordingly, 40 per cent of respondents said that they feel negatively about GM food, whilst only 15 per cent said that they feel positively. A similar pattern emerged when people were asked whether GM food is a good or a bad thing: with 40 per cent saying it is a 'bad thing', and 14 per cent said it is a 'good thing'. Perhaps more interestingly, these two questions also show that a sizeable minority could be found in the middle of the scales used (35 and 40 per cent, respectively). These results may suggest an underlying *ambivalence* about this issue, something which has been identified in previous research on this matter (see Grove-White *et al.* 1997; Gaskell *et al.* 1997) and to which the discussion returns later.

The results with regard to behavioural intentions were also broadly in line with the overall 'affective' responses towards GM: a sizeable minority (28 per cent) were happy to eat GM food, whilst almost half of the sample (46 per cent) disagreed with this. Similarly, 50 per cent of the sample said that they would try to avoid purchasing GM food products.

In order to further explore public attitudes towards agricultural biotechnology, the UEA/MORI survey contained statements about specific risks and benefits, this time in relation to both GM food *and* crops (see Table 2.2). Most of the questions were adapted for the present survey from statements administered as part of the GM Nation? quantitative feedback forms (see PDSB 2003; Pidgeon *et al.* 2005). Table 2.2 shows that an overwhelming majority (85 per cent) thought that 'we don't know enough about the long-term effects of GM food on our health', while only 9 per cent agreed with the statement that 'GM crops are safer than traditional crops because they have been more thoroughly tested'. Next to uncertainties about the health impacts of GM food, 63 per cent of the sample were concerned about the 'potential negative impact of GM crops on the environment'. A large majority also (68 per cent) agreed with the statement that 'I am worried that if GM crops are introduced it will be very difficult to ensure that other crops are GM free'. It appeared also that a majority (56 per cent) agreed that 'GM food could make farmers dependent on big companies that have patents on GM crops'. Finally, as also shown in Table 2.2, three out of four were worried that 'this new technology is being driven more by profit than by the public interest', while 69 per cent agreed that GM crops would 'mainly benefit the producers and not ordinary people'. The results presented above clearly show that a range of concerns exist about the risks of GM food and crops, across a range of health, environmental and governance issues, congruent with that found in the GM Nation? debate.

Distrust: 'There was widespread mistrust of government and multi-national companies' (Steering Board Finding 4)

The Steering Board's report highlights that, alongside concern about possible risks, there was also a strong degree of suspicion amongst debate participants about the motives of those taking decisions about GM (especially governments and multi-national companies), expressed as a lack of trust. The relationship between risk perceptions of new technologies and (dis)trust is of course a well researched area, and several models and approaches to this issue have been proposed by researchers (for general discussions of this issue see Wynne 1992; Cvetkovich and Löfstedt 1999; Poortinga and Pidgeon 2003b). In the case of GM Nation?, the Steering Board's report noted suspicions that the government has already made up its mind about GM, alongside a more general distrust of the motives and agendas of modern government. The report also highlighted unease over the perceived power of multinational companies, which were seen as being motivated by profit rather than to meet society's needs.

The UEA/MORI survey contained general questions about trust in information sources to tell the truth about GM Food (see Table 2.3). Here both national/regional government and the EU, alongside food manufacturers and the biotechnology industry, were trusted far less than other actors (doctors, environmental organisations). Interestingly however, the

Table 2.2 Specific risks and benefits of GM food and crops

	Strongly disagree	Tend to disagree	Neither/ Nor	Tend to agree	Strongly agree	Don't know
Specific risks						
I don't think we know enough about the long-term effects of GM food on our health[1]	1	3	8	33	52	2
GM crops are safer than traditional crops because they have been more thoroughly tested	21	27	32	8	1	9
I am concerned about the potential negative impact of GM crops on the environment[1]	2	8	22	37	26	5
I am worried that if GM crops are introduced it will be very difficult to ensure that other crops are GM free[1]	2	5	18	35	33	6
GM food will make farmers dependent on big companies that have patents on GM crops	2	5	26	36	20	10
I am worried that this new technology is being driven more by profit than by the public interest[1]	1	6	13	38	37	4
I think GM crops would mainly benefit the producers and not ordinary people[1]	1	8	21	33	31	4
Specific benefits						
I believe GM crops could help to provide cheaper food for consumers in the UK[1]	9	14	23	39	6	8
I think that some GM crops could benefit the environment by using less pesticides and chemical fertilisers than traditional crops[1]	9	11	26	37	7	9
I believe that GM crops could improve the prospects of British farmers by helping them to compete with farmers around the world[1]	10	16	32	26	5	11
I believe that GM crops could benefit people in developing countries[1]	7	10	19	38	18	6
I believe that some GM non-food crops could have useful medical benefits[1]	5	6	34	34	6	15

Source: UEA/MORI GM Survey 2003 (weighted dataset, n=1363).

Note:
[1] Statement adapted from the GM Nation? public debate feedback form (see PDSB 2003).

Table 2.3 Trust in stakeholders: responses to the question 'To what extent do you trust the following organisations and people to tell the truth about GM food' (%)

	Distrust a lot	Distrust a little	Neither/ Nor	Trust a little	Trust a lot	No opinion
Doctors	1	2	14	42	39	2
Consumer rights organisations (e.g. Consumers Association)	2	5	13	43	33	4
Environmental organisations	2	6	14	45	31	3
Scientists working for universities	2	4	17	46	29	2
Scientists working for environmental groups	2	7	15	48	25	2
The Food Standards Agency (FSA)	3	6	14	45	26	5
Friends and family	1	3	23	30	40	2
Department of Environment, Food and Rural Affairs (DEFRA)	5	7	18	44	20	6
People from your local community	2	5	33	42	14	4
Farmers	5	10	26	36	19	3
Scientists working for government	15	21	19	34	8	3
Scientists working for the biotech industry	14	21	23	28	9	5
Local authorities	10	17	33	31	5	3
Biotechnology industry	15	20	25	27	8	5
Food manufacturers	17	26	21	29	5	3
The European Union (EU)	20	19	25	25	7	4
The national government	23	25	19	25	5	3
The Welsh Assembly (n=235)	15	12	27	29	11	6
The Scottish Parliament and its Executive (n=265)	14	23	31	24	5	3

Source: UEA/MORI GM Food Survey 2003 (weighted dataset, n=1,363).

government agency charged with regulating food safety in the UK, the Food Standards Agency, was relatively *highly* ranked in this list, suggesting at least some degree of differentiation in people's understandings of the agencies of risk governance (see also Walls *et al.* 2004).

Regarding the specific dimensions of trust in government, a range of factors appear to influence trust in risk managing institutions, which may be summarised under the rubrics of *competence, care* and *consensual values* (Johnson 1999). In this study, respondents were asked to evaluate government policy on GM food using items designed to measure *competence, credibility, reliability, integrity* (vested interests), *care, fairness,* and *openness* (see Table 2.4). The results show that most respondents were indeed critical about the government and its perceived handling of GM, across the full range of items used. A principal components analysis was then conducted in order to examine whether the evaluation of government could be described by a number of underlying dimensions. The trust statements could be described by two main factors, together accounting for 68 per cent of the variance of the original variables. Most items loaded high on the first factor, which accounted for 43 per cent of the variance. This factor was concerned with the items aimed at measuring competence, care, fairness and openness, and can be interpreted as a *general trust* factor. That is, it represents a general evaluation of government policy on GM food. The second factor accounted for 25 per cent of the original variance and contained the items 'the government distorts facts in its favour regarding GM food', 'the government changes policies regarding GM food without good reasons', and 'the government is too influenced by industry regarding GM food'. This factor reflects a sceptical view of how government GM policies are brought about, and can be labelled as *scepticism*. These results are comparable to similar analyses conducted on five risk cases in earlier research (see Poortinga and Pidgeon 2003b). However, it may well be the case here that people hold very limited knowledge of the precise functions of risk regulation (see also Walls *et al.* 2004), and that such judgements reflect more general beliefs about the role and legitimacy of 'government' rather than the handling of this particular risk issue *per se*.

Knowledge: 'There a broad desire to know more and for further research to be done' (Steering Board Finding 5)

The Steering Board reported that there was a strong desire, from all participants to the GM Nation? debate, to be better informed about GM from sources that they could trust. GM debate participants were reported as wanting agreed 'facts' that were accepted by all organisations and interests, and confidence in the independence and integrity of information about GM, such that an individual could resolve the various contested claims about the technology. There was also a view expressed that, because the science was uncertain, more research was needed on GM.

Table 2.4 Evaluation of government (%)

	Strongly disagree	Tend to disagree	Neither/ Nor	Tend to agree	Strongly agree	No opinion
Competence						
The government is doing a good job with regard to GM food	24	25	28	11	1	11
The government is competent enough to deal with GM food	27	27	19	19	2	6
Credibility						
The government distorts facts in its favour regarding GM food	3	9	23	34	22	9
Reliability						
The government changes policies regarding GM food without good reasons	3	9	30	30	17	11
Integrity (vested interests)						
The government is too influenced by the biotechnology industry regarding GM food	2	7	25	35	20	10
Care						
The government listens to concerns about GM food raised by the public	19	33	20	20	2	5
The government listens to what ordinary people think about GM food	31	35	15	11	2	6
Fairness						
I feel that the way the government makes decisions about GM food is fair	23	25	31	11	1	9
Openness						
The government provides all relevant information about GM food to the public	35	33	17	7	2	7
Bias of government						
The government wants to promote GM food	2	6	23	39	21	9
The government is not in favour of GM food	24	33	26	4	2	11

Source: UEA/MORI GM Survey 2003 (weighted dataset, *n*=1,363).

Table 2.5 presents the UEA/MORI findings on a range of direct measures of ambivalence, questions about the need for information, and about independence of regulation, although, as in the case of the trust items, all of the questions were restricted to the question of GM food.[1] A large proportion of our sample appeared ambivalent about GM food when asked directly in this way. That is, 56 per cent agreed with the statement that 'I have mixed feelings about GM food', while 31 per cent agreed with the statement that 'There are so many arguments for and against GM food. I could be persuaded by any of them'. We also found that many people feel that they do not have enough information to form a clear opinion about GM food. Only 19 per cent indicated they felt they were well informed about GM food. Probably as a result, and entirely congruent with the Steering Board's findings regarding the wider issue of GM agriculture, most people felt a strong need for information, with more than four out of five (84 per cent) saying that they needed more information to form a clear opinion about GM food. These findings are in line with other qualitative and survey findings too, and may reflect a wider sentiment than just specifically about GM, since many people endorse the need for more information about a range of science and technology issues when asked (see, e.g., DTI/MORI 2005). The survey also contained items about the need for independent regulation, and here a very high proportion agreed that there was a need for independence from government (79 per cent) and from industry (80 per cent). However, we should also note here that these percentages showed a marked increase over the identical questions asked twelve months earlier, unlike other measures of perceptions of GM food itself which had themselves shown little change. As the second survey took place after the controversy in the UK over Britain's participation in the spring 2003 Iraq war, and at the time of the events surrounding the death of the scientist David Kelly, we interpreted this as reflecting, in part, a more general disaffection and disillusionment at that point in time with central government.

Benefits: 'Developing countries have special interests' (Steering Board Finding 6)

The Steering Board's report noted that there was a 'debate-within-the-debate' on the possible role GM might play in developing countries, with at least an initial assumption that GM agriculture might contribute future benefits in terms of food production or medical, social and economic benefits. Again, differences were reported between the open participants (who rejected this proposition) and the 'narrow-but-deep' groups who did not. The former tended to reject *any* possible benefits, while the latter were more prepared to endorse these. In the UEA/MORI survey, a clear majority (56 per cent) did indeed feel that 'GM food crops could benefit people in developing countries' (see Table 2.2). However, it appears that, at the same

Table 2.5 Ambivalence, need for information and independent regulation (%)

	Strongly disagree	Tend to disagree	Neither/ Nor	Tend to agree	Strongly agree	No opinion
Ambivalence v Certainty						
I have mixed feelings about GM food	10	12	18	40	16	4
There are so many arguments for and against GM food. I could be persuaded by any of them	16	24	24	26	5	4
Need for information						
I am well informed about GM food	28	32	16	15	4	5
I need more information to form a clear opinion about GM food	3	5	7	33	51	2
Independent regulatory organisations						
Organisations separate from government are needed to regulate GM food	2	4	10	39	40	5
Organisations separate from industry are needed to regulate GM food	1	3	10	38	42	6

Source: UEA/MORI GM Survey 2003 (weighted dataset, *n*=1,363).

time, a substantial proportion of our sample agreed that there could be other (potential) benefits of GM food and crops. Almost half of the sample (45 per cent) believed that GM crops 'could help to provide cheaper food for consumers in the UK', while 44 per cent agreed that 'some GM crops could benefit the environment by using less pesticides and chemical fertilisers than traditional crops'. People's responses to the statement that 'I believe that GM crops could improve the prospects of British farmers by helping them to compete with farmers around the world' were fairly equally distributed: 31 per cent agreed, 26 per cent disagreed, and 32 per cent neither agreed nor disagreed. Finally, a sizeable minority (41 per cent) believed that 'some GM non-food crops could have useful medical benefits'. It is worth nothing that a relatively large number of people (34 per cent) neither agreed nor disagreed with the latter statement, while 15 per cent had no opinion or did not respond to the question.

In further analysing the risks and benefits items shown in Table 2.2, Pidgeon *et al.* (2005) have suggested that it provides further evidence of considerable levels of *ambivalence* about GM agriculture amongst a large proportion (over 50 per cent) of the UEA/MORI sample. Such individuals believed that there were risks from GM *alongside* possible benefits. Prior qualitative studies have also demonstrated that the general public have detailed but often conflicting views on agricultural biotechnology (Grove-White *et al.* 1997; Marris *et al.* 2001). In these studies many people expressed arguments both for and against agricultural biotechnology. Such ambivalence can arise in a number of ways. For example, people's cognitive and affective responses to a certain facet of biotechnology might be in conflict. Or they might believe agricultural biotechnology could indeed bring useful environmental benefits but do not trust corporations or the authorities to invest in delivering those particular benefits (as opposed to attributes, such as a longer shelf-life for products, which more directly benefit producers). Equally, both perceived benefits and perceived risks may be viewed as highly significant, and possibly difficult to directly trade off; for example, environmental or other benefits set against a perceived long-term health risk to one's family from eating GM food.

The Value of Public Engagement: 'The debate was welcomed and valued' (Steering Board Finding 7)

The Steering Board concluded that, across all parts of the debate (open-access and closed meetings, as well as the website), people were glad that it had happened, both as a platform for expressing their own views and as a way of hearing those of others. Echoing the comments about trust noted above, however, there was widespread suspicion that government would ignore the results, although people nevertheless also expressed a hope that their contribution would make a difference (otherwise, why turn up to a meeting?). In all some 20,000 individuals across the UK were estimated to have taken part in the various open meetings (PDSB 2003), while the website recorded over 24,000 unique visitors during the course of the debate.

While it was clear that the debate process motivated a very large number of people to take part, as well as to organise their own local meetings in what was a very short period of time (six weeks), what is less clear from the debate account is its wider impact across British society. Although a large number of participants were both informed and motivated enough to engage through the open meeting and website, this is an insufficient basis to estimate the levels of awareness amongst the population at large. As one of the explicit objectives of the GM Nation? debate had been to create widespread awareness amongst the population of the programme of debate activities, we included questions on our survey on people's awareness and evaluation of the debate process.

In the UEA/MORI survey a great majority (71 per cent) of those interviewed had never heard of the GM Nation? debate (even though fieldwork was completed in the six weeks immediately following the debate). In addition, 13 per cent had heard of the debate but knew nothing about it. Only 15 per cent of survey respondents reported that they were to some extent aware of the debate: specifically, about one in eight indicated that they knew a little about the debate, while only 3 per cent said they knew a fair amount or a lot about it. These figures can be interpreted in at least two ways. On the one hand, this finding does suggest that a small proportion of the British adult population had been made aware of its existence. Given the esoteric nature of the subject, and the relative lack of advertising, tabloid or mainstream television coverage of the debate,[2] this figure might be regarded as representing a modest success. On the other hand, the bulk of the sample had not heard of the debate at all, suggesting that GM Nation? failed to meet one of its objectives, of creating widespread awareness of the programme of debate.

Anticipating that awareness of the debate among at least a significant proportion of the sample would be non-existent, our survey respondents were then provided with information by interviewers about the background and the process of GM Nation? (see box), following which they were asked to evaluate the debate on a range of measures.

Information provided to survey respondents on GM Nation?

As you may know, 'GM Nation? The Public Debate' is a nationwide discussion of issues related to genetic modification (GM) of crops and food. It is sponsored by the government, and managed by an independent board of people representing diverse views on GM. During June and July 2003 a series of regional and local meetings were organised to allow people to have their say about the role of GM in the UK. 'GM Nation? The Public Debate' is organised to involve the public in the important decision as to whether or not GM foods and crops should be grown commercially in Britain. The findings from the meetings will be fed back to the Government to help inform their policy-making on GM foods and crops.

Views on the value of the debate when described in this way to respondents appeared to be mixed (see Table 2.6). On the one hand, a clear majority (66 per cent) felt that *in principle* a debate was a good way for the public to get more involved in making decisions about GM foods and crops, something which echoes the Steering Board's own observations, as well as other more recent research on public engagement with science (DTI/MORI 2005). However, concerning the precise *substance* of the process, people were far less clear, something which is hardly surprising given that most respondents knew nothing at all about the debate. For example, people's responses to the statement 'Organising the debate is a waste of taxpayers' money' were fairly equally distributed (37 per cent disagreed, 35 per cent agreed and 20 per cent neither agreed nor disagreed), while the great majority (42 per cent) neither agreed nor disagreed with the statement 'the debate has been run fairly, without promoting any specific views on GM foods and crops', with an additional 30 per cent responding with no opinion to this latter question.

Despite clear support for the *principle* of a debate on GM foods and crops, there remained scepticism about the *impacts* of the debate. This again echoes the substance of the Steering Board's comments on debate participants' mistrust of government motives. More than half (59 per cent) the UEA/MORI sample agreed with the statement 'The debate will make no difference, because the government has already made its mind up on GM foods and crops'. Equally, almost half of the sample (45 per cent) disagreed, while about one in four (23 per cent) agreed that 'the debate shows that the government listens to what normal people think about GM foods and crops', while fully 68 per cent agreed with the statement that it 'does not matter whether there is a debate on GM or not. In the end European and International laws will determine what will happen'.

In the final substantive question on the survey, we asked respondents about the usefulness of other public involvement exercises, such as GM Nation? in the future. Here, a large majority (77 per cent) felt that it was indeed important to have public debates on other important new developments in science and technology.

Concluding comments

In seeking to compare the outcomes of the GM Debate? process with those from our own nationally representative survey, it should be clearly borne in mind that these were two rather different types of process, whose participants were engaged in very different ways, and whose objectives only partially overlapped. However, we do believe that the empirical comparison

Table 2.6 GM Nation? evaluation statements (%)

	Strongly disagree	Tend to disagree	Neither/ Nor	Tend to agree	Strongly agree	No opinion
Evaluation of the debate						
The debate is a good way for the public to get more involved in making decisions about GM foods and crops	3	9	16	53	13	7
It is unclear to me what the debate is about	8	23	19	31	14	6
The debate shows that the government listens to what normal people think about GM foods and crops	16	29	26	20	3	7
Organising the debate is a waste of taxpayers' money	7	30	20	25	10	8
The debate has been run fairly, without promoting any specific views on GM foods and crops	5	9	42	13	1	30
Impact of the debate						
The debate will have an influence on government's policies on GM foods and crops	11	25	27	25	3	10
Because of the debate I trust the government more to make the right decisions about GM foods and crops	18	27	25	18	3	9
The debate will make no difference, because the government has already made its mind up on GM foods and crops	2	13	19	38	21	7
It does not matter whether there is a debate on GM or not. In the end European and international laws will determine what will happen	3	8	15	43	25	5

Source: UEA/MORI GM Survey 2003 (weighted dataset, n=1,363).

between GM Nation? and the UEA/MORI survey findings is sufficiently robust to lend at least some additional corroboration to the wider out-comes of the debate process itself – in particular, in relation to widespread concern about GM, mistrust of government and business, people's desire for further information, and the value placed on the principles of public engagement and deliberation. However, where our results diverge from the Steering Board's report is our finding of widespread ambivalence about agricultural GM amongst our sample across a range of measures (and a lack of commentary on this issue in the Steering Board's report is somewhat surprising, given established research findings also showing this to be the case for agricultural GM). Our findings are important for GM policy in Britain, because, as noted above, a number of academic commentators (e.g. Campbell and Townsend 2003; Toke 2005), as well as some stakeholders in the debate, have claimed that GM Nation? was *so* methodologically flawed as to have rendered little of useful value for policy. While the outcomes of any form of public debate around such a politically sensitive issue are always likely to be the subject of claim and counter-claim, our analysis suggests a more complex interpretation, with a number of the more fine-grained details of the GM Nation? findings clearly borne out in our survey data. In addi-tion to this, Grove-White (2003) has argued that GM Nation? must be judged as a deliberative engagement rather than a research exercise, and as such was not designed as an opinion poll or a psychometric survey. One can also add here that deliberative processes may fall short of the ideal in research methodology because of intense budgetary, political and time pres-sures (and GM Nation? was no exception to this rule; see Horlick-Jones *et al.* 2004).

A methodological criticism of the analysis presented here might be that neither method (our quantitative survey nor the mixed-methodology GM Nation? design) was appropriate to delivering a robust mapping of beliefs and discourses about GM, as envisaged in the original objectives of the GM Nation? debate. Certainly, it can be problematic to use a survey to elicit opinion on issues where people have low prior knowledge and awareness, where multiple framings and discourses around the issue exist, and for which people require extended deliberation in order to construct any form of position or preference at all (see, e.g., Fischhoff and Fischhoff 2001). And, while our survey results do map onto the GM Nation? findings, it is often only at a very coarse level of analysis. Hence such surveys are unlikely ever to be a sole substitute for properly planned and conducted dialogue and deliberative exchange. Equally, and with the full benefit of hindsight, one would not necessarily choose the GM Nation? process as a sole means of accessing public discourse on such a socially contested issue. The lesson from our analysis is that properly conducted and integrated social science surveys can provide useful complementary evidence – and in some cases critical evidence – if used as one part of large-scale public engagement exercises.

Acknowledgements

Work reported in this chapter was supported by grants from the Leverhulme Trust under the 'Understanding Risk' programme, and the Economic and Social Research Council (including its Science in Society programme). The authors wish to thank in particular the GM Nation? Debate Steering Board and the COI for their assistance in the work, and Michele Corrado and Claire O'Dell of MORI.

Notes

1 It is a matter of conjecture as to whether different responses would have been obtained had the questions read 'Food and Crops' rather than just 'Food', as we have here. We suspect the difference would have been only marginal for these types of items.
2 For an analysis of the media coverage of the debate process see Horlick-Jones *et al.* (2004).

References

AEBC (2001) *Crops on Trial*. London: Agriculture and Environment Biotechnology Commission.

Beierle, T. C. and Cayford, J. (2002) *Democracy in Practice: public participation in environmental decisions*. Washington, DC: Resources for the Future.

Campbell, S. and Townsend, E. (2003) 'Flaws undermine results of UK biotech debate', *Nature*, 425: 559.

Chess, C. and Purcell, K. (1999) 'Public participation and the environment: do we know what works?', *Environmental Science and Technology*, 33: 2685–92.

Cvetkovich, G. T. and Löfstedt, R. E. (eds) (1999) *Social Trust and the Management of Risk*. London: Earthscan.

DEFRA (2004) *The GM Dialogue: government response*. London: Department for Environment, Food and Rural Affairs.

DTI/MORI (2005) *Science in Society: findings of qualitative and quantitative research conducted for the Office of Science and Technology, Department for Trade and Industry*. London: DTI.

Finucane, M. L., Alhakami, A. S., Slovic, P. and Johnson, S. M. (2000) 'The affect heuristic in the judgement of risks and benefits', *Journal of Behavioral Decision Making*, 13(1): 1–17.

Fiorino, D. J. (1990) 'Citizen participation and environmental risk: a survey of institutional mechanisms', *Science, Technology and Human Values*, 15: 226–43.

Fischhoff, B. and Fischhoff, I. (2001) 'Public's opinions about biotechnologies', *AgBioForum*, 4(3/4): 155–62.

Gaskell, G., Durant, J., Wagner, W., Torgerson, H., Einsiedel, E., Jelsoe, E. *et al.* (1997) 'Europe ambivalent on biotechnology', *Nature*, 387: 845–7.

Glass, J. J. (1979) 'Citizen participation in planning: the relationship between objectives and techniques', *Journal of the American Planning Association*, 452: 180–9.

Grove-White, R. (2003) 'GM-debate methodology works in the real world', *Nature*, 426: 495.

Grove-White, R., McNaghten, P., Mayer, S. and Wynne, B. (1997) *Uncertain World: genetically modified organisms, food and public attitudes in Britain.* University of Lancaster: Centre for the Study of Environmental Change.

Gundry, K. G and Heberlein, T. A. (1984) 'Do public meetings represent the public?', *Journal of the American Planning Association*, 50: 175–82.

Heberlein, T. A. (1976) 'Some observations on alternative mechanisms for public involvement: the hearing, public opinion poll, the workshop and the quasi-experiment', *Natural Resources Journal*, 16: 197–212.

Horlick-Jones, T., Walls, J., Rowe, G., Pidgeon, N. F., Poortinga, W. and O'Riordan, T. (2004) 'A deliberative future? An independent evaluation of the GM Nation? public debate about the possible commercialization of transgenic crops in Britain, 2003', (Working Paper 04–02). Norwich: Centre for Environmental Risk: University of East Anglia.

—— (in press) 'On evaluating the *GM Nation?* public debate about the commercialization of transgenic crops in Britain', *New Genetics and Society.*

House of Lords Select Committee on Science and Technology (2000) *Science and Society 3rd Report, HL Paper 38.* London: HMSO.

Jackson, J. S. and Shade, W. L. (1973) 'Citizen participation, democratic representation and survey research', *Urban Affairs Quarterly*, 9 (September): 57–89.

Johnson, B. B. (1999) 'Exploring dimensionality in the origins of hazard related trust', *Journal of Risk Research*, 2(4): 325–54.

Joss, S. (1998) 'Danish consensus conferences as a model of participatory technology assessment: an impact study of consensus conferences on Danish Parliament and Danish public debate', *Science and Public Policy*, 25: 2–22.

Langford, I. H. (2002) 'An existential approach to risk perception', *Risk Analysis*, 22(1): 101–20.

Loewenstein, G. F., Weber, E. U., Hsee, C. K. and Welch, N. (2001) 'Risk as feelings', *Psychological Bulletin*, 127(2): 267–86.

McComas, K. A. (2001) 'Public meetings about local waste management: comparing participants to nonparticipants', *Environmental Management*, 27: 135–47.

Marris, C., Wynne, B., Simmons, P. and Weldon, S. (2001) *Public Perceptions of Agricultural Biotechnologies in Europe (FAIR CT98–3844).* University of Lancaster: Centre for the Study of Environmental Change.

OST/Wellcome (2000) *Science and the Public: a review of science communication and public attitudes to science in Britain.* London: Office for Science and Technology/Wellcome Trust.

PDSB (2003) 'GM Nation? the findings of the public debate (report by the Public Debate Steering Board)', London: Department of Trade and Industry. Available online at www.gmnation.org.uk (accessed 2 June 2006).

Phillips, Lord, Bridgeman, J. and Ferguson-Smith, M. (2000) *The Report of the Inquiry into BSE and Variant CJD in the UK.* London: The Stationery Office.

Pidgeon, N. F. (1998) 'Risk assessment, risk values and the social science programme: why we do need risk perception research', *Reliability Engineering and System Safety*, 59: 5–15.

Pidgeon, N. F., Hood, C., Jones, D., Turner, B. and Gibson, R. (1992) 'Risk perception', in *Risk – Analysis, Perception and Management: report of a Royal Society study group* (89–134). London: The Royal Society.

Pidgeon, N. F., Poortinga, W., Rowe, G., Horlick-Jones, T., Walls, J. and O'Riordan, T. (2005) 'Using surveys in public participation processes for risk decision-making: the case of the 2003 British GM Nation? public debate', *Risk Analysis*, 25(2): 467–80.

Poortinga, W. and Pidgeon, N. F. (2003a) *Public Perceptions of Risk, Science and Governance: main findings of a British survey on five risk cases (Technical Report)*. Norwich: Centre for Environmental Risk, University of East Anglia.

—— (2003b) 'Exploring the dimensionality of trust in risk regulation', *Risk Analysis*, 23: 961–72.

—— (2004a) 'Public perceptions of genetically modified food and crops, and the GM Nation? public debate on the commercialization of agricultural biotechnology in the UK: main findings of a British survey' (Working Paper 04–01). Centre for Environmental Risk: University of East Anglia, Norwich.

—— (2004b) 'Trust, the asymmetry principle, and the role of prior beliefs', *Risk Analysis*, 24(6): 1475–86.

—— (2005) 'Trust in risk regulation: cause or consequence of the acceptability of GM food?', *Risk Analysis*, 25: 199–209.

POST (1995) *Plant Biotechnology: a consensus* (Report 56). London: Parliamentary Office of Science and Technology.

—— (2001)*Open Channels: public dialogue in science and technology* (Report 153). London: Parliamentary Office of Science and Technology.

Renn, O. (1999) 'A model for an analytic-deliberative process in risk management', *Environmental Science and Technology*, 33: 3049–55.

Renn, O., Webler, T. and Wiedermann, P. (eds) (1995) *Fairness and Competence in Citizen Participation: evaluating models for environmental discourse*. Dordrecht: Kluwer.

Rowe, G. and Frewer, L. J. (2000) 'Public participation methods: a framework for evaluation', *Science Technology and Human Values*, 25(1): 3–29.

Rowe, G., Horlick-Jones, T., Walls, J. and Pidgeon, N. F. (2005) 'Difficulties in evaluating public engagement initiatives: reflections on an evaluation of the UK GM Nation? public debate about transgenic crops', *Public Understanding of Science*, 14: 331–52.

Rowe, G., Poortinga, W. and Pidgeon, N. F. (2006) 'A comparison of responses to internet and postal surveys in a public engagement context', *Science Communication*, 27: 352–75.

Royal Society (1985) *The Public Understanding of Science*. London: The Royal Society.

Slovic, P., Finucane, M. L., Peters, E. and MacGregor, D. G. (2004) 'Risk as analysis and risk as feelings: some thoughts about affect, reason, risk and rationality', *Risk Analysis*, 24: 311–22.

Stern, P. C. and Fineberg, H. C. (1996) *Understanding Risk: informing decisions in a democratic society*. Washington, DC: US National Research Council.

Stirling, A. (2005) 'Opening up or closing down? Analysis, participation and power in the social appraisal of technology', in M. Leach, I. Scoones and B. Wynne (eds) *Science and Citizens: globalization and the challenge of engagement*. London: Zed Press.

Toke, D. (2005) *The Politics of GM Food*. London: Routledge.

UKCEED (1999) *UK National Consensus Conference on Radioactive Waste (Final Report)*. Peterborough: UK Centre for Economic and Environmental Development.

Walls, J., Pidgeon, N. F., Weyman, A. and Horlick-Jones, T. (2004) 'Critical trust: understanding lay perceptions of health and safety risk regulation', *Health, Risk and Society*, 6(2): 133–50.

Wynne, B. (1992) 'Risk and social learning: reification to engagement', in S. Krimsky and D. Golding (eds) *Social Theories of Risk*, Westport, CT: Praeger.

3 The UK stem cell bank
Creating safe stem cell lines and public support?

Loes Kater

Introduction

The last few decades have shown a growth in collections of human biological material. These large collections of samples, stored in biobanks, play an important role in biomedical research (Lewis 2004; Holm and Bennett 2001). Research from biobanks is thought to lead to new treatments for disease. Recently a new sort of biobank has been set up: the UK Stem Cell Bank, the first of its kind in the world. Stem cells, especially those derived from embryos, are able to differentiate into a wide range of somatic cell types. It is because of their pluripotent character that stem cells are expected to influence modern medicine profoundly by introducing cell replacement therapies and immunological compatible replacement tissues. Various diseases could benefit from cell-based therapies; Parkinson's disease, diabetes, heart failure and spinal cord lesions, amongst others (Wert and Mummery 2003). The UK Stem Cell Bank (SCB) was set up to ensure the quality of human stem cell lines used in research and therapy. However, at this stage it is still uncertain how stem cells could be successfully applied in tissue engineering and cell therapies. There are still several techno-scientific and social issues to be resolved. Therefore the governance of the SCB is a manifold challenge.

One of the challenges for the management of the bank was how to involve the public. Widespread public concern about biomedical developments was raised after the cloning of Dolly the sheep in 1998. This initiated a public consultation on cloning (Wellcome Trust 1998) in which participants expressed concerns relating to biomedical subjects such as the use of embryos for scientific experimentation. Public anxieties around cloning developments were translated into an emerging anti-science climate in the UK (Parry 2003: 148). The public consultation on cloning indicated that stem cell technology was being introduced at a time when science–society relations were under strain. In response to this evolving climate there has been a call for increased public participation in the governance of science and technology. The notion of governance refers to the idea that governing is a multi-actor process instead of a top-down activity. However, the precise role of the public in the new governance of science and technology is still the subject of much discussion (Kirejczyk 2003).

In examining the role of the public in the organisation of the UK Stem Cell Bank, this chapter contributes to the growing interest in the governance of life-sciences and public involvement (Jasanoff 2004; Jones and Salter 2003; Frewer and Salter 2003; Fuller 2000). In this analysis I will draw upon insights developed within the field of science and technology studies (Latour 1987; Callon *et al.* 1986). According to Latour science is politics by other means: doing science is doing politics. Within science and technology studies actor-network theory has been successfully used to analyse the development of scientific practices. According to this theory new scientific practices develop by translating current connections between objects and humans. When new technologies are introduced, links are redefined, and this is called 'translation'. This can be briefly illustrated by Callon's study on the scallops of St Brieuc Bay (Callon 1986). In the 1970s the production of scallops in St Brieuc Bay declined because of several factors (marine predators, hard winters, over-consumption). Three marine biologists developed a conservation strategy for the scallop population. Callon's analysis identified four moments of translation in the attempts of the researchers to impose their definition of the situation on others. First (problematisation): the researchers sought to become necessary to other actors in the scene by defining the nature of the problems facing them and then suggesting that these problems would be resolved if the actors negotiated the 'obligatory passage point' of the researchers' programme of investigation. Second (*interessement*): a series of processes by which the researchers sought to lock the other actors into the roles that had been proposed for them within that programme. Third (enrolment): a set of strategies in which the researchers sought to define and interrelate the various roles they had allocated to others. Fourth (mobilisation): a set of methods used by the researchers to ensure that supposed spokesmen for various relevant collectivities were properly able to represent those collectivities.

The goal of stem cell therapy is to repair a damaged tissue that cannot heal itself in order to cure diseases like Alzheimer's and Parkinson's disease. Scientists believe that stem cell therapies could become a clinical reality within five to ten years from the time of writing this chapter, but a huge research effort, and a great deal of what Latour would call 'politics and negotiation', is still needed to achieve this goal. The concept of translation can be applied in the analysis of the development of stem cell research and stem cell therapy, particularly regarding the enrolment of important actors. For example: a biologist concludes in a scientific article, based on a list of several tests, that embryonic stem cells are more flexible than adult stem cells. This conclusion may be used in a government report to argue in favour of embryonic stem cell research. In turn this would imply a need for more embryos, and therefore donors of spare embryos would have to be sought. This process is called a 'translation', as there has been a shift from a scientific article to an advisory report to a call for donors. A condition for successful translations is an infrastructure of humans, institutes and materials (cell cultures, computers, reports, tests, labs), in short a heterogeneous network. The United Kingdom aims to be the world leader in stem cell research and therapy, and the British initiative to set up the first European stem cell bank is an important link in the regulatory infrastructure to make stem cell therapy possible in the future. The main focus in this chapter is on the efforts to enrol important actors in this network, specifically the public.

Key figures involved with the setting up and running of the SCB were interviewed, and this information was supported by an analysis of the SCB's policy documents. The central conclusion of this study is that the public has to be enrolled in the network of the bank, just like other actors such as the stem cell community and regulatory authorities. These alliances are essential for the success of the SCB. The SCB should embody trust and safety in at least two ways. First, it should ensure the development of safe stem cell products by re-culturing and testing stem cells for use in research and therapy. Second, the success of stem cell therapy not only relies on the production of safe stem cell lines, but depends on public support for, and trust in, stem cell therapy, and in the ethically sound production of stem cell lines. These issues receive as much attention as techno-scientific issues. I argue that the public is enrolled in two different ways: the general public as a rhetorical actor, and a specific public as a consulted actor. I will argue in favour of more small-scale interactions with a specific public as a new form of governance.

'Quality control is really important ... in making it happen'

The SCB was set up with the full support of the UK government. It is a top-down initiative led by government authorities, primarily the Medical Research Council (MRC). According to the government's 2002 spending review (SR 2002) stem cell research was the second largest investment made by the MRC, with a budget of £26 million. The National Institute for Biological Standards and Control (NIBSC) was chosen to host the SCB. The UK SCB is thus located in South Mimms in Hertfordshire, a few miles north of London. For many years NIBSC had been involved with the production of new medicines, particularly biological medicines. The combination of the cell experience plus its background in ensuring the quality of human medicines was the main reason that the NIBSC was chosen to host the SCB.

> There are three main words or descriptors if you like for what we do at the institute here and they are: quality, safety and standardisation. And anything we do in the stem cell bank will reflect those three key elements.
> (Interview with Glyn Stacey, Stem Cell Bank director, 19 January 2004)

According to the director of NIBSC, safeguarding the quality of the process in the present is crucial for the future development of stem cell therapy.

> [T]he second point which is very important is that aside from the research that the bank will support there is also an intention to bank cells that ultimately could be used for human treatment, the clinical grade cells. In order for that to happen you need to be able to convince the regulatory authorities, the patients and everyone involved in the chain that what you are doing is of good quality. And the way to be able to do that is to trace back the materials that you have used with information that says that they have been handled properly all the way through. That they have not been exposed to nasty viruses and they haven't had medium constituents that might have mad cow disease and all sorts of things like that. By thinking about that from the very beginning of the process when you get down the line maybe ten or fifteen years and people start to want to use cells that are there for clinical therapy. The trail is laid, everything is in place. You do not have to go back to square one. So the quality control is really important to push the field forward in making it happen.
>
> (Interview with Stephen Inglis, NIBSC director, 19 January 2004)

The focus on the elements of quality and safety is inextricably bound up with the recent crisis of public trust in science. Safety and quality have always been important, but what is new is how these issues are communicated to the public and the role the public has in considering these issues. The relationship between science and the public, more specifically public trust in science, has a high priority on the UK government's agenda. There has been a steady decline in the public's trust in science in Britain due to events surrounding GM crops, BSE, gene therapy, and organ transplantation practices such as the Alder Hey incident. Creating public trust is now considered as a key measure of political success or failure in developing science (Jones and Salter 2003). Events like GM crops and BSE play an important role as 'horror' stories and examples of *what* should be avoided and *how* it can be avoided. The early development of gene therapy and how it entered the public consciousness is a prime example.

Gene therapy is about injecting genes or gene products into body tissues in the hope of correcting abnormal genes. The first clinical gene transfer experiment was launched in 1990, and since then a few hundred clinical trials have been launched. However, the majority of these trials have not been successful. Moreover, in 1999 a young patient died in the United States as a direct consequence of gene treatment. Although this occurred in the United States, the impact of this incident on the development of gene therapy was worldwide because of the response of the public. This incident undermined the idea of gene therapy and made it difficult to get research back on track again. Policymakers and others involved are anxious to avoid a similar scenario in stem cell therapy research. The gene therapy incident made policymakers realise that the speed of developments in stem cell research is something they must pay careful attention to.

> The other thing that we have to avoid at all costs is that gene therapy has been damaged considerably by the fact that clinical trials in America and France caused the death of young patients. You have to be a bit careful if you say that because the one trial actually cured eleven children before one died. I mean it is not fair to say that it was either a failure or irresponsible. But the public reacted very badly, or the media reacted very badly, to the ones that went wrong. So one of the things that we have got to do is to balance the fact that government, and the public and in America the charities in particularly, are saying we want results quickly. We want children to be cured. But at the same time an enormous amount of damage could be done if clinical trials are undertaken without them being as safe as we can make them.
>
> (Alf Game, BBSRC, 21 January 2004)

The public response to the gene therapy incident was mild at first, but public perception took an abrupt turn after the Federal Drug Authority (FDA) cited the researchers in December 1999 for multiple protocol violations in the trial. In addition the National Institutes of Health (NIH) found out that only 6 per cent of all serious adverse effect observed in patients during clinical gene transfer studies were reported to the NIH. A further blow to the public's confidence in clinical gene transfer studies was the revelation that some investigators had significant financial interests in the outcome of the clinical research they were conducting. What was first perceived of as an ordinary risk in medical research turned out to be more than just a tragic incident. This raised public concern over genetic engineering and had serious consequences for the future development of gene therapy.

Without proper coordination, similar problems could affect stem cell research. For example, if a virus contaminated a final product the potential for a series of incidents would be quite high. To avoid similar disasters in the development of stem cell products a Steering Committee was put in place to oversee and manage the activities of the bank. One of the activities of the Steering Committee was to develop a code of practice for the bank and for the use of stem cell lines. In August 2003 a draft Code of Practice for the UK SCB was published.[1] The purpose of the Code is to assure professionals and the public that the UK SCB is operating ethically, safely and reliably. The Steering Committee wanted to ensure that the Codes of Practice were clear for all who use them, and to that end submitted the drafts for consultation. The Committee invited many organisations, centres and individuals to comment.

A second perspective within the Steering Committee is represented by the presence of two lay members. These lay members are important in communication with the broader public for two reasons; to provide information, and to spread information.

> Because we need to understand the public perception of what we are doing. But also to inform the public – the lay members will go out, because most of them belong to other consumer networks, and they disseminate the information outwards. So it needs to be explained to the public exactly what the MRC is doing.
> (Interview with Diane McLaren, MRC chief executive,
> 19 January 2004)

Creating trust is an important premise for the way decisions are made in setting up the SCB. How is trust created? Quality and safety are not merely about decisions on the sort of materials you will use but also about how you communicate these and other decisions to customers and the outside world – to the stem cell community and the public. These latter have no access to the particular labs in which the work is being done. They have no control over the actual work, and so they have to engage with the SCB project by establishing certain relationships. Trust has to be built into the image of the SCB, hence, for example, its location at NIBSC with its longstanding history and reputation in the production of biological medicines.

'You have to work hard at it and the bank has worked hard'

To ensure that the SCB will provide safe products in the future, the bank had to establish relationships with the stem cell community. Because there was no experience with stem cell research at the NIBSC, knowledge and experience had to be gained from stem cell researchers in the wider community. The bank also had to gain the trust of the stem cell community. The SCB invested in working with stem cell researchers. They went to the laboratories to meet with researchers, to discuss stem cell lines in general and various aspects of growing cell lines. Growing and maintaining stem cell lines is still a very difficult activity. Therefore it is important to have a close interaction with stem cell laboratories. But establishing a relationship of trust is equally important.

> There was a feeling in a number of labs that I went to that the bank would take cells and run away and makes lots of money. And we had to try and win those people over. And that has been a joint activity with the MRC, what they do and what we do as a bank, heading away and getting them on board and persuading them.
> (Interview with Glyn Stacey, UK SCB director, 19 January 2004)

But I think it is important to stress this is an active process and you have to work hard at it; and the bank has worked very hard and the Research Councils have worked very hard to bring the community on board.

(Interview with Diane McLaren, MRC chief executive, 19 January 2004)

In order to create a stem cell bank within the NIBSC – an institute for biological standards and control – personal contacts between the staff of the bank and the stem cell community had to be established. Contacts, knowledge and expertise are all important, but the need to control – where possible – the development of stem cell research is equally important. Within the UK the development of genetically modified food made policy-makers realise that the *direction* of stem cell research required careful consideration and control.

We are worried about a lot of things. Research on GM plants was almost destroyed because of a very foolish approach to marketing by a company, Monsanto. Monsanto did not realise that the public would not see the benefits in producing crops that were herbicide resistant. I mean they basically just did not realise that people would not see this is a good thing. So I could draw an analogy with stem cells and say that if the first applications of stem cells would appear as improved collagen in lips or so, which is not impossible actually, that is not going to go down well, is it? So one of the things that we have been very clear on is the idea that a lot of progress has to be made in areas that are going to been seen as very important. This is why the MRC is focusing on heart disease, Parkinson's disease and diabetes. Because they are all achievable targets and they are the ones where you can make a real impact quickly. It is important that the first impacts that come out are high medical things that people are going to see as having benefits to sick people and children. ... Purely the fact that this research will be seen as being worthy, in other words morally appropriate. It is all very well using embryos to cure diabetes. It is probably not a good idea [using them] to make people's lips look bigger.

(Alf Game, BBSRC, 21 January 2004)

The fear that the development of stem cell technology may head in the wrong direction is ever-present – for instance, if technology took a turn towards cosmetics before producing medical benefits. This fear is linked to a perception that the public response would be extremely negative, that such applications are not considered by the public to be a legitimate use of embryos. The use of embryos in the development of stem cell therapy is highly controversial and therefore the results or benefits of this research have to legitimise their use. Moreover, one of the arguments presented in the House of Lords in 2002 in support of setting up a stem cell bank was that an end result would be a reduction in the use of embryos.

When the SCB was set up, one of the expectations, or rather a claim, was that in the longer term fewer embryos would be destroyed as research becomes more centralised. A reduction in the use of embryos is not just a morally desirable goal, it is equally important in order to counter, to a certain extent, the moral views of pro-life movements. In the parliamentary debate on stem cell research a growing anti-research lobby focused on the ethical problems raised by embryo research. This anti-research lobby, according to the Medical Research Council, was a threat to the future of stem cell research (Parry 2003: 149). At the first conference on stem cell research, organised by the MRC in 2002, pro-life groups were demon-strating outside the conference against the use of embryos as a source in stem cell research. Over time the MRC developed a strategy to *involve* them instead of leaving them aside.

> So they [pro-life groups] are quite active, but we try to involve them in everything, we invited them to comment on the guidance documents [the previously mentioned draft codes] for example. Nothing happened, but we did ask them to. ... But that's quite good isn't it? – because we can assume they don't think there is anything wrong. ... And we also invited them to the annual conferences and they come to those and they ask a lot of questions. So again it is all-encompassing, trying to ensure that you keep people inside the tent, rather than outside.
>
> (Interview with Diane McLaren, MRC chief executive,
> 19 January 2004)

The pro-life movement is a specific public, and it was considered important to enrol the group in the network. There are other specific publics that are addressed.

In the report 'Public consultation on the Stem Cell Bank' (People Science & Policy Ltd 2003), an initiative of the Medical Research Council, another specific public is defined and addressed. The report describes a consultation forum with experts and a series of focus groups with members of the public. Views were sought on making donations and the types of structures and safeguards that people wish to see built into the management frame-work and operation of the bank. The focus groups were composed of blood donors or potential organ donors, non-donors, men and women who had successfully received IVF treatment, and couples who were undergoing IVF treatment. In the groups it was explained that the SCB would be built up by researchers donating the stem cell lines they create for their research. Other researchers would request stem cell lines donated by patients or friends and relatives of patients to derive the lines for their research. The main benefit of the SCB for participants was the potential to control access to stem cell lines and the uses to which stem cell lines were put (People Science & Policy Ltd 2003: 26). In addition, the issue of consent for donating to the SCB was discussed and several participants expressed the wish to be able to specify diseases for which their donation would be used. How this could or should be organised by the SCB was not specified.

If surplus IVF embryos remain the main source of embryonic stem cells, then women undergoing IVF treatment or couples with frozen embryos are defined as an important group of potential donors. The focus group interviews indicated that people who had received IVF treatment had quite specific views about embryos. This had implications for dealing with this particular group on donation issues in the future. This report identified a number of implications for communication with the public. In general the focus groups revealed that members of the public were not likely to be stem cell donors unless they found themselves in particular circumstances. This led to a distinction made between *potential donors* (which could include everyone) and *likely donors* (those who find themselves being asked to donate). Therefore separate communication activities were suggested for likely donors and other members of the public (People Science & Policy Ltd 2003: 4).

The recommendations for communication with likely donors illustrate the previous argument – that, in order to make the SCB a successful project, alliances have to be created with several actors: that is, with the stem cell community and the pro-life movement, but also with potential donors and the public in general. This requires ongoing hard work in establishing and building the relationships that are important for the bank's network.

Conclusions

Since the late 1990s there have been frequent calls for transparency and increased public involvement in the field of biotechnology. Public involvement, in particular the organising of public debates, is presented as the key solution to the troubled relationship between science and society. Inspired by the premise of the democratic ideal, public debates are supposed to reduce public concern over possible dangerous developments in biotechnology, and to create support for its potential benefits. However, which role the public can or should play is still the subject of discussion.

People disagree about the ethical boundaries and the public policies for governing stem cell research. In a pluralistic society different moral beliefs are a regulatory challenge to public policy. To increase public involvement, public debates are organised to make an inventory of the different opinions and to mirror shared visions in the policy. Despite the democratic basis of public debate, there are several reasons why it is an illusion to imagine that this type of forum will resolve the moral and policy issues in stem cell research. First, it is difficult enough to organise large public debates, but it is even more difficult to adequately filter shared visions and to capture the diversity of the debate. Second, even after shared visions are analysed, there will always be new issues arising as technological developments proceed. And third, the extent to which these debates are genuinely 'open' and 'public' is highly questionable.

Despite the general impression that there has been an open public dialogue, the debate on human embryonic stem cells in the US for example has been characterised as a debate among elites and experts. The debate was a consensus amongst formal bodies that are closely connected to the research federations and institutions pressing for the acceptability of stem cell research (Root Wolpe 2001: 185). In general, public debates are a struggle over definitions, and experts are usually the most successful players in making their perspective dominant in the debate. In other words, for several reasons public debates are not socially robust.

In science and technology studies, the work of scientists and policy-makers has been described as a process of the creation of networks (Latour 1987; Callon *et al.* 1986). Translations and alliances are essential to make (scientific) projects successful. At a time when science–society relationships are under strain, the public has to be convinced and moved into the developing network in order to create broad support for the development of future stem cell therapy. In the emerging network of the SCB the public is clearly an ally that has to be won over, for at least three reasons: to create public support, to deal with ethically complicated issues, and to gain information. In examining the establishment of the SCB, two roles for the public have been identified and analysed: one for the general public, and one for specific publics.

As we have seen, the general public or public perception of the development of future stem cell therapy was taken into account in the strategy of the SCB and the development of stem cell research. This was done by focusing on quality and safety. Public perception was also considered in decisions on how quickly results should be achieved. Results that were not supported by well controlled clinical trials could lead to a disastrous public response similar to that of the gene therapy incident. The public *does* want to see results, but these results have to be associated with the right applications. That is within the medical area and not the cosmetic area. Public distrust in science has created a trajectory in which 'the public' in general has a role, in issues of public concern. Thus the public is represented as a relevant actor in the development of stem cell therapy. However, public involvement is restricted to a *representation* of public expectations, concerns and beliefs. There is no genuine open dialogue with the public; the communication can be characterised as a one-way communication process. Those involved with setting up the SCB take *public perception* instead of the *public* into account when decisions are made.

In the second instance a *specific public* was consulted – for example, as with the report 'Public consultation on the Stem Cell Bank' commissioned by the MRC. In the focus groups, several groups of people such as blood donors and couples that had successfully received IVF-treatment were invited to give their views on the subject of donating embryos. The main aim of the focus groups was to find out what safeguards people wanted to see built into the management framework and operation of the bank. Though maybe not wholly a two-way communication process, this consultation procedure at least had some possibilities for deliberation and meaningful dialogue. Second, a dialogue was also sought in which pro-life groups were asked to comment on the SCB's guidance documents. The pro-life groups are known for their objections against the use of embryos for stem cell research and therapy. Although this is a strategy to keep people 'inside the tent' rather than outside, it creates room for the development of new notions. Third, lay members were enrolled on the SCB's Steering Committee in order to provide information and to spread information. This involvement is a two-way communication process, as meaningful dialogue can be set up during meetings of the Steering Committee. In the meeting there is the possibility to deliberate, to express doubts and to discuss uncertainties and new developments.

A policy framework to deal with public involvement needs to be elaborated in more detail. Based on this study of the organisation of the SCB I argue that small-scale interactions and consultations with the public in the field of stem cell research are a more effective means of public involvement. Deliberation with specific publics will give an immediate opportunity to communicate over different views, instead of merely making an inventory of them. Moreover, it will provide an opportunity to express uncertainties. Small-scale interactions and consultations are not based on rather confusing notions such as transparency and control. Further research on alternative two-way communication processes is required to contribute to the socially robust governance of stem cells.

Acknowledgements

I would like to thank colleagues at the University of Twente and colleagues involved in the NWO project for their input and helpful comments.

Notes

1 The Code of Practice for the Use of Human Stem Cell Lines was the subject of a consultation in 2004.

References

European Commission (2001) 'Enhancing democracy: a white paper on governance in the European Union'. Brussels, 25 July 2001.
Callon, M. (1986) 'Some elements of a sociology of translation: domestication of the scallops and the fishermen of St Brieuc Bay', in John Law (ed.), *Power, Action and Belief: a new sociology of knowledge*. London: Routledge and Kegan Paul.

Callon, M., Law, J. and Rip, A. (1986) *Mapping the Dynamics of Science and Technology: sociology of science in the real world*. Basingstoke: Macmillan.

Frewer, L. and Salter, B. (2003) 'The changing governance of biotechnology: the politics of public trust in the agri-food sector', *Applied Biotechnology, Food Science and Policy*, 1(4): 199–211.

Fuller, S. (2000) *The Governance of Science*. Buckingham: Open University Press.

Holm, S. and Bennett, R. (2001) 'Genetic research on tissues stored in tissue banks', *Canadian Journal of Policy Research*, 2(3): 26–37.

Jasanoff, S. (2004) 'Science and citizenship: a new synergy', *Science and Public Policy*, 31(2): 90–4.

Jones, M. and Salter, B. (2003) 'The governance of human genetics: policy discourse and constructions of public trust', *New Genetics and Society*, 22(1): 21–42.

Kirejczyk, M. (2003) 'Citizens and experts in the embryonic stem cells debates. Dutch and British policies compared', paper prepared for the ECPR Second General Conference, Marburg, Germany.

Latour, B. (1987) *Science in Action: how to follow scientists and engineers through society*. Cambridge, MA: Harvard University Press.

Lewis, G. (2004) 'Tissue collection and the pharmaceutical industry: investigating corporate biobanks', in R. Tutton and O. Corrigan, *Genetic Databases: socio-ethical issues in the collection and use of DNA*. London: Routledge.

Parry, S. (2003) 'The politics of cloning: mapping the rhetorical convergence of embryos and stem cells in Parliamentary debates', *New Genetics and Society*, 22(2): 177–200.

People Science & Policy Ltd (2003) *Public Consultation on the Stem Cell Bank*. London: Medical Research Council.

Root Wolpe, P. M. G. (2001) 'Expert bioethics as professional discourse: the case of stem cells', in S. Holland, L. Lebacqz and L. Zoloth *The Human Embryonic Stem Cell Debate. Science, Ethics and Public Policy*. Cambridge, MA: MIT Press.

Wellcome Trust (1998) *Public Perspectives on Human Cloning: a social research study*. London: The Wellcome Trust.

Wert, G. d. and Mummery, C. (2003) 'Human embryonic stem cells: research, ethics and policy', *Human Reproduction*, 18(4): 672–82.

4 Public biotechnology inquiries

From rationality to reflexivity

Tee Rogers-Hayden and Mavis Jones

Introduction

The use of public engagement instruments in the governance of complex science and technology matters is now *de rigeur* in liberal democratic nations, tempering the combined drives for scientific progress and market competitiveness with often more precautionary human values approaches. This has not always been the case. Where previously many states depended almost solely on expert opinion to inform and legitimate policy decisions, increasingly there is evidence of a reliance on public input in policy development. The reasons for this shift lie within the context of modernity. Interrogating the relationship in wealthy liberal democracies between modernity, governance and public inquiries, this chapter uses a comparative approach located in two liberal democracies: Canada and New Zealand (also known by the Māori name 'Aotearoa'). The Canadian Royal Commission on New Reproductive Technologies (RCNRT) and Aotearoa New Zealand's Royal Commission on Genetic Modification (RCGM) operated in different national contexts, in different policy arenas; through comparison of their processes and outcomes, patterns emerge which offer insights towards a more reflexive design for public engagement mechanisms.

Modernity and governance

To understand the expression of modernity in current political practices such as the ones we describe here, it is useful to consider first how the evolution of modernity has been theorised. Modernity is an important influence on liberal democracies described by both Ulrich Beck and Anthony Giddens (1990) as a transitional period. Beck (1992a) conceptualises three periods of societal transition: modernity, second modernity, and reflexive modernisation (or risk society). Modernity refers to industrialisation, or more precisely several interrelated historical processes, events, ideas and periods that reinforced each other – the Renaissance, the Reformation, the Enlightenment, and the democratic, industrial and scientific revolutions (Oelschlaeger 1991). A key element of modernity was the transformation of structures of knowledge. Reliance on tradition (religion) and organicism as the foundation of knowledge was replaced by a new dependence on Rationality, 'objectivity', and an 'atomistic' portrayal of reality. The modern reductionist scientific method that developed is based on dualisms: 'objective' instead of 'subjective', 'facts' not 'values', 'Rational' not 'Romantic'. Romanticism was a historical movement characterised by a reaction against the new faith in Progress through Rationality and modern Science. The operating principle of Romanticism was self-reflection. To this end, Romanticism's practitioners, such as the poets Coleridge and Wordsworth, communicated in an indirect, imperfect, dialectical way, in which their works were not presented as the 'final word' but attempted to engage the readers in their creation (Jones *et al.* 1993). The role of language in upholding and challenging dominant knowledge systems, as articulated in the Romantic response to modernity, is central to this chapter.

Second modernity, the phase in which, according to Beck, industrialised societies currently operate, forms the transitional stage between industrial society and reflexive modernity. In second modernity society becomes aware that the industrialisation created through the ideologies and practices of modernity has brought with it new hazards and, despite this awareness, still attempts to manage these hazards with modernist processes (Beck 1992a, 1995). Environmental and value-based controversies, such as that over biotechnologies, are seen by Beck as part of a societal progression from industrialisation to a period of proactive reflection; they critique modernity by drawing attention to the costs of industrial progress and technological development (Beck 1992a). For Beck, second modernity describes societies in which modernist, Rationalist approaches (such as those informing science and law) begin to be seen as incapable of managing new mega-hazards; and therefore, incapable of achieving progress (Beck 1992b).

The last stage is an alternative modernity – an 'ideal type' of future – in which society comprehensively and proactively reflects upon and addresses the processes, ideals and failings of modernity. Reflexive modernity, Beck (1999) argues, follows second modernity with the developments forming second modernity as a necessary, rather than an optional, part of the process that society will go through before becoming self-confronting. Environmental problems will then not be seen as external but will be considered as central to institutions.

Likewise, Anthony Giddens (1990) depicts a shifting political paradigm: from a modern Rationalist philosophy of government to the conditions of late modernity – characterised by complex relationships between citizens, the state and experts who both define issues and give policy advice. He locates this shift within societies exhibiting features of capitalism and industrialisation. As Giddens asserts, the competitive nature of capitalism supports a culture of constant technological innovation and unfettered discovery that is allowed to thrive without significant interference from political institutions. This has resulted in societies where the scope and pace of change are expanded drastically in comparison with those in historical record.

The effect of the shift from modernity to late modernity is observable in the tensions of governance in liberal democracies. Simply put, legitimacy in a liberal democracy requires at least the appearance of doing what the citizens want it to do. If the state *is* responding to its constituents' interests but does not appear to be (due to closed processes, for example) then it may not enjoy legitimacy; conversely, if the state *is not* responding to public interests but appears to be doing so, it may still be seen as legitimate. In conditions of modernity, the complexity of the social order requires that the certainty an individual has about a policy issue is based on trust in authorities instead of on that individual's knowledge or familiarity with the issue. Therefore, the situation is tenable only in so far as the institutions in question are able to command public trust.

Taken together, the work of Beck and of Giddens describes conditions suffered by liberal democracies in approaching the governance of controversial technologies. Political pressures are emerging which challenge the conventional approach to policy development based primarily on expert models. No longer able to achieve policy based on (Rational) expert authority, such states must find a means to incorporate value-based (Romantic) inputs in order both to progress and to retain political control. Increasingly the institutions of the state are turning to new, or updated, measures to demonstrate their commitment to listen to the concerns of their citizenry and, implicitly, a promise to act on these concerns. One such approach is the *commission of inquiry*, a model with a long history which, due to its loosely defined structure, continues to be popular as demands for public involvement in policy decisions escalate in second modernity.

Public engagement: commissions of inquiry

The commission of inquiry is the highest form of inquiry, providing non-legally binding advice to government. These commissions are popular for several reasons, including tradition and flexibility; but more significantly, the resources available to them (as official state consultation exercises) allow them to maintain a high profile, disseminate their findings widely, and have at least the appearance of rigour through the use of multiple methodologies.

Despite the popularity of official inquiries they have repeatedly been criticised for their role in actually creating obstacles to public participation. A key criticism is that they perform merely a rhetorical function, that of providing legally non-binding advice. Thus, while giving the public perception of review and control, the real effect is to postpone governmental action (Burton and Carlen 1979). Similarly, it is suggested that inquiries are carried out to subdue powerful pressure groups, when a possibility exists that those groups threaten the continuation of state-condoned activities (Doyle and McEachern 2001). Further, official inquiries may be employed as a way of addressing a complex multifaceted issue when a bureaucracy's resources are stretched, or to maximise opportunities for a specific and usually elite group of stakeholders when they are under a controversial cloud (Ashforth 1990). Part of maximising those opportunities is to structure inquiries in ways that create hurdles to broad participation by constraining input and objections. Thus, many criticisms focus on that construction of dominant power/knowledge systems to structure inquiries (see Rogers-Hayden and Hindmarsh 2002; Walls *et al.* 2005).

The effect of this is that, despite apparent attempts in contemporary liberal democracies to access human values approaches towards producing legitimate policy outcomes, these engagement exercises are constructed within modernist governance worldviews that limit the ability to produce transformational outcomes, and essentially reproduce these worldviews instead of liberating them.

In Commonwealth countries, the commission of inquiry often takes the form of a Royal Commission. We explore our thesis via two cases. The first is the Canadian Royal Commission on New Reproductive Technologies (RCNRT) (1989–93) which more than a decade after its report has seen a policy outcome – the Assisted Human Reproduction Act (2004). The second is Aotearoa New Zealand's Royal Commission on Genetic Modification (RCGM) (2000–1), for which the policy outcome is the lifting of the moratorium on the field trails and releases of GM crops, and the introduction of the New Organisms and Other Matters Bill 2003 which makes provision for the 'conditional release'[1] of GMOs. Within these cases we critique such features as the disciplinary orientation of the Commissioners; their methods of public engagement; the presence and role of hegemonic and counter-hegemonic discourses; and the translation of the process to the policy outcome.

The Canadian Royal Commission on New Reproductive Technologies

Critiques

Designing and conducting robust public consultation exercises is no easy task; different national and cultural contexts present different problems. Canada's small population is spread across one of the largest geographies in the world. Though still part of the British Commonwealth, it is a nation settled by both English and French, attempting to balance the tensions between its two national languages (and its significant diasporic immigrant populations speaking languages other than English or French), and of course coming to post-colonial terms with its indigenous First Nations population. As a result, consultation efforts must address linguistic and regional differences, not to mention travel logistics. These challenges combine to form a uniquely problematic landscape for the effective conduct of a commission of inquiry. Nevertheless, a robust template exists in living memory for a Canadian Royal Commission which has met with broad public approval: the internationally recognised Mackenzie Valley Pipeline Inquiry (or the Berger Inquiry).[2] Very few commissions, the RCNRT included, have been able to reproduce the success of Berger. In the case of the RCNRT, it faced criticism regarding the impact of its internal difficulties, membership, flaws in its process of inquiry, and the deliberately vague caution that permeated its recommendations (Anonymous 1993; Eichler 1993; Massey 1993; Vandelac 1993). These all compromised its ability to be regarded as a clear success. These critiques will be discussed in order.

The Royal Commission on New Reproductive Technologies (or the Baird Commission) was appointed after extended lobbying by the Canadian Coalition for a Royal Commission on New Reproductive Technologies, a collection of women's groups, health advocates and academics (Eichler 1993). The Commission's mandate was to 'examine current and potential scientific and medical developments related to reproductive technologies, but also to go beyond them to consider: the impact of the technologies on society as a whole; their impact on identified groups in society, specifically women, children, and families; and the ethical, legal, social, economic, and health implications of these technologies' (RCNRT 1993: 2). From the government's perspective, the creation of the RCNRT was strategically timed in the midst of a parliamentary debate on abortion and a federal ban on funding research using foetal tissue. The RCNRT's deputy director of research and evaluation states that there was 'a perceived need to identify a women's issue that would help to deflect the dissent around the issue of abortion' (Hatcher Roberts 1999: 16).

Membership

The original Commissioners were Dr Patricia Baird (Chair), a paediatrician and geneticist; Dr Bruce Hatfield, a private practitioner in internal medicine; Martin Hébert, a lawyer; Dr Grace Marion Jantzen, a lecturer in Religion; Maureen McTeer, a lawyer/politician; Suzanne Rozell Scorsone, from a Catholic Archdiocese; and Dr Louise Vandelac, a sociology professor. The Commission had been struck after a prolonged lobbying effort by the Coalition for a Royal Commission on New Reproductive Technologies (of which McTeer and Vandelac were a part). The Coalition made recommendations for appointments, some of which were accepted by the Prime Minister's Office; others, most notably the Chair, were rejected in the interests of 'balance' (Hatcher Roberts 1999).

It is common for governments to seek expert opinions in policy development. So, why should the commissioners' professions be a contentious issue? As is typical in appointed expert committees, RCNRT membership was dominated by individuals whose professional discipline falls within the modernist purview (for example, scientists and lawyers). Although this does not necessarily mean that these members espoused or identified with modernist viewpoints, professional training in such fields is designed to limit problem definition to the Rational. Disciplinary orientation, then, could dictate limits to the ability to imagine solutions sympathetic to marginalised alternatives (for example, social or holistic solutions). Similarly, some feminist critiques have focused on how the problematic medical model of women's reproductive health is reflected in policy. For example, though the Commission's final report did call for public education campaigns on issues such as infertility risks associated with STDs, it featured no similar call regarding risks associated with IVF (including death, via ovarian hyperstimulation syndrome). The implicit assumption – that the preferred alternative to involuntary childlessness is the medical treatment of infertility with its accompanying risks, rather than acceptance of involuntary childlessness – reflects and reproduces the priorities of the modernist worldview which characterises biomedical approaches to human reproduction.

However, as suggested above, the disciplinary orientation did not necessarily mean that all Commissioners were firmly embedded in a modernist policy paradigm. In 1991 four Commissioners (Hatfield, Hébert, McTeer and Vandelac) publicly expressed unhappiness with the exclusionary design of the public participation programme and the RCNRT's internal politics, citing 'an undemocratic internal structure' (Eichler 1993: 197). They attempted to obstruct the quorum but the government responded by appointing two more Commissioners (Bartha Maria Knoppers, a Montreal law professor, and Susan McCutcheon, a lay member). The situation culminated in a lawsuit filed by the dissenting Commissioners, who were eventually fired. In the aftermath the Coalition called for the RCNRT to be disbanded, but to no avail. The Commission lived out its life short of four members but secure in having delivered its remit. Nevertheless, the cumulative effect of the lawsuit and sackings was ironic: the very body charged with the task of de-politicising reproductive technologies succeeded in, if anything, rendering them more highly politicised than before.

Public engagement

The RCNRT held eighteen public meetings: one in each of the ten provincial and two territorial capitals; one each in large urban centres with significant academic and health research contingents (Montreal, Vancouver, Calgary, London and Toronto); and an extra hearing in Toronto, the largest city. It also held over fifty colloquia and conferences on specific issues, such as surrogacy, IVF, professional training, and ethical and moral aspects.[3] Massey (1993, 1994) conducted an analysis of the RCNRT's public participation process. She found a variety of procedural flaws, including poor design of public hearings (which were held in large, urban hotel ballrooms, with no provisions for travel allowance or child care), and an intimidating and dismissive hearing atmosphere. Most important of these was the invisible public information campaign – resulting in poor public awareness that the Commission even existed. In fact, a further irony of the Commission's 'troubles' lay in the excess of media attention devoted to the legal battle – for it was at this point that many Canadians first became aware that there *was* a Royal Commission on New Reproductive Technologies. In addition, the RCNRT's public engagement efforts appeared to be characterised by a dominantly neo-liberal approach, proving especially problematic considering the well-known politicisation of these technologies regarding effects on vulnerable groups (the disabled, minorities, the poor). An RCNRT internal memo pointed to the lack of representation from a variety of communities, including francophones, ethnocultural groups, religious groups and aboriginals, suggesting that lack of input from these groups was 'because they had nothing to say on our mandate or because they did not understand the issues or our process' (Massey 1993: 248). The RCNRT noted that some volunteer groups, who participated and complained about the process, would have other opportunities to participate; however, the way in which these 'other opportunities' eventually materialised did not allow for this. The RCNRT held a second set of hearings, known as 'selected stakeholder' consultations, to which they invited participants judged to 'carry significant weight in the public policy realm' (Massey 1994). These were primarily pharmaceutical firms and national professional organisations – not the volunteer groups who had been promised further input. Criticism of the RCNRT by a former Commissioner suggests that shortcomings in the public engagement programme might be attributable to the lack of a clear research agenda for the first several months of the Commission's tenure (Vandelac 1993).

Discourse

A thread running through the RCNRT's report and subsequent events (including a voluntary moratorium and the policy outcome, described below) was reference to regard for some variant of 'Canadian values' (or ethos, or ethics); a vague but undisputed determination which carries forward an interpretation of survey and consultation findings in the Baird Commission's report. This reference to *values* is primarily associated with the feminism-informed discourse permeating the lobbying efforts of the Coalition. It could be characterised as the strongest *counter-hegemonic* discourse in the sense that, in order to spur the creation of the Commission, it positioned itself in opposition to the government's lack of activity, to that point, in the regulation of assisted human reproduction. In the political climate at the time of the Commission's report and hearings, however, certain feminist discourses could no longer be realistically characterised as counter-hegemonic. The policy culture was already enacting other measures to demonstrate a commitment to many initiatives associated with feminist policy discourses: for example, its embrace of gender-neutral language, and measures in the public sector to bring gendered professions in line in terms of equal pay for work of equal value. It was in the state's interest to co-opt feminist ideology and discourses; feminist critiques of dominant ideologies had in fact gained a sufficient foothold in political consciousness that 'feminism' had virtually become black-boxed and subsequently mobilised with the power to legitimate policy actions.

But what of the discourses which 'had nothing to say' on the RCNRT mandate? First Nations attempts to engage with the Commission proved to be somewhat fruitless, as values and worldviews structured the mode of engagement as well as its context (Massey 1994). Perhaps even more dangerous, in terms of the success of the policy outcome, is the policy process's failure to sufficiently account for the francophone response to the RCNRT and the AHR Act in particular. This marginalisation has had political consequences, as the next section will discuss.

Policy outcome

Two years after the publication of the Commission's final report, *Proceed with Care,* the then Minister of Health Diane Marleau announced a voluntary moratorium on several reproductive technologies deemed contrary to the Canadian ethos. The next year, 1996, the moratorium evolved into Bill C-47, The Human Reproductive and Genetic Technologies Act (HRGTA). The Bill lived many lives through several health ministers until it passed Parliament as Bill C-13 on 28 October 2003. Its final incarnation was Bill C6, An Act Respecting Assisted Human Reproduction and Related Research (or the AHR Act), the guise under which it received Royal Assent on 31 March 2004.

Technologies banned under the original Act included germ-line genetic alteration, human embryo cloning, creation of animal–human hybrids, retrieval of sperm or eggs from cadavers or foetuses for reproduction, ectogenesis, sex selection, transfer of embryos between humans and other species, research on human embryos later than fourteen days after conception, and creation of embryos for research purposes only. As the policy developed, definitions were fine-tuned for prohibited acts, and controlled practices (including IVF and gamete storage) designated to be monitored by an Assisted Human Reproduction Agency of Canada (AHRAC). AHRAC has positioned itself to be formed in the image of the UK Human Fertilisation and Embryology Authority (HFEA). Established in 1990 via the Human Fertilisation and Embryology Act, the HFEA has been an internationally influential model for reproductive regulation. The AHR Act, however, has now fashioned itself as 'one of the world's most comprehensive legislative initiatives in the area of AHR' (Health Canada

2006). Interestingly, the authors of the Canadian framework also intend to establish, concurrent with AHRAC, an instrument similar to the UK Human Genetics Commission (Nisker 2004, personal communication) – a body which has lent considerable legitimacy to the UK biotechnology governance framework (Jones and Salter 2003; Salter and Jones 2006).

It is worth noting that the committee which developed the regulatory framework was comprised of members with a selection of worldviews distinctly different from earlier approaches. At least four of the members were ethicists who were primarily interested in feminist/social justice perspectives. Among the others – including those belonging to conventionally modernist professions – almost all had expressed associations with Romantic worldviews.[4] The minds behind the AHR Act have had the luxury of the last decade, while developing a framework based on their interpretations of the Canadian context, to observe and learn from the politicisation of technological issues elsewhere. They have thus been able to anticipate some of the governance issues for which their framework must account. In its current incarnation, the Bill's language is a strategic mix of the overcautiously precise (in terms of defining the practices it regulates) and the deliberately vague (in terms of how the framework will be executed). However, it has been successful in responding to the discursive construction of certain practices of commodification, such as surrogacy and the sale of gametes, as holding the potential for coercion based on relationships of authority or socioeconomic class (Nisker 1997).

In the end, this has translated into a policy outcome which reflects a concern with discrimination. It insistently rejects practices which could result in the commodification of reproductive tissue and processes, in line with arguments lodged by prominent Canadian feminist philosophers (some of whom were involved in the advisory committee which hammered out the terms of the framework) that such practices are coercive and devaluing. In many ways the new policy framework represents a departure from the neo-liberal, industry-influenced approach which has characterised technology policy in other policy arenas in Canada. However, some elements of the philosophy/worldviews which the RCNRT's critics would have preferred to be acknowledged in the policy outcome – regarding a Canadian society increasingly comprised of non-traditional families, the rising popularity of holistic health models, and a turning away from the biomedical model – were sidelined in favour of a policy discourse which leaves unquestioned some aspects of the biomedical model, and the assumption that the family is defined by biological relations over social ones.

Interestingly, the tendency during the Commission's progress to elide some of the complexities of Canadian popular opinion may in fact return to haunt it. As noted above, the francophone response to the AHR Act has not been overwhelmingly positive.

In fact, two specific actions were taken in the province of Québec – where the majority of Canadian francophones live– to regain control over the political disposition of reproductive technologies. The first was the action taken by the province to contest parts of the federal law on the basis of provincial jurisdiction. The second was a *projet du loi* to govern assistedreproduction in Québec. The outcomes of these are still pending, as a change in government has delayed policy movement, possibly until 2007. These actions represent the concern of a significant proportion of the francophone population with the treatment of the family and the embryo in the federal regulation. More importantly, this concern is embedded in a longstanding battle between representations of francophone popular opinion against the federal government which, they often charge, has failed to politically account for the distinct nature of francophone culture within an essentially anglophone dominance. As with many aspects of the Canadian reproductive technologies policy framework, it is likely that these events will have effects in the tenure of the AHR Act – for example, when the AHRAC comes up for parliamentary review three years after its establishment.

Concluding remarks

Considering the RCNRT's troubled existence – internal problems, a lawsuit, calls for dissolution by the group that called for its creation – it is of little surprise that the path to policy should be far from smooth. None the less, the final report was not rejected by its critics. It did, after all, condemn the commercialisation of human tissues and processes – a major concern of Coalition members. There was pessimism at the report's release that its positive potential would end up buried in bureaucracy and never reach a policy outcome.

Eventually, after years of recognising the pressure to regulate, the AHR Act provided closure on the process begun with the Royal Commission. Regardless of the critiques discussed here, a policy outcome influenced by the RCNRT's final report should not be characterised as a bad move. Although flawed, *Proceed With Care* did manage to distil an image of Canadian public opinion on the issues which was, overall, acceptable to the majority of those interested (in particular, the NAC). And it did reflect some perspectives which could be characterised as 'radical' – and have been carried through to the policy outcome. Its stand against commercialisation, and its rejection of the conventional definition of the family espoused by its more conservative critics, are both evident in the AHR Act. If this seems relatively radical, it is not necessarily transformative in Canadian political culture, which periodically seeks to distance itself from US regulatory approaches (or, in this case, the lack thereof) and re-affirm its commitment to socialist values. A truly radical view might have acknowledged alternatives to medicalised reproduction, including other definitions of family or fulfilment. For example, the RCNRT supported the provision of assisted reproduction for single women and same-sex couples – perhaps a non-traditional approach, but one that supports the goals of the medical–industrial establishment; yet there was little acknowledgement of the possibility of a fulfilled life without children or as the caretaker for non-biologically-related children.

If anything, the Canadian process to design a framework for assisted human reproduction is illustrative of the social and ethical complexity of biotechnology governance in late modernity. From a flawed consultative process driven by modernist-defined expertise, to a government advisory committee dominated by precautionary ethical approaches, an outcome has been achieved which makes concessions to the latter without rejecting the former. Acknowledging the Romantic while operating within the Rational appears, after all, to be the only solution currently available to those hoping to revolutionise governance in neo-liberal democracies. Interestingly, the provisions of the AHR Act have been implemented but the composition of AHRAC (originally scheduled to begin its tenure in Vancouver in early 2006) is yet to be announced; again, a change in government has delayed policy progress on this front. The Act states that membership 'must reflect a range of backgrounds and disciplines relevant to the Agency's objectives'. This reflects back to the Coalition's response to the RCNRT's report. At the centre of the Coalition was the powerful National Action Committee for the Status of Women (NAC) which strategically maintained a media presence throughout the Commission's tenure. NAC spokesperson Gwynne Basen was quoted on the national news regarding the licensing body proposed by the Commission's report: 'it would be very important that it not be dominated by professionals' (CBC Archives 2004).

The New Zealand Royal Commission on Genetic Modification

Critiques

Unlike the case of the RCNRT there is a dearth of published critiques of the RCGM. It is, however, noteworthy that during the RCGM many of the same environmentalists who had called for the Royal Commission started to question if they had done the right thing as the Commission's processes began unfolding. There were numerous issues, including the appointed Commissioners and the processes surrounding participating interest groups of 'interested persons'.

The Commission consisted of a chair and three other Commissioners. Similar to the RCNRT, the Commissioners were all professionals. The Commissioners appointed to the RCGM were the Right Honourable Sir Thomas Eichelbaum, Chief Justice of New Zealand from 1989 to 1999; Dr Jean Fleming, a molecular reproduction and endocrinology researcher; Dr Jacqueline Allen, a general practitioner in south Auckland with community and Māori health expertise; and the Right Reverend Richard Randerson, Bishop of the Anglican Church.

Membership

The extent of discontent felt by some about the Commissioners appointed for the RCGM was indicated when the umbrella group for many environmental and conservation groups throughout the country – the Environment Conservation Organisation (ECO) – initially contemplated boycotting the RCGM due to its perceived bias (interview with Berylla of the Environment Conservation Organisation, 2001). A number of groups highlighted the lack of environmental or ecological focuses in the backgrounds of the Commissioners. Rogers-Hayden and Hindmarsh (2002) analysed the composition of the Commission, concluding that the backgrounds of the Commissioners reflected modernist-embedded fields. They commented that three of the four commissioners had either legal or scientific credentials, while the fourth was from the Anglican Church – an organisation that could be identified as possessing an overly anthropocentric (rather than ecologically focused) position. Thus, although the credentials of the Commissioners did not necessarily determine their positions on the RCGM, they were from backgrounds that generally relied upon Enlightenment worldviews. This meant that any environmental inclinations the Commissioners held would arguably tend towards reaffirmation of modernist ideals, being reformist (and shallow green) rather than recognising the interconnectedness of ecosystems (deep green) (see Dryzek 1997). In this sense the Commission could be seen as biased in its interpretive framework to view and judge genetic modification (Rogers-Hayden and Hindmarsh 2002).

Public engagement

Large-scale public participation did occur in the RCGM. Approximately 11,000 people from the public out of a population of just under four million, at that time, made submissions. As previously noted (Rogers-Hayden 2004; Rogers-Hayden and Hindmarsh 2002), approximately 10,000 out of those 11,000 submissions were against or tending to be against GM. but this pubic opposition to GM was dismissed by the Commission who stated that it was not a referendum. These were all written submissions, and in order for people to present oral submission and participate in the cross-examination process they had to gain 'interested persons' (IP) status. Rogers-Hayden and Hindmarsh (2002) have elsewhere detailed some of the issues confronting the public attempting to gain IP status.

Discourse

Some of the influences of the modernist worldview on the discourse of the RCGM can be seen as affecting both the discourses they entertained and that which they reproduced (see Rogers-Hayden 2004; Rogers-Hayden and

Hindmarsh 2002). With regard to the RCGM's elicitation of discourses, all interest groups (groups of IPs) were required to send their written submissions to the Commission as responses to sixteen set questions. Their answers had to 'stand alone', meaning they could not cross-reference to other answers. This 'parts approach' is characteristic of modernity as it portrays issues as existing in isolation from one another. Although accommodating of science, as it reflects the scientific method, it disadvantaged holistic approaches (see Rogers-Hayden and Hindmarsh 2002). Although the environmental groups did use science (see Rogers-Hayden and Campbell 2003), the environmental groups in general found that the template restricted them from contextualising their answers within their worldviews, and in portraying the interconnectedness of the issues. Obviously an alternative, or optional, template may have been offered in the interest of equity to all parties, whereby those groups adopting holistic approaches could have appropriately represented their arguments. Thus, like the dominance of the biomedical discourse in the RCNRT, the RCGM encouraged another modernist discourse through a template which acted to marginalise environmental and holistic input.

The RCGM findings, like the submission template, did not overtly appeal to the Rational. Instead, in their summation they discussed Te Ao Māori (the traditional Māori worldview), the ecological worldview, and the Judeo-Christian worldview. But the RCGM did not discuss its own operational worldview, the hegemonic worldview of progress and science. Instead, the RCGM looked for shared values between the divergent worldviews. The RCGM listed seven core values which they stated were common to all New Zealanders, and which formed the basis of their decision-making. These were: the uniqueness of Aotearoa/New Zealand; the uniqueness of the cultural heritage; being part of a global family; the well-being of all; freedom of choice; participation and sustainability (RCGM 2001). These core values are of course upheld by both the bioproponents and the environmental groups, as it is highly unlikely that an organisation or person would come out against something such as 'the well being of all'. However, on closer inspection, contrasting discourses are revealed between bioproponents and environmentalists. An example of this can be seen in the uses of the term 'sustainability', which, for GM proponents, means using GM, whilst for environmentalists, this is instead achieved by organics (see Rogers-Hayden and Hindmarsh 2002).

Policy outcome

In contrast to the RCNRT, Aotearoa New Zealand made comparatively quick legislative changes in what continues to be a highly volatile political environment. Since the release of the RCGM's report in July 2001, the government has been working though the recommendations of the Commission. Although, as mentioned, official inquiries have often been characterised as being undertaken to subdue powerful pressure groups, and thus portray continued opposition to an issue as unreasonable, the release of the RCGM's findings have not subdued the resistance against GM in Aotearoa New Zealand.

On 29 October 2003 the government lifted the moratorium on GMO field releases as part of its biotechnology strategy (although there have been no GMO releases to date). Along with this plan the government has created Toi te Taiao, the Bioethics Council (see Toi te Taiao: the Bioethics Council 2004), and funded research into social, economic, ethical, environmental and agricultural issues related to GM. A further part of this is changing legislation to deal with GMO releases. The New Organisms and Other Matters Bill 2003 passed its third and final reading on 14 October 2003 (Fitzsimons 2003a), after returning from a Parliamentary Select Committee (Hobbs 2003). The Bill includes alterations to the Hazardous Substances and New Organisms Act to make legislative provision for 'conditional' (traceable) releases of GMOs and reviewing liability arrangements surrounding GMOs. The Education and Science Select Committee, which considered the Bill and reported back to Parliament in September 2003, has like the RCGM itself been criticised as predetermining the outcome as it 'blatantly and deviously referred the Bill governing GMO releases to a select committee with no Green MPs' (Fitzsimons 2003b).

As the government continues to work though the RCGM's recommendations, public opposition to GM also continues. Since the RCGM released its findings, protests have included the uprooting of a GM potato trial at Lincoln University Crop and Food Research Institute Laboratory, the occupation of the offices of the ERMA by concerned Māori, and 3,500 'ordinary citizens' signing a pledge to take direct action against GM (Genus and Rogers-Hayden 2005).

Concluding remarks

This path of development leading to the RCGM's findings can be seen as attempting to legitimate the development of GM in Aotearoa New Zealand. It involves a series of co-optations. The counter-hegemonic discourses of the environmentalists, which can be seen as offering a challenge to those of the bioproponents, were appropriated by the bioproponents. This process partly explains, for example, the co-optation of the notion of a sustainable nation by bioproponents, and later the RCGM's notions of sustainable process (Rogers-Hayden and Hindmarsh 2003). Furthermore, the government's use of the report could be seen as attempting to legitimate their action through the RCGM. The dominant governmental discourse on the recent legislative changes has been described as following the Commission's recommendations. This is, however, consistently disputed by the Green Party of Aotearoa New Zealand.

Discussion: reproducing modernity in public engagement

Despite differences in their timescales, geographic contexts and subject matter, there are a number of similarities between the two Commissions. Both created a panel of Commissioners whose backgrounds reflected a particular definition of relevant expertise, and represented modernist embedded professions. They can be seen as technocratic forms of governance in their reproduction of Rational operations. Both created processes which marginalised public engagement. In the case of the Royal Commission on New Reproductive Technologies, this took the form of a biomedical discourse which, although still operating within a dominant discourse of feminism, managed to marginalise First Nation, disabled and pro-life discourses. As Massey (1994) notes, the title of the RCNRT's final report is telling: although voices opposing wholesale technological progress are acknowledged by the words 'with care', the overall message is to 'proceed'. Reductionist discourses such as scientific discourses were also privileged in the Royal Commission on Genetic Modification, marginalising holistic discourses such as those of the environmentalists. Interestingly, to overcome these disparities both Commissions chose discourses of commonality ignoring the variations and nuances of those who submitted. Thus, while the RCNRT argued for recognising the Canadian ethos, the RCGM argued for New Zealander's core values, as the basis for decision-making. Both Commissions therefore acted to transform Romantic oppositional discourse into an instrument of legitimation.

Implications for public engagement in biotechnology policy

By way of conclusion, we wish to offer some thoughts on the implications of our analysis for future public engagement exercises. A common theme between the two case studies is the need to clearly articulate the goal of a public engagement exercise and plan how to best achieve this, from the initial research agenda through to the outcome. Both examples revealed to varying extents an *ad hoc* process which lacked clarity. Addressing this problem involves asking what the point of the exercise is and how the findings should be used. In the cases we have explored in this chapter, the RCGM and the RCNRT, there is evidence that the flexible nature of Royal Commissions can in fact work to their detriment as effective investigative mechanisms for public input. Critiques of these Commissions highlight the need to consider ahead of time how the non-Rational discourses, once elicited, are to be represented with integrity in the policy process. There are obvious difficulties, which can be anticipated, around the translation of human values into policy outcomes – a process which by its very nature is reductionist and tends to elide the philosophical nuances of which pluralist systems are comprised. Both Commissions also pose the question of representation – if it is representation that is important, why rely on processes that invite public submissions? The follow-on from this is also how to value, and elicit, the views of those engaged and those not engaged in the issue.

This point is entwined with a second – that there is a substantial body of literature offering critiques, suggestions and discussion on public inquiries/ participation and deliberation which appears to have not been consulted despite the vast amounts of energy spent on the public participation exercises. For example, a reading of existing literature would highlight the issue of 'debate Rationalisation', public participation vs. public sampling, and participation, learning and deliberation exercises (Andersen and Jaeger 1999; Hamstra 1995; Irwin 1995; Joss and Durant 1995; OECD 1979; Rifkin 1998; Schott 1993; Sclove 1995). A literature review would also have enabled the Commissions to conceive of the fourth point – resource inequity. They could thus have considered plans for overcoming these differences within the inquiries. In the case of the RCNRT, it would not be unreasonable to expect that lessons could have been drawn from the success of the Berger Inquiry's public engagement programme, especially with more than a decade in between the two exercises.

Finally, we recommend that, for these exercises to be learning exercises in themselves and not merely replicate previous mistakes, it is essential to incorporate a reflective evaluation into the inquiry/public forum (see Pidgeon and Rogers-Hayden 2005). This level of reflection and learning is necessary to free public participation from the technocratic processes which reproduce modernist outcomes and open them up to forms where publics have genuine opportunities for participation.

All of the above suggestions – clearly articulated goals based on learning from past experience (both within and outside national contexts and policy arenas); a strategy for incorporating a plurality of (non-dominant) discourses; and a robust mechanism for evaluation – could be addressed at the planning stage of an inquiry. However, we should be clear that we understand that these will not be *easily* addressed. Confronting the preoccupations of modernity with a reflexive, human values-based approach to policy design struggles against an entrenched dominance of the Rational in neo-liberal governments. This is especially true as long as sources of expertise in policy development tend to be confined to those perceived as legitimate within the Rational paradigm, a situation which makes it more difficult for those involved in policy deliberation to expand the limits of imagination outside their disciplinary orientation or worldview. Happily, with the move in many nations to extend policy committee membership to lay people and non-professional forms of expertise, a broadened variety of worldviews is slowly becoming recognised in technology policy development. While this goes far towards incorporating the Romantic in policy discourses, there are still persistent problems associated with large-scale public engagement and consultation exercises. An attempt to elicit the perspectives of publics is ill-equipped to maintain any philosophical reflexivity unless that reflexivity is clearly articulated as one of its goals. The work towards such a reflexive design may be incremental, but should it eventually accomplish a mechanism for public engagement in technology policy that merits the descriptor *democratic*, it might well be worth the wait.

Notes

1 A conditional release is the release of a GMO according to certain criteria established by the Environmental Risk Management Authority (ERMA). For example, this may include continued monitoring of the organism once it is released or that only sterile plants are released (Hobbs 2003).
2 For more detail on the Berger Inquiry, see the Canadian Broadcasting Corporation's online archives at http://archives.cbc.ca/300c.asp?id=1-73-295 (accessed 25 April 2004).
3 Colloquia often involved participation of prominent academic feminists, bioethicists and technology critics, including Ursula Franklin, author of *The Real World of Technology* (1990); Heather Menzies, author of *Fast Forward and Out of Control: how technology is changing your life* (1989); Christine Overall, author of *The Future of Human Reproduction* (1992); and Susan Sherwin, author of *No Longer Patient: feminist ethics and health care* (1992).
4 Per findings, currently unpublished, from the ESRC grant 'Policy Learning in Risk Governance' (M. Jones, B. Salter and N. Pidgeon, investigators).

References

Andersen, I.-E. and Jaeger, B. (1999) 'Scenario workshops and consensus conferences: towards more democratic decision-making', *Science and Public Policy*, 26(5): 331–40.

Anonymous (1993) 'Inside the Royal Commission', in G. Basen, M. Eichler and A. Lippman (eds), *Misconceptions: the social construction of choice and the new reproductive and genetic technologies*, vol. 1. Hull: Voyageur.

Ashforth, A. (1990) 'Reckoning schemes of legitimation: on commissions of inquiry as power/knowledge forms', *Journal of Historical Sociology*, 3(1): 1–22.

Beck, U. (1992a) *Risk Society: towards a new modernity*. London: Sage.

—— (1992b) 'From industrial society to the risk society: questions of survival, social structure and ecological enlightenment', *Theory, Culture and Society*, 9: 97–123.

—— (1995) *Ecological Politics in an Age of Risk*. Oxford: Blackwell.

—— (1999) *World Risk Society*. Malden, MA: Blackwell.

Burton, F. and Carlen, P. (1979) *Official Discourse: on discourse analysis, government publications, ideology and the state*. London: Routledge and Kegan Paul.

Cameron, B. (1993) 'Brave new worlds for women: NAFTA and new reproductive technologies', in J. Brodie (ed.), *Women and Canadian Public Policy*. Toronto: Harcourt Brace.

Canadian Broadcasting Corporation Archives (2004) The Royal Commission Reports. Available online at http://archives.cbc.ca/IDC-1-75-610-3370/science_technology/infertility/clip10 (accessed 15 June 2006).

Doyle, T. and McEachern, D. (2001) *Environment and Politics*. London: Routledge.

Dryzek, J. (1997) *The Politics of the Earth: environmental discourses*. Oxford: Oxford University Press.

Eichler, M. (1993) 'Frankenstein meets Kafka: the Royal Commission on new reproductive technologies', in G. Basen, M. Eichler and A. Lippman (eds), *Misconceptions: the social construction of choice and the new reproductive and genetic technologies*, vol. 1. Hull: Voyageur.

Fitzsimons, J. (2003a) 'New organisms and other matters bill'. Available online at http://www.greens.org.nz/searchdocs/speech6825.html (accessed 31 January 2004).

—— (2003b) 'Govt. shafts Greens over GE committee', Press Release, 5 July. Green Party of Aotearoa New Zealand.

Franklin, U. (1990) *The Real World of Technology*. Concord, Ontario: Anansi.

Genus, A. and Rogers-Hayden, T. (2005) 'Genetic engineering in Aotearoa New Zealand: opening up or closing down debate?', in M. Leach and I. Scoones (eds), *Science, Citizenship and Globalisation*, DRC Series. London: Zed Books.

Giddens, A. (1990) *The Consequences of Modernity*. Stanford, CA: Stanford University Press.

Hamstra, A. (1995) 'The role of the public in instruments of constructive technology assessment', in S. Joss and J. Durant (eds), *Public Participation in Science: the role of consensus conferences in Europe*. London: Science Museum.

Hatcher Roberts, J. (1999) 'Coalition building and public opinion: new reproductive technologies and Canadian civil society', *International Journal of Technology Assessment in Health Care*, 15(1): 15–21.

Health Canada (2006) Assisted human reproduction internationally. Available online at http://www.hc-sc.gc.ca/hl-vs/reprod/hc-sc/general/international_e.html (accessed 15 June 2006).

Irwin, A. (1995) *Citizen Science: a study of people, expertise and sustainable development*. London: Routledge.

Jones, J. P. III, Natter, W. and Schatzki, T. R. (1993) '"Post"-ing modernity', in J. P. Jones III, W. Natter and T. R. Schatzki (eds), *Postmodern Contentions: epochs, politics, space*. New York: Guilford.

Jones, M. and Salter, B. (2003) 'The governance of human genetics: policy discourse and constructions of public trust', *New Genetics and Society*, 22(1): 21–41.

Joss, S. and Durant, J. (1995) 'Introduction', in S. Joss and J. Durant (eds), *Public Participation in Science: the role of consensus conferences in Europe*. London: Science Museum.

Massey, C. (1993) 'The public hearings of the Royal Commission on New Reproductive Technologies: an evaluation', in G. Basen, M. Eichler and A. Lippman (eds), *Misconceptions: the social construction of choice and the new reproductive and genetic technologies*, vol. 1. Hull: Voyageur.

—— (1994) *The Public Participation Program of the Royal Commission on New Reproductive Technologies: an evaluation*. Unpublished thesis, Simon Fraser University.

Menzies, H. (1989) *Fast Forward and Out of Control: how technology is changing your life*. Toronto: Macmillan.

Nisker, J. (1997) 'In quest of the perfect analogy for using *in vitro* fertilization patients as oocyte donors', *Women's Health Issues*, 7(4): 241–7.

Oelschlaeger, M. (1991) *The Idea of Wilderness*. London: Yale University Press.

Organization for Economic Co-operation and Development. (1979) *Technology on Trial: public participation in decision-making related to science and technology.* Paris: Organization for Economic Co-operation and Development.

Overall, C. (1992) *The Future of Human Reproduction*. Toronto: University of Toronto Press.

Pidgeon, N. and Rogers-Hayden, T. (2005) 'Public engagements on GM and nano-technology', *Science and Public Affairs*. June, 14–15.

Rifkin, J. (1998) *The Biotech Century: harnessing the gene and remaking the world*. New York: Tarcher/Putnam.

Rogers-Hayden, T. (2004) *Commissioning Genetic Modification: the marginalisation of dissent in the Royal Commission on Genetic Modification*. Unpublished thesis, University of Waikato.

Rogers-Hayden, T. and Campbell, J. R. (2003) 'Moving beyond science? Environmentalists' submissions to New Zealand's Royal Commission on Genetic Modification', *Environmental Values* 12(4): 515–34.

Rogers-Hayden, T. and Hindmarsh, R. (2002) 'Modernity contextualises New Zealand's Royal Commission on Genetic Modification: a discourse analysis', *Journal of New Zealand Studies*, 1(1): 41–61.

—— (2003) 'Sustaining progress or progressing sustainability? Reading the report of Aotearoa New Zealand's Royal Commission on Genetic Modification'. Unpublished conference paper presented at the Ecopolitics XIV Conference. Sustainability: a new goal for human endeavour? RMIT Melbourne, 27–30 November 2003.

Rose, H. (2004) 'Beware the cowboy cloners', *Guardian* online, 16 February. Available online at http://www.guardian.co.uk/print/0,3858,4859426-103677,00.html (accessed 15 June 2006).

Royal Commission on Genetic Modification (2001) *Report of the Royal Commission on Genetic Modification*, Appendix 2, Wellington, 2001.

Royal Commission on New Reproductive Technologies (1993) *Proceed with care: Final report of the Royal Commission on New Reproductive Technologies*. Ottawa: Canada Communications Group.

Salter, B. and Jones, M. (2003) 'The governance of human genetics: policy discourse and constructions of public trust', *New Genetics & Society*, 22(1): 21-42.

—— (2006) 'Change in the policy community of human genetics. A pragmatic approach to open governance', *Policy & Politics*, 34(2): 347–66.

Schott, T. (1993) 'World science: globalization of institutions and participation', *Science, Technology & Human Values*, 18(2): 196–208.

Sclove, R. E. (1995) *Democracy and Technology*. New York: The Guilford Press.

Sherwin, S. (1992) *No Longer Patient: feminist ethics and health care*. Philadelphia, PA: Temple University Press.

The Knowledge Basket (2006) Legislation NZ, Bills & SOPs for 2004/5/6. Available online at http://www.knowledge-basket.co.nz/gpprint/docs/welcome.html (accessed 20 June 2006).

Toi te Taiao: the Bioethics Council (2004) 'Welcome'. Available online at http://www.bioethics.org.nz/index.html (accessed 20 June 2006).

Vandelac, L. (1993) 'The Baird Commission: from "access" to "reproductive technologies" to the "excesses" of practitioners of the art of diversion and relentless pursuit', in G. Basen, M. Eichler and A. Lippman (eds), *Misconceptions: the social construction of choice and the new reproductive and genetic technologies,* vol. 1. Hull: Voyageur.

Walls, J., Rogers-Hayden, T., Mohr, A. and O'Riordan, T. (2005) 'Seeking citizens' views on GM crops: experiences from the United Kingdom, Australia, and New Zealand', *Environment*, September, 22–36.

5 The precautionary principle on trial

The construction and transformation of the precautionary principle in the UK court context

Chie Ujita, Liz Sharp and Peter Hopkinson

Introduction

The concept of the precautionary principle (PP) has emerged as a major concept in the arena of environmental policy (for example, Principle 15 in the Rio Declaration on Environment and Development: the United Nations Conference on Environment and Development (UNCED), 1992), and is especially significant in discussions and practices concerning the risks from genetically modified (GM) crops. Despite this, the concept of the PP has no universal definition and is sometimes criticised (Alder 2000; Morris 2000) because of the vagueness of its definition and usage. Public debates about genetic modification usually revolve around the extent to which policies and practices should exercise precaution. Academic discussion frames such debates as considering the extent to which the 'precautionary principle' should be exercised. This latter is the principle which states that society should seek to avoid environmental and social damage by careful forward planning, blocking the flow of potentially harmful activities.

The PP has come into stark focus in the GM debate through the UK government funded Farm Scale Evaluation (FSE) Programme. This programme was set up to test some of the effects of growing GM in external environments. It has been developed and conducted by groups or representative bodies, including the government, the biotechnology industry, some scientists and some farmers, whilst others, such as pressure groups, organic farmers and many lay people, have been strongly opposed. The FSE has therefore created a situation where the risks, costs and benefits of GM are fiercely contested. Furthermore, the FSE programme has been frequently disrupted through the direct action of protestors who have attempted to prevent or damage individual trials, most notoriously through uprooting crops. A number of these actions have led to arrests and charges, and a few to subsequent prosecutions in the criminal courts. The PP has featured prominently in documentation, discussion and as a basis or justification for action in some of these trials.

The court represents a core social setting where society deals with problems. It is a highly controlled environment in which judge and jury hear arguments constructed by the Prosecution and the Defence, before a judgment is reached. The apparent equity of this process provides a potential route for society to clarify difficult concepts. In this light, it seems highly appropriate to examine how the concept of the PP fares in the courtroom setting. To our knowledge, no other studies have attempted a similar investigation of the precautionary principle in UK court cases.

Foucauldian discourse analysis and the PP

This chapter uses Foucauldian discourse analysis to explore how the PP has been represented and developed in one case-study criminal prosecution. Foucauldian discourse analysis is an analytical approach based on perspectives introduced by Foucault (1972, 1977, 1980), followed and applied by other thinkers (Hajer 1995; Sharp and Richardson 2001). The advantage of Foucauldian discourse analysis is that it allows us to 'think of discourses as constantly being contested and challenged and therefore not necessarily always omnipotent' (Carabine 2001: 273). This approach draws attention to how the meanings of specific issues are constituted and developed through time. As Taylor writes, 'controversy is basic ... [to] discourse analysis because it involves the study of power and resistance, contests and struggles' (Taylor 2001: 9). The Foucauldian perspective on discourse is appropriate for understanding the court context under the criminal proceeding, where the Defence and the Prosecution promote the different and competing views in order to convince the jurors and the judge.

Applying discourse analysis to the investigation of a courtroom setting is not new. For example, in the field of discursive psychology, many researchers investigated legal discourses in courtroom (e.g. Edwards and Potter 1992; Drew 1990). However, these researchers deemed the context to be studied – such as the court – in a very restricted way. Foucault (1977: 276), on the other hand, conceived of the court as a place where 'society as a whole does not judge one of its members, but that a social category with an interest in order judges another that is dedicated to disorder'. From this perspective, the investigation of the discourses within the court context throws light on the wider social context.

The term 'discourse' is understood as 'multiple and competing sets of ideas and metaphors embracing both *text* and *practice*' (Sharp and Richardson 2001: 196). The category of '*text*' incorporates any written materials and speech produced in any context. Examples include a company policy statement in its environmental report, witness statements for court case proceedings, and speech recorded as a transcript of the court case. It includes both text produced for the courtroom and text produced for wider purposes, such as distribution to the public. In contrast, '*practice*' is defined here as actions which have material effects on the policy-making processes and on society as a whole. A farmer's choice to participate in the FSE programme and to grow GM crops on his land can be seen as an example of practice.

In order to study discourses about the precautionary principle, we need to examine how the text and practice of discourses are manifested. Academic writing is one domain in which discourses about the precautionary principle are manifested, and another domain is in the courtroom. Each of these different societal contexts imposes different rules and norms. The academic domain can be understood as a relatively 'open' forum in which, as long as particular rules of argumentation are followed, a wide range of arguments can be expressed on a variety of topics. The courtroom is a more closed setting in which permissible topics and arguments are tightly restricted.

The aim of this chapter is to compare manifestations of the PP in academia with those in the courtroom. In order to do this, we begin by drawing on the literature to define an academic 'precautionary discourse' and a 'science-based' academic discourse. The chapter then turns to the courtroom context. Having set the background to the trials, the analysis uses courtroom documents, other literature, and the actions of those in the courtroom to identify Prosecution and Defence discourses. In particular, we focus on the more precautionary Defence discourse, tracing how it is differently represented by different actors, and demonstrating which elements of this discourse are later reproduced by the judge. A critical evaluation then compares the representations of the PP in court with those found in academia. We conclude by reflecting on the implications of the findings for the development of the PP.

For ease of expression, we use the term 'discourse' to refer to the two academic discourses as well as the two courtroom discourses. Our conceptualisation, however, is more fluid than this implies. This is partly because there are variations and differences within the academic discourses themselves. Further, our expectation is that the academic precautionary discourse will have many elements in common with the Defence discourse. Hence, we might suggest that they are both manifestations of a larger ('societal') precautionary discourse. Our purpose in tracing the similarities and differences between these manifestations is to consider the role that the courts are playing in influencing and developing our conceptualisation of the precautionary principle.

Table 5.1 illustrates the relation of two sets of discourses in the different domains.

Background

Theoretical framework for two academic discourses

The PP argues for policy-makers to anticipate and avoid irreversible risks that may arise from the limits of knowledge. In academic debates, the PP usually covers the risks of a new technology with respect to at least some of the following: the environmental impacts, the issues of human and animal health, the potential economic costs and benefits, short-term, long-term and

Table 5.1 Summary of two sets of discourses and their main advocates from the court context

	Academia	
	Academic precautionary discourse	Science-based academic discourse
Courtroom		
Defence discourse	Activists Expert witnesses for the defence Defence lawyers	
Prosecution discourse		Biotech company Contracted farmer The UK government Prosecution lawyers Police officers

cumulative impacts, regional, national and global scales of consideration, risks as understood through a combination of an expert's and a lay-person's knowledge, as well as dialogue with the public. The PP has been increasingly recognised and applied in practice as a part of the environmental policies (e.g. EC 2000). In academic debates, many different conceptualisations of the PP have been evolved. These can be mapped broadly into two different representations.

The 'science-based' academic discourse

Proponents of the 'science-based' academic discourse have typically constructed legitimacy of GM on the basis of techno-scientific arguments. Risk is understood as the amalgam of the probability of the occurrence of an unwanted event and the magnitude of the consequence of the occurrence (RCEP 1998). Within the realm of the 'science-based' academic discourse, risk can be quantified. Advocates of the 'science-based' academic discourse generally take a critical stance against the PP. However, some of them accept the PP with highly limited interpretations (see, e.g., Morris 2000). This narrowly defined precaution is sometimes called 'weak precaution', and is often associated with the combination of probabilistic risk assessment and cost–benefit analysis. In our interpretation, the 'science-based' academic discourse is used to strengthen the environmental discourses promoted by the biotechnology industry, as well as the current GMO policy and regulatory systems in the UK and EU.

The academic precautionary discourse

Proponents of the academic precautionary discourse (e.g., O'Riordan and Jordan 1995; Stirling and Mayer 2000; Wynne 2001; Irwin 2004), on the other hand, have constructed their rationale on the basis of the importance of socio-cultural 'value'. The importance of a dialogue with multiple stakeholders is stressed because it enables lay knowledge and a wider range of views to be brought into the risk-assessment and the policy-making processes.

For proponents of the academic precautionary discourse, the term 'value' does not mean 'emotion' against science, but implies a measurement tool developed by the society through its history and geographic traits. Many different precautionary approaches for the management of GM issues can be identified (e.g., Grove-White *et al.* 1997; Stirling and Mayer 1999; Grove-White *et al.* 2000). Two points should be noted here. First, none of these academics reject the importance of science. Contributors to the academic precautionary discourse understand the concept of the PP as assisting in the development of both science and policy. Second, because of its diversity, some academics draw on elements of the PP without making explicit use of the term. This diversity also provided a space for the recognition of 'Strong Precaution', which was often associated with the demand for 'zero risk', and which is widely advocated by environmentalists.

Some precautionary commentators have argued that, while the concept of the PP has begun to be frequently mentioned in environmental policy, including GM issues, the PP is still difficult to bring into the legal context (e.g., Cameron and Wade-Gery 1995; Jordan and O'Riordan 1999; Royal Society of Canada 2001). Others have argued that science alone cannot meet the legal standards in court where scientific, socio-economic criteria and layperson's knowledge are required (e.g., Salter 1988; Fisher 1999; McEldowney and McEldowney 2001; Belt and Gremmen 2002; Wilkinson 2002; Ho and Saunders 2003). In their view, the PP is appropriately given some space in the legal context. This chapter also aims to examine whether and in what way the PP is given a role in the UK courtroom.

Background to the case study

In October 2003, the UK government-funded FSE programme finally released its results (DEFRA 2003). The FSE programme was the world's largest field trial of GM crops, took over three years to complete, and compared biodiversity in fields growing genetically modified (GM) crops with that in fields growing conventional crops. The FSE results have drawn massive media attention, but there has been little consensus over their implications. For example, the proponents of GM said that the result of FSE shows GM crop 'is good for farmers and better for biodiversity ...' (CropGen 2003), while the opponents of GM highlighted the scientific methodology of the FSE as 'too narrow, but even so the results show that GM crops would damage the environment' (Soil Association 2003). Still others argue that the method of the FSE were flawed and were of little scientific value at all (ENDS 2003).

The opponents of GM have been critical of the FSE since the programme was announced. They have focused on the uncertainties and risks of GM crops over and above those investigated by the FSE, and have argued that the FSE itself posed risk, and that the trials should not be permitted. During the years 1999 to 2003, some of these opponents went into the FSE trial sites and uprooted GM crops. Figures from 2001 indicate the scale of this activity: of over 100 FSE trial sites, twenty were damaged by anti-GM activists (GeneWatch UK 2002). It is important to note that not all activists were caught and charged by the police. Even if they were charged, the case was sometimes dropped (for detail see Stallworthy 2000; Genetix Engineering Network 2001). However, over the past few years at least three GM court cases have been brought to the Crown Courts, and many others – usually those where the GM crops are of lower value – have appeared in the lower courts. The charges imposed on these activists could be a variety of criminal charges, such as aggravated trespass (e.g. '*Tilly v DPP, DPP v Tilly and others*': *The Times* 2001), theft (e.g. the first trial of 'the Norwich case': Hall 2000), and criminal damage (e.g. 'the Worcester case': Shrimsley 2001). In addition to these criminal cases, uprooting of GM crops also led to some civil cases (e.g. '*Monsanto plc v Tilly and others*': *The Times* 1999).

From interviews with the anti-GM activists and reading their texts, it appears that they understood their activities as falling within the category of what Thoreau (1966) called 'civil disobedience'. In Thoreau's view on civil disobedience, the law needs to be broken if 'the remedy will not be worse than the evil; but if it is of such a nature that it requires you to be the agent of injustice to another …' (Thoreau 1966: 231). There are several academics (e.g., Rawls 1972; Elliott and Quinn 2002; Carter 2005) who also address the theory of civil disobedience, and who say that civil disobedience could possibly be justified under certain circumstance. In the cases examined in this chapter, the activists who were charged with criminal damage decided to defend themselves with a plea of not guilty, through a defence of 'lawful excuse'. The statute that introduced the defence of lawful excuse is set out in the box.

Lawful excuse

Lawful excuse is stated under the Criminal Damage Act 1971

Section 1:

1 A person who without lawful excuse destroys or damages any property belonging to another intending to destroy or damage any such property or being reckless as to whether any such property would be destroyed or damaged shall be guilty of an offence.

Section 5:

1 This section applies to any offence under section 1(1) …

2 A person charged with an offence to which this section applies
   shall, whether or not he would be treated for the purposes of this
   Act as having a lawful excuse apart from this subsection, be treated
   for those purposes as having a lawful excuse

   (a) ... or
   (b) if he destroyed or damaged or threatened to destroy or damage
        the property in question or, in the case of a charge of an
        offence under section 3 above, intended to use or cause or
        permit the use of something to destroy or damage it, in order
        to **protect property** belonging to himself or another or a right
        or interest in property which was or which he believed to be
        vested in himself or another, and at the time of the act or acts
        alleged to constitute the offence he believed –;
        (i) that the property, right or interest was in **immediate need of
             protection**; and
        (ii) that the **means** of protection adopted or proposed to be
             adopted were or would be **reasonable** having regard to all
             the circumstances.

3 For the purposes of this section it is immaterial whether a **belief** is
   justified or not if it is **honestly** held.

4 ...

5 This section shall not be constructed as casting doubt on any
   defence recognised by law as a defence to criminal charged.

(N.B. Emphasis added by the present authors.)

In our study, the defendants, namely the activists in court, rely on section
5(2)(b) and 5(3), which require four conditions:

1 The purpose to protect property (see section 5(2)(b)).
2 The property is in imminent danger (section 5(2)(b)(i)).
3 Reasonable means (section 5(2)(b)(ii)).
4 Honest belief (section 5(3)).

In the GM court cases, the defence of lawful excuse will only be relevant if
the activists can show that their purpose in damaging GM crops was to
protect property vested in themselves or another (1). Moreover, at the time
of damaging GM crops they must believe the property that they intended
to protect is in imminent danger (2). They must also believe that damaging
GM crops was a reasonable means of protecting the property (3). Finally,

it is important to note that lawful excuse does not require the activists' beliefs to be reasonable (4); it is their 'honest belief' with respect to points (2) and (3) which is required. (This interpretation is based on a general explanation of lawful excuse presented by Smith 1996.)

The Mold case

This section focuses on the recent jury trial in the UK of two anti-GM activists who destroyed GM crops at the FSE site in Wales, were charged with criminal damage, and went to Mold Crown Court in 2003 ('*R v Tilly and Davies*' in 2003 – hereafter, we call this GM court case 'the Mold case'). The court heard that the two defendants did not deny the facts, but pleaded not guilty. The main defence in the Mold case was that the defendants had a lawful excuse. This is the third time that activists had been brought before a jury for criminal damage to GM crops.

In the first jury trial at Norwich Crown Court in 2000 twenty-eight members of Greenpeace, including Lord Melchett (a former chairman of Greenpeace), were the defendants ('the Norwich case': Kelso 2000), while two local residents who do not belong to any organisation were the defendants in the second jury trial at Worcester Crown Court in 2001 ('the Worcester case': *The Guardian* 2001). According to Stallworthy (2000: 729), 'when criminal damage charges are brought against GMO protesters, the subjective mental element for the defence of lawful excuse may result in a sympathetic view on the part of jurors'. In practice, the defendants of these two previous GM court cases were acquitted, and these verdicts consequently strengthened other activists' views that trial by jury tends to be more sympathetic to the defendants (e.g., Randle 1995; Genetix Engineering Network 2002).

Unlike in the two previous jury trials, the defendants of the Mold case were found guilty. It must be noted here that this guilty verdict is not this chapter's primary interest; it is, however, intriguing, and we will speculate on the reasons for this verdict later in the chapter. Any conclusions on this point are necessarily incomplete, however, as the Contempt of Court Act 1981 prohibits observation or exposure of the process through which the jurors reached their verdict.

The Mold case was selected for investigation for three reasons: (i) the relevance of the case in terms of activities and the legal process; (ii) access to some of the legal documents; and (iii) the timing, which made it practical to observe the criminal proceeding. Evidence about the court process is comprised of the authors' written notes from observation of the trial, interviews with key actors for the defence, a transcript of the court case, four expert witness statements for the defence, and two proofs of evidence (often called 'defence statements') produced by the activists. Additionally, texts from a non-legal context have helped elucidate the background position taken by different individuals involved in the Mold case.

It should be noted that the time and the cost of obtaining this essential information were serious obstacles to the investigation. At the time of writing this chapter in 2004, transcribing a court case costs around £450 for each day of the trial. Access to expert witness statements and proofs of evidence depends on the goodwill of individuals. The issue of confidentiality was also important for almost all legal materials obtained from the primary sources.

The story presented by the prosecution

During his opening address, the prosecution barrister showed a video tape recorded by the police both from ground level and from a helicopter, showing the process of activists uprooting GM crops on the FSE site. The video showed activists arriving at the site dressed in white suits and their movements into and through the fields, accompanied by police requests for the protesters to leave the site and the eventual arrest of several protesters. The video provided a powerful piece of visual evidence demonstrating a well-organised protest and apparently pre-meditated plans to damage the crops. This video was shown very early in the prosecution case, thereby creating a very strong (and likely negative) image for the jury of the protest action for the remainder of the trial.

At the beginning of the trial, the prosecution barrister emphasised that 'GM crops are not on trial' and thus jurors were asked not to consider whether GM science is good or bad. This was the reason why the prosecution did not seek to challenge the scientific evidence on GM which the defence submitted to the court.

The Prosecution barrister constructed his speech on the basis of the prosecution witness statements, given by:

1 The contracted farmer under the FSE programme.
2 The product development manager of Bayer (then Aventis).
3 Police officers.

All the prosecution statements included a description of the incident in July 2001 to explain to jurors what happened at the site on that specific day. However, among these witnesses, statements written by (1) and (2) particularly explained about the FSE and GM crops growing at that site. They described the FSE as a programme conducted on the basis of an agreement between the government and the body representing the farming and biotechnology industry, the Supply Chain Initiative on Modified Agricultural Crops (SCIMAC). The company, Aventis, was under the umbrella of SCIMAC and provided GM seeds, herbicides and agronomic advice to the contracted farmer. This means that the company did not conduct the research itself, and that it was the contracted farmer's role to cultivate the crops for the sake of the research. The prosecution stressed that GM crops in the FSE site were lawfully growing under licence from the EU and the UK government. The prosecution implicitly referred to the licence as the evidence of safety, which provided sufficient precaution. SCIMAC Codes of Practice and Guideline were also mentioned as the agronomic advice, which was given to the contracted farmer and implemented throughout the FSE trial.

SCIMAC rules included a recommended separation distance between GM crops and non-GM crops. But the court did not hear any detail of SCIMAC and its rules. Instead of an oral explanation, the judge, the jury and the defence were able to see the exhibit submitted by the prosecution. The exhibit included an experimental permit issued to Aventis CropScience, providing it with Part C consent (required under the EU regulatory frameworks for GMOs) given to GM maize in 1998, and SCIMAC Codes of Practice and guideline.

In his closing speech, the prosecution barrister argued that the two activists damaged GM crops on that day because they aimed to get extra publicity to make political points. This was the reason why they did not destroy all GM crops at the FSE trial site. He stated that the activists did not honestly believe there was property at imminent risk from the FSE trial. This was demonstrated because they took actions at the weekend, a more convenient time to rally and to reach more people than any other day. Finally, the prosecution barrister argued that the defendants did not honestly believe their actions were the last resort because there were many other reasonable means to protect property, for example by contacting a Welsh Assembly Member for the farm growing GM crops. Therefore, the prosecution argued that the two defendants had no right to rely on the defence of lawful excuse.

The case presented by the defence

The defence barristers emphasised that the issue in this court case was 'not about what was done, it's about why it was done ...' (the Mold case, 2 April 2003, p. 9, para. E). The defence therefore accepted that '[t]he question [of this case] isn't whether the release of these [GM] genes at this stage into the environment is dangerous. It's whether the defendant honestly believed it was dangerous' (the Mold case, 2 April 2003, p. 17, para. A).

During their direct examination of the defendants, the defence barristers' questioning emphasised the activists' concerns about the scientific uncertainties and risks of GM crops, which could have local and global impacts, and which could cause short-term, long-term and cumulative adverse effects. For example, one of the defendants described GM technology as 'young science', containing a considerable amount of unpredictability and uncontrollability. Among many critiques of the FSE programme presented by the defence, the lack of practicability of 'the separation distance' was one of the focal points. The defence insisted that the separation distance defined by SCIMAC and applied for the FSE was totally insufficient to protect non-GM plants from the risk of cross-pollination by GM pollen. Moreover, public distrust about the safety of GM foods and crops was also stressed in their argument as a core element that shaped their beliefs.

The defendants frequently cited the scientific evidence provided by expert witnesses in order to strengthen jurors' knowledge about GM and the current regulatory systems. Experts in the courtroom do not necessarily mean scientists from the academic arena, but people who are 'knowledgeable in the field to fully understand the complexities involved' (Freiman and Berenblut 1997: 264–5). These experts do not only provide evidence, but they also comment on the specific issues about a case, which cannot be covered by lay persons' common sense and experience alone. Their knowledge obviously helps the courts to understand complexities involved in environmental issues, though some critics note that the close relationship between lawyers and their experts can lead to bias (Wilkinson 2002; Freiman and Berenblut 1997).

In the Mold case, the defence had four expert witnesses, each with distinct areas of expertise as follows:

1 Cross-pollination by bees.
2 Cross-pollination by wind.
3 Regulatory failure around GM technology.
4 Risks and uncertainties around GM technology.

Three of these four expert witnesses had an academic background.

The court heard that the two defendants honestly believed their purpose was to protect property from GM risks. The property at risk included other farmers' conventional and organic crops, honeybees, honey products, the environment (including soil and wildlife), and human and animal health. The activists honestly believed these properties were in imminent danger due to GM contamination arising from cross-pollination. This was because at the time of the incident GM crops on the site were just about to flower. Finally, they honestly believed that damaging GM crops was a reasonable means of protecting property because they had sought other ways to avoid the risks (for example, they had written to the Government, DEFRA, and the Farmers Union of Wales, arguing that the licence was insufficient protection). Therefore, the defence barristers argued that the damage done to GM crops by the two activists should be treated as within the realm of lawful excuse.

The judge's view

During the Mold case, the judge stated that 'The cross-examination [by the prosecution] did not test the defendants' knowledge or views about the use of GM crops' (the Mold case, 1 April 2003, p. 40: para. F); and he went on to say: 'The test under section 5 [of the Criminal Damage Act] is whether the defendant has or might have a belief, and it doesn't matter ... whether that belief is true or not as long as it is genuinely held' (the Mold case, 1 April 2003, p. 41: para. C). Therefore, he said, he wasn't 'going to deal with cross pollination. It doesn't matter how it is likely to happen, if they honestly believed [that it would]' (the Mold case, 2 April 2003, p. 2: para. F). For these reasons, scientific evidence provided by the expert witnesses was not examined during the court proceeding.

After the jury had heard all the evidence, the judge summed up the case. In the first part of his summing-up, the nature of the law which the jury had to consider was explained. This means that the judge required jurors to consider: first, whether the activists aimed to protect other maize crops; second, whether these crops were in imminent need of protection; third, whether damaging GM crops was a reasonable means in all the circumstances. Finally, jurors needed to consider whether they thought the activists honestly believed other maize crops were in imminent risk, and honestly believed the means were reasonable. The judge summarised the arguments presented by both sides as follows: 'There are two arguments, put shortly. The prosecution really say this is publicity. The defence say no' (the Mold case, 2 April 2003, p. 24, para. F).

In the second half of his summing-up, the judge summarised the evidence. He referred to the licence given to the GM crops owned by the prosecution witness, and he described the day of the incident in 2001. He then moved on to the summary of the defendants' views of GM crops and the FSE site, which sought to explain to the jurors why the defendants pulled up GM crops at that site on that day. At the end of his summing-up, the judge asked the jury to reach a unanimous verdict.

Analysis of legal discourses

Looking at the court transcript and the documents submitted to the court helps to understand the legal discourses put forward by the defence and prosecution lawyers. This court-based textual information alone, however, is not sufficient to fully understand the ideas and rationales that contribute to each of these discourses. Instead, it is important to supplement the court information with evidence from the non-legal context in which the different court actors are engaged. For example, for a fuller understanding of Aventis's view of GM and anti-GM campaigning, it is essential to investigate not only the legal documents, but also other materials such as the company reports, its website, and activities taken by the company and by the pro-GM lobbying group to which the company subscribed. Moreover, by considering these non-legal documents, differences between the views of different actors begin to emerge. Hence, this analysis now supplements the information obtained from the court documents with information drawn from these non-legal contexts. In doing so, we begin to highlight some similar features between the legal discourses and corresponding academic discourses. The section begins with an analysis of selected contextual factors for each legal discourse. A more detailed focus on different elements of the defence discourse is then developed.

The prosecution discourse

The prosecution in the Mold case implicitly argued about safety, the legal justification for the planting, and the implementation of sufficient precaution, through referring to the licence given to the biotechnology company Aventis and its GM crops. Focusing on safety (rather than risk) is one of the key characteristics of the 'science-based' academic discourse. The actors associated with the prosecution, such as the biotechnology industry including SCIMAC and Aventis, a farmer growing GM crops for the FSE, the current GMO policy and law in the UK and EU, are understood to be proponents of the 'science-based' academic discourse. This means that, as long as the prosecution attempted to justify its arguments by relying on the licence authorised *by* and *to* these actors, the prosecution discourse should be understood as one specific legal manifestation of a wider 'science-based' discourse. According to Burns (1999: 50), the opening speech is 'not simply an argument about "what happened". It is a battle about the frameworks within which events should be understood ...'

As we have seen, the prosecution barrister cited the names of SCIMAC, Aventis, the contracted farmer, the Directives framed in EU law, and the UK government, but the court did not hear these actors' detailed views of GM or other key issues. The closest these actors' views came to the court was that SCIMAC was quoted in documents submitted by the prosecution barrister. In this document to the court, SCIMAC stated that in its view GM crops were beneficial to humans and the environment because GM technology ensured safety and efficacy in use (SCIMAC 1999). Information about the separation distance by SCIMAC was also stated in the document, but was not cited in the prosecution barrister's speech. These detailed judgements underlying the prosecution discourse were not discussed in the court because the prosecution barrister and the judge regarded such information as unnecessary in terms of the court proceedings and their aims.

When the opening speech by the prosecution is viewed in this way, the purpose of the prosecution discourse could be understood as persuading the jurors that the safety of GM crops at the FSE site was substantially equivalent to that of non-GM crops; and therefore that there was no need for protection against imminent danger.

The defence discourse

According to Burns (1999: 57), 'direct examination ... conveys the witness's understanding of the meaning of a past event, embedded in the perceptual judgements he [sic] makes'. In the Mold case, direct examination presented the two defendants' views of the GM risks that led them to their direct actions for the protection of non-GM crops. Although the PP is not directly mentioned by the two defendants in court, it is implicitly done in their court arguments about their views on risks and scientific uncertainties of biotechnology. For example, one of the defendants understood genetic engineering as 'a very unpredictable science, so little is known about it. It is also a highly random science' (the Mold case, 1 April 2003, p. 10: para. B); and therefore, the activists aimed to protect people or animals from the risks of eating foods/feeds containing GM materials 'where we don't know what effect it would have on them' (the Mold case, 1 April 2003, p. 21: para. C).

The PP implicitly manifested in the defence discourse is very similar to those argued by advocates of the academic precautionary discourse. This means that the defence argued that the two defendants took the action of damaging GM crops as a reasonable last resort, because they believed there was a need for anticipatory action to prevent other property from risks. The property at risk included: other farmers' conventional and organic crops, the environment including soil and biodiversity, and human and animal health. The risks from GM contamination were believed to be unpredictable, uncontrollable and irreversible. The activists also believed that the adverse effects could derive from two main sources, namely (i) the limits of knowledge about GM technology, and (ii) the insufficient protection measures taken by the government and the company through the licensing system. Undoubtedly, pulling up GM crops was the most critical action which led to the emergence of the defence discourse.

Detailed analysis of the defence discourse

In the defence discourse, all actors for the defence seemed to be integrated under the single conception of GM issues. However, detailed investigation of the defence discourse reveals a considerable degree of difference between the activists, lawyers and expert witnesses.

Lawyers and the Activists

Through their defence statements and subsequent interviews, the activists expressed a wish to bring the wider range of GM issues to the court. This included the association of GM with intensive agriculture and developing countries, other activities they took in the past, experience of other GM court cases, the civil rights of freedom to choose, and their concerns about all GM crops at all FSE sites throughout the period of the FSE. On the other hand, the defence lawyers suggested a focus on the specific GM crop at the specific site on the specific day. From the legal point of view, the other issues were unhelpful and legally irrelevant. In this light, the activists appeared to feel a degree of pressure. However, in a subsequent interview the defendants explained that they understood the lawyers' advice as demonstrating which of their concerns would be admissible in the courtroom context (Tilly 2003b). The statements produced by the activists as a preparation for their court case contained explicitly referred to the PP (Davies 2003a; Tilly 2003a), but the two defendants did not use the term in the courtroom. The decision not to use the PP in court was made by the defendants themselves. They said that they thought 'the concept might be too technical for the jurors' (Davies 2003b).

Lawyers and expert witnesses

Those of the expert witnesses for the defence who have been interviewed said that they did not feel pressure from the lawyers (Mayer 2003; Hopkinson 2003). However, an investigation of their statements found that their arguments were highly focused. This was in contrast to their academic writings and/or other texts produced by the same authors for the non-legal domain. Moreover, the experts' own conceptualisations of the PP, as well as their own primary concern with the GM issue, were largely excluded from the expert witness statements (e.g., Wynne and Mayer 1993; Mayer and Stirling 2002; Hopkinson 2002, 2004). These differences seemed to have stemmed from (i) the experts' roles in the court case, which was decided by the lawyers, (ii) a lack of space in the witness statement to state their opinion, and (iii) the experts' preference to make the statements absolutely clear.

The case study found that there were several elements in common between the defence discourse in the court context and the academic precautionary discourse in the academic context. However, these discourses were not exactly the same as each other. This was because the view introduced by the defence discourse was narrower than the academic precautionary discourse. For example, the potential advantages of GM in terms of health, economic and environmental aspects are not referred to by the defence discourse. This was partly because the defendants regarded GM as an 'unnecessary' technology and adopted the PP with its very strong interpretation. But, more importantly, the difference emerged from the criterion applied by lawyers in order to construct legally relevant arguments.

The judge and the defence discourse

The most distinctive difference between these two discourses was the gap between the defence lawyers' view of property and the judge's definitions of the term 'property' in his summing-up. This means that, while the defence intended to establish the meaning of property to be a broad set of issues pertaining to environment, health and conventional and organic farming, the judge used the term 'property' referring only to non-GM maize crops which could be potentially damaged through cross-pollination.

In the judge's interpretation of lawful excuse, the jurors were only asked to take account of the imminent risks at the specific place. This means that long-term and cumulative risks to the environment and commercial crops, and the risks to biodiversity, were excluded from the range of protection. The national and the global scale of impacts were also deemed as beyond jurors' concerns. The process of narrowing the definition of lawful excuse can be understood as a result of the judge's view constructed in accordance with the legal criteria of the Act (for the detail of legal criteria, see Lord Hailsham of St Marylebone 1990).

It is interesting to compare the definition of property in the Mold case to other similar GM court cases in the past. For example, in the summing-up of the first trial, 'the Norwich case' ('*R v Bellotti, Melchett and Others*', 2000), the judge did not specify the property to mean only maize. In the Norwich case, the property included 'maize crops liable to cross-pollinate the pollen from the trial crop and any other property susceptible to direct damage by that pollen, but [did] not include anything growing wild on any land or any wild creature'. In short, the definition in the Norwich case included honeybees and honey products as part of the 'property' covered by lawful excuse.

The current English law defines property under lawful excuse to mean 'a tangible nature ... including wild creatures which have been tamed ...' (Lord Hailsham of St Marylebone 1990: 449). A significant difference between the Mold and the Norwich cases emerges from the judge's summing-up in the Mold case. While the judge had definitely heard the defence arguments about the implication of bees and GM pollen, he seemed to regard the issues on bees/honey as unimportant, or unnecessary to his summing-up. This is a surprising omission, as bees clearly fit into the category of 'wild things that have been tamed'. The reasons for these differences are not immediately clear and require further investigation.

In conclusion, while the PP and the defence discourse intended to protect the wider range of property regardless of whether it belonged to individuals or society, lawful excuse protected only liable property that belonged to individuals and was threatened by imminent danger.

Table 5.2 shows different elements of discourse (including actions taken, and statements/comments made both in and out of court) taken by different groups of key actors in the GM court case.

Conclusions

The critical focus of this chapter has been the construction processes of two discourses in the UK courtroom. We found that the prosecution and defence discourses established competing but not contradictory views of a specific event. This means that both discourses accepted that the activists damaged GM crops growing at the FSE site in Wales in July 2001.

Table 5.2 Actions and statements by the key actors

Main actors	Actions (e.g.)	Out of court statements and speech (e.g.)	Statements for court case and speech in court (e.g.)
Activists	Opposing GM crops/foods Pulling up GM crops at 'the site' Pulling up GM crops at other sites Appearing in other GM court case Petition against having GM crops and the sites for GM Taking other anti-GM campaigning	Statements in publication by own NGO/on own website Comments to the media Opinion at the public hearing	Proof of evidence for the defence Speech in court
Bayer (Aventis)	Promoting GM in UK Cooperating with the government for the FSE • Finding contracted farmers • Providing seed and advice to the farmers	Statements in company report/own website/related lobby group Comments in the advisory bodies for policymakers Comments on other governmental reports Comments to the public hearing Comments to NGOs	Prosecution witness statement
Farmer	Growing GM crop for the FSE	Dialogue with the activists and other general public Comments to the media	Prosecution witness statement

(continued on next page)

Table 5.2 (*continued*)

Main actors	Actions (e.g.)	Out of court statements and speech (e.g.)	Statements for court case and speech in court (e.g.)
Experts witness for the defence	Providing knowledge for defence	Academic paper Comments to the media Comments in the advisory bodies for policymakers Comments on other governmental reports Comments to the public hearing Comments to NGOs	Experts witness statements for the defence
Defence lawyer	Taking a role as the defence in court	Comments to the media	Letters to experts and defendants Documents for legal passage Speech in court
Prosecution lawyer	Bringing the case to the criminal proceeding	None	Indictment Other documents for legal passage Speech in court
Judge	None	None	Words during the trial Summing-up Sentencing
Jury	Finding the verdict	None	None

However, the prosecution interpreted this activity as contravening the Criminal Damage Act 1971 for purposes of obtaining publicity, while the defence argued that the activists took this action with a lawful excuse for the protection of the environment, non-GM crops, human health and other property.

This analysis has found that the term PP was not directly used in the speech by the defence, and only some aspects of the PP were brought into the court context through the plea for a lawful excuse. However, these aspects were important, for they formed the kernel of the defence argument – that the activists had a lawful excuse for their actions. During the court proceedings, the judge did not interrupt the defence lawyers while they presented their arguments and narratives. This can be understood to mean that the court did not reject the presence of the PP in the courtroom as long as the concept of the PP was used as a foundation of the defendants' belief. For the activists, belief is essential. This is because belief leads the activists to conceive GM risks in a particular way, which in turn underpins their direct actions. Under lawful excuse, the issue of belief was one of the critical points that determined the final verdict.

Why is it difficult to present the PP in the courts?

In this chapter we have made a case endorsing the claim that the PP is difficult to bring into the current UK court context. The case study showed three main reasons for this difficulty.

First, the Criminal Damage Act 1971 does not require the defendant's belief to be reasonable, and therefore any expert witness who provided scientific evidence was not called to the witness box. These experts implicitly and explicitly incorporated the PP with science in their legal statements, and each expert had their own conception of the PP and their own concerns about GM risks. Some experts were willing to appear in court. But the case study showed that there was little place for the scientific evidence underpinning the PP in the GM court cases addressing lawful excuse. The issue of reasonableness becomes important when we consider the policy implications of legal discourse. This point will be discussed later.

Second, the defence discourse is produced before the trial by different sets of actors with subtly different interpretations. Some of these are seen by the defence lawyers as irrelevant in the court context. Thus the activists and expert witnesses who affirm an explicitly precautionary principle outside the court present a muted form of precaution within the court.

Finally, property which the activists tried to defend was not seen by the judge in the Mold case as property under the defence of lawful excuse. In the Mold case, the jury was told that it could consider the evidence which was excluded from the judge's summing-up if the jurors understood these issues were relevant. However, the judge's selection of the evidence (stated in his summing-up) seemed to have some impact on the jury's final views on the defence discourse, including the range of the PP implicitly brought into the defendants' arguments in court.

These findings enabled us to illustrate a 'filtering process' in the construction of the PP in the court context, shown in Table 5.3. This table shows, first, how the defence lawyers, and second, how the judge, act as filters to limit and narrow the range of the defence arguments and the meaning of the PP allowed to be presented in court.

As Table 5.3 illustrates, the activists and the experts held an interpretation of the PP that is close to the academic precautionary discourse. Indeed, some of the expert witnesses are important contributors to the academic debate. However, the legal setting modified the way in which the actors shape their views and arguments, limiting the extent of precaution that can be expressed in debates on GM crops. This happened before and during the court proceeding. The filtering process effectively demonstrates the progressively narrower definitions of 'property' that were allowable in the legal context. Some remaining questions are: (i) why the defence lawyers accepted the activists' preference for the wider views of property; and (ii) why the property defined by the judge in the Mold case is narrower than that defined by the judge in the Norwich case. We are currently pursuing these issues.

Does the guilty verdict rule the PP out of the courts?

In the Mold case, the verdict was guilty. However, this does not mean that the PP was excluded from the court. The verdict in this case showed the prosecution discourse won the contest in the eyes of the jury. The justification contained in the prosecution discourse heavily relied on the licence being legally authorised. The prosecution implicitly argued that the actors on the prosecution side had provided sufficient precaution. In essence, while the defence include the wider perspectives of the PP, the prosecution only covers weak precaution and relies heavily on scientific rationale.

If the verdict had been not guilty, this would imply that the interpretation of the PP expressed in the defence discourse was deemed by the jury to be more convincing. In such a case, the verdict would have given some positive recognition to the PP, in recognising that the defendants held honest beliefs founded on the PP. Such an outcome would, however, merely confirm that they held this belief, rather than making any comment on whether this belief was appropriate.

Policy implication of legal discourses

Whilst the policy implication of legal discourses is not our primary aim, it is worth considering briefly. First, we should stress that the defence discourse in the Mold case did not cover the full range of the key aspects of the PP that are important for policy formation. Even if the defendants and

Table 5.3 The filtering processes in the Mold case

The key aspects of the PP		Activists and Experts	Defence lawyers	LE defined by judge
Information resources	Layperson's knowledge	▓	▓	▓
	Expert's (scientific) knowledge	▓	▓	▓
Time	Short-term	▓	▓	▓
	Long-term and cumulative	▓	▓	
Zone	Regional	▓	▓	▓
	National	▓	▓	
	International	▓	▓	
Economy	Costs — Specific crops (e.g. maize)	▓	▓	
	Any non-GM crops	▓	▓	
	Other products (e.g. honey)	▓	▓	
	Benefits	▓		
Environment	Risks	▓	▓	
	Advantages	▓		
Human/Animal health	Risks	▓	▓	
	Advantages			

the defence experts had been able to bring their arguments to the court without any modification by lawyers, there are many other issues that need to be covered. For example, these would include the issue of patents, labelling, ethical and religious concerns, and the political implications with other nations. Second, and as we have seen, because of the nature of lawful excuse that only requires honesty, the criminal court for addressing lawful excuse will not be a forum in which the reasonableness of different scientific beliefs about the PP can be tested.

Hence, whatever the verdict, the limited nature of precaution allowable in the court context is not sufficient to contribute to policy development. Viewing the GM court case in this way, it is clear that legal discourses are independent from policy-making processes. The activists themselves were aware that they were unable to bring their political intention to the criminal court proceedings, and this was indeed the image that the prosecution lawyer intended to establish in the court in order to invalidate the defence discourse. In essence, this perspective demonstrates that the legal setting is simply inadequate for policy development.

There is another perspective which indicates some paradoxical effects of the GM court cases which have an indirect influence on society and on policy formation processes. Direct actions and subsequent court processes are widely reported through the mass media (e.g. BBC 2003b). Photographic images of the activists pulling up GM crops are often used by the mass media as symbols of anti-GM perspectives (e.g. BBC 2003a). Therefore, while it is impossible at present to identify direct impacts of the GM court cases on policy, these cases are symbolically important, not least for anti-GM activists. For activists, each GM court case is understood as a part of their sequential actions for social change, and therefore has implications far beyond the courtroom. An example of their impact can be identified by the decisions of major biotechnology companies (including Bayer) to withdraw their research facilities from the UK partly because of their consideration of anti-GM activists (Vidal and Sample 2003; Mason 2004). Understanding the GM court cases as a symbolic forum assists us in understanding the implications of the court context for the wider social context in the UK.

It is clear that the problem of the society over the GM issues is closely linked to GMO policy formation, and both of these require consideration of the issue for the future. But the criminal proceeding for lawful excuse does not allow the issue of the future to be brought into the court, and it also excludes many other important aspects of the PP. Consequently, the precautionary principle in court is narrower and weaker than the same concept applied in the non-legal context. Therefore, we conclude that the courtroom setting for criminal proceeding is not an ideal forum for society to debate the GM controversy. However, the visual images of anti-GM campaigning activities are important for UK society as symbols of anti-GM perspectives, and the courtroom provides us with an interesting context where the two societal discourses take distinctive shapes.

References

Alder, J. (2000) 'More sorry than safe: assessing the precautionary principle and the proposed International Biosafety Protocol', *Texas International Law Journal*, 35: 173–205.

BBC (2003a) 'Doctors review GM crop evidence'. Available online at http://news.bbc.co.uk/1/hi/sci/tech/2711801.stm (accessesed 1 November 2003).

—— (2003b) 'Women deny wrecking GM crop', Available online at http://news.bbc.co.uk/1/hi/sci/tech/3007573.stm (accessesed 1 November 2003).

Belt, H. V. D. and Gremmen, B. (2002) 'Between precautionary principle and "sound science": distributing the burdens of proof', *Journal of Agricultural and Environmental Ethics*, 15: 103–22.

Burns, R. P. (1999) *A Theory of the Trial*. Princeton, NJ: Princeton University Press.

Cameron, J. and Wade-Gery, W. (1995) 'Addressing uncertainty: law, policy and the development of the precautionary principle', in B. Dente (ed.), *Environmental Policy in Search of New Instruments*. Dordrecht: Kluwer.

Carabine, J. (2001) 'Unmarried motherhood 1830–1990: a genealogical analysis', in M. Wetherell, S. Taylor and S. J. Yates (eds), *Discourse as Data: a guide for analysis*. London: Sage.

Carter, A. (2005) *Direct Action and Democracy Today*. Cambridge: Polity Press.

CropGen (2003) 'Momentous day for British agriculture'. Available online at http://www.cropgen.org (accessesed 20 October 2003).

Davies, Y. (2003a) 'R v Tilly and Davies: draft proof of Yvonne Davies as at 20.03.2003'.

—— (2003b) Interview with Mrs Davies, Carmarthen, 3–4 September.

Department for Environment, Food and Rural Affairs (DEFRA) (2003) 'GM crop farm Scale Evaluation results published today'. Available online at http://www.defra.gov.uk/news/2003/031016b.htm (accessesed 16 October 2003).

Drew, P. (1990) 'Strategies in the contest between lawyer and witness in cross-examination', in J. Levi and A. Walker (eds), *Language in the judicial process*. New York: Plenum.

Edwards, D. and Potter, J. (1992) *Discursive Psychology*. London: Sage.

Elliott, C. and Quinn, F. (2002) *English Legal System, fourth edition*. Essex: Pearson Education.

ENDS (2003) 'Sound science has its say on GM crops and biodiversity', *ENDS Report*, no. 345. 27–31.

European Commission (EC) (2000) *Communication from the Commission on the Precautionary Principle*. Brussels: COM(2000)1 final of 2.2.2000.

Fisher, E. C. (1999) 'The precautionary principle as a legal standard for public decision-making: the role of judicial and merits review in ensuring reasoned deliberation', in R. Harding and E. Fisher (eds), *Perspectives on the Precautionary Principle*. Sydney: The Federation Press.

Foucault, M. (1972) *The Archaeology of Knowledge*. London: Tavistock.

—— (1977) *Discipline and Punish: the birth of prison*. London: Penguin.

—— (1980) *Power/Knowledge: selected interviews and other writings 1972–1977*. New York: Harvester Wheatsheaf.

Freiman, M. J. and Berenblut, M. L. (1997) *The Litigator's Guide to Expert Witnesses*. Canada: Canada Law Book Inc.

Genetix Engineering Network (2001) *Genetix Update*, 20, Winter 2001.

—— (2002) *Genetix Update*, 22, Summer 2002.

GeneWatch UK (2002) 'Progress of the 2001 trials and plans for 2002', Available online at http://www.genewatch.org/CropsAndFood/Crop%20Trials/Upd2002.htm (accessesed 12 February 2003).

Grove-White, R., Macnaghten, P., Mayer, S. and Wynne, B. (1997) *Uncertain World*. Lancaster: CSEC.

Grove-White, R., Macnaghten, P. and Wynne, B. (2000) *Wising Up*. Lancaster: CSEC.

Guardian (2001) 'Part cleared over GM crop attack', *The Guardian*, 20 November.

Hajer, M. A. (1995) *The Politics of Environmental Discourse: ecological modernisation and the policy process*. Oxford: Clarendon Press.

Hall, S. (2000) 'Jury split over GM crop destroyers', *The Guardian*, 20 April.

Ho, M. and Saunders, P. (2003) 'Precautionary principle is science-based', *Science in Society*, 18: 35–7.

Hopkinson, G. (2002) 'The GM debate', *The Beekeepers Quarterly*, 71.

—— (2003) Interview with Mr Geoff Hopkinson, Stafford, 30 July.

—— (2004) 'Help', Available by e-mail from C.Ujita@bradford.ac.uk (accessesed 5 November 2003).

Irwin, A. (2004) 'Fearing the unknown', Available online at http://www.spiked-online.com/Articles/0000000CA375.htm (accessesed 11 February 2004).

Jordan, A. and O'Riordan, T. (1999) 'The precautionary principle in contemporary environmental policy and politics', in C. Raffensperger and J. Tickner (eds), *Protecting Public Health & The Environment: implementing precautionary principle*. Washington, DC: Island Press.

Kelso, P. (2000) 'Greenpeace wins key GM case', *The Guardian*, 21 September.

Lord Hailsham of St Marylebone (ed.) (1990) *Halsbury's Laws of England, fourth edition, Vol. II (I)*. London: Butterworth.

Mason, J. (2004), 'Syngenta shuts GM labs in UK', *Financial Times*, 1 July.

Mayer, S. and Stirling, A. (2002) 'Finding a precautionary approach to technological developments – lessons for the evaluation of GM crops', *Journal of Agricultural and Environmental Ethics*, 15: 57–71.

Mayer, S. (2003) Interview with Dr Mayer, Tideswell, 31 October.

McEldowney, J. F. and McEldowney, S. (2001) *Environmental Law and Regulation*. London: Blackstone Press.

Morris, J. (2000) 'Define the precautionary principle', in J. Morris (ed.), *Rethinking risk and the Precautionary Principle*. Oxford: Butterworth-Heinemann.

O'Riordan, T. and Jordan, A. (1995) 'The precautionary principle in contemporary environmental politics', *Environmental Values*, 4: 191–212.

R v Bellotti, Melchett and Others (2000) The Norwich Crown Court.

R v Tilly and Davies (2003) The Mold Crown Court.

Randle, M. (1995) *How to Defend Yourself in Court*. London: The Civil Liberties Trust.

Rawls, J. (1972) *A Theory of Justice*. Oxford: Clarendon Press.

Royal Commission on Environmental Pollution (1998) *Twenty-first Report Setting Environmental Standards*. London: The Stationery Office.

Royal Society of Canada (2001) *Elements of precaution: recommendations for the regulation of food biotechnology in Canada*. Ottawa: The Royal Society of Canada.

Salter, L. (1988) *Mandated Science*. Dordrecht: Kluwer.

Sharp, L. and Richardson, T. (2001) 'Reflecting on Foucauldian discourse analysis in planning and environmental policy research', *Journal of Environmental Policy and Planning*, 3: 193–209.

Shrimsley, R. (2001) 'Britain urged to boost GM crops', *Financial Times*, 20 November.

Smith, J. (ed.) (1996) *Smith & Hogan Criminal Law, eighth edition*. London: Butterworth.

Soil Association (2003) 'Soil Association comments on farm-scale trials'. Available online at http://www.soilassociation.org (accessesed 23 October 2003).

Stalloworthy, M. (2000) 'Damage to crops – Part 1'. *New Law Journal*, 19: 728–9.

Stirling, A. and Mayer, S. (1999) *Rethinking Risk: a pilot multi-criteria mapping of agenetically modified crop in agricultural systems in the UK*. Brighton: Science and Technology Policy Research Unit (SPRU).

Stirling, A. and Mayer, S. (2000) 'Precautionary approaches to the appraisal of risk: a case study of a genetically modified crop', *International Journal of Occupational Health and Environmental Medicine*, 6(4): 296–311.

Supply Chain Initiative on Modified Agricultural Crops (SCIMAC) (1999) *Guidelines for Growing Newly Developed Herbicide Tolerant Crops*. (n.p.): SCIMAC.

Taylor, S. (2001) 'Locating and conducting discourse analytic research', in M. Wetherell, S. Taylor and S. J. Yates (eds), *Discourse as Data: a guide for analysis*, London: Sage.

The Times (1999) 'Monsanto plc v Tilly and others'. *The Times*, 30 November.

The Times (2001) 'Tilly v DPP. DPP v Tilly and others'. *The Times*, 27 November.

Thoreau, H. D. (1966) 'Civil disobedience', in O. Thomas (ed.), *Henry Davis Thoreau: Walden and civil disobedience*. New York: W. W. Norton.

Tilly, R. (2003a) *R v Tilly and Davies: Proof of Rowan Tilly as at 24.03.2003*.

Tilly, R. (2003b) Interview with Ms Tilly, York, 27 December.

United Nations Conference on Environment and Development (UNCED) (1992) *Rio Declaration on Environment and Development*. New York: UNCED.

Vidal, J. and Sample, I. (2003), '5 to 1 against GM crops in biggest ever public survey'. *The Guardian*, 25 September.

Wilkinson, D. (2002) *Environment and Law*. London: Routledge.

Wynne, B. (2001) 'Expert discourses of risk and ethics on genetically modified organisms: the weaving of public alienation', *Notizie di Politeia*, XVII, 62: 51–76.

Wynne, B. and Mayer, S. (1993) 'How science fails the environment', *New Scientist*, 138: 33–5.

6   The social construction of the biotech industry

Kean Birch

Introduction

Because the biotech industry is still in its infancy – having 'as a whole ... never been profitable' (Ernst & Young 2003: 5) – it represents an ideal research subject in both economic sociology and science and technology studies (STS) for those wishing to study the development of a technology market. The current cross-fertilisation of ideas between these two disciplines makes it a particularly germane topic (see Callon 1998a; MacKenzie 2003; Woolgar *et al.* 2005; Barry and Slater 2005). It has also been of relevance to United Kingdom and European policy-makers concerned with the development of the *knowledge economy*, exemplified by hi-tech industries like biotechnology (OECD 1996). However, the current policy emphasis on biotechnology as a saviour of our economy and the planet (see BIGT 2003 foreword by Tony Blair) also makes such research an increasingly important activity because this policy discourse tends to disguise the political and economic motivations for such policy changes and instead concentrates, contradictorily, on the *natural* innovativeness of hi-tech as a potential boost to competitiveness. Thus it deliberately ignores the effects of national political decisions on economic activity, highlighted by Laura Tyson (1992), and their economic and social implications. Furthermore, according to Paul Krugman (1996) the very concept of competitiveness leads to a biased industrial and trade policy focused on export-driven manufacturing sectors – which can represent a small proportion of a country's overall production – to the detriment of service sectors.

Herbert Gottweis (1998: 159) argues that the specific European concern with competitiveness in biotechnology has developed because the United States biotech industry 'attain[ed] a mythical status in the European policy discourse', a status that drove subsequent policy.[1] This myth predated the emergence of the biotech industry by some years, with the West German federal government even changing its constitution in 1969 so that it could support large-scale research (Gottweis 1998: 183). More recently, European Union (EU) drug regulation has been deliberately reoriented to promote European R&D and competitiveness (Abraham and Lewis 2000), which has had an important impact on the wider global, not just the European, regulation of medicines (Abraham and Reed 2002, 2003). In the UK the USA has also been characterised as a threat to UK biotechnology competitiveness over several decades, starting with the 1980 *Spinks Report* (ACARD 1980). The rhetoric of the 'threat' to the UK then persisted, being reiterated in a 1993 House of Lords Select Committee on Science and Technology report and again in the report from the government's most recent initiative, the 2003 Bioscience Innovation and Growth Team (BIGT).

Whilst the USA has always dominated the biotech industry, illustrated by the annual Ernst & Young reports (2000, 2001, 2003), the reasons for this are not always as clear-cut as mainstream accounts contend. According to these accounts, US dominance was achieved through several factors particular to the biotech industry (see Sharp 1996; van Reenen 2002). First, the US benefited from first-mover advantage. For example, Celltech (the first UK biotech firm) was founded in 1980 when Genentech (founded in 1976) went public and doubled its share value in one day (Owen 2001: 6). Second, the US had a more entrepreneurial environment that encouraged the formation of 'dedicated biotech firms' (DBFs) keen to exploit the new technological opportunities (see ACARD 1980; Walsh *et al.* 1995; Acharya *et al.* 1998). Third, the US makes a larger investment in the biological sciences and has done so from an earlier time than other countries (Walsh *et al.* 1995). The US public funding agencies spend nearly half of the total public science budget on biotech research, amounting to $21.3 billion at their disposal in 2003 (Cooke 2003), creating a virtuous circle of inward investment as foreign firms sought to benefit from US capabilities by investing in US firms (Buctuanon 2001: 29). Consequently, European pharmaceutical firms have collaborated with US biotech firms more than with their national biotech firms (Sharp 1996). Fourth, US DBFs engaged in a larger and denser series of collaborative arrangements – networks of complementary firms and public research organisations (Acharya *et al.* 1998; Acharya 1999). Finally, European financial markets were and are more stringent in their regulatory requirements stifling investment in DBFs (Acharya *et al.* 1998; Prevezer 2003).

The following discussion has been designed to illustrate how these differences in the biotech industry arose as a result of deliberate legal and political institutional change in the USA, motivated by a fear of lost competitiveness. As such I will first seek to expound a social explanation of the biotech market by highlighting several interconnections between science and technology studies (STS) and economic sociology that problematise economic theories of technological change. This will, in particular, draw upon the work of Karl Polanyi (1957) in his discussion of markets as 'instituted process'. Second, I will detail a series of deliberate changes to law and industrial policy in the USA designed to benefit the biotech industry and thereby construct a specific technological market. The focus is centred on the USA because these changes have been instituted most clearly there as part of a political programme of national interest. In the final section, I will explore what the construction of the biotech industry means for global distributive justice, especially in relation to US government policy and its impact on foreign countries.

Old and new connections between science studies and economic sociology

In classical economic models of economic growth, production is characterised by two factors of production: capital (i.e. machinery) and labour (i.e. skills). As a model it assumes that technology is first an exogenous public good and second that it is neutral, merely shifting the production function (Nelson and Winter 1974; Coombs *et al.* 1987). Later economic theories, such as those of Joseph Schumpeter (1939), problematised this rather simple formulation by arguing that knowledge and technological change play a vital part in growth, although arguing again that basic science is located outside of the firm itself. However, the more recent work of Nelson and Winter (1982) and others in evolutionary economics has sought to reposition these aspects of growth in the firm, although, once again, not without several problematic issues (see van den Belt and Rip 1987). It is useful to draw upon work in science and technology studies (STS) and economic sociology to explore these limitations in economics, especially in the creation and development of a particular market and the technologies that both enable it to function and are influenced by it.

Science and technology studies

Science and technology studies (STS) has always had a close affinity with research in both economics and topics related to the economy, as evident in early STS analyses of production technology (see MacKenzie and Wajcman 1999) and considerations of economic influences on technological change (see Misa 1997). Outside of these specific STS approaches there has been an array of associated research of particular relevance to biotechnology, such as innovation studies (see Dosi 1988), that developed around the Schumpeterian and firm-based traditions – in other words, evolutionary economics – in economic theory, and largely based on Freeman's (1982) concept of 'systems of innovation'. Just some examples of this research include Coombs and Metcalfe (2002) on firm capabilities, Woiceshyn (1995) on product innovation, and della Valle and Gambardella (1993) on strategies for innovation. Alongside these approaches there has also been an institutionalist tradition of research on biotechnology, both from an economic (that is, transaction costs) and sociological (that is, varieties of capitalism) perspective. Again, just a few examples include the work of Casper and Kettler (2001) on the institutional differences between German and UK biotechnology sectors, Malinowski (2000) on US 'responsive regulation', and Coriat *et al.* (2003) on a 'new science-based innovation regime'.

One commonality shared by all this research is the general treatment of the economy or markets as exogenous phenomena, whether in terms of an external influence on technological trajectories or development, or as an environment in which firms operate. It is only with the institutionalist approaches that the economy or market is considered in any way a contingent artefact, although this tends to apply only to the sociological arguments. In contrast, economic theory is largely based on the work on transaction costs by Ronald Coase and Oliver Williamson and presumes that institutional arrangements will automatically assume the most efficient form; i.e. either market, firm, or, in later formulations, network. In sociological approach there is no such supposition. However, some of the recent research (Callon 1998a; Barry and Slater 2005) in STS on markets and the economy has sought to problematise the very treatment of markets in economic sociology that criticised these economic theories.

In this new approach, strongly centred on work by Michel Callon (1998a, 1998b, 1999), no distinction is made between the economic and the social because both are based, in some sense, upon the 'calculativeness' of human activity. As a consequence, there is no distinction between economic and social worlds, which presents a particular critique of the perspective that positions the latter as an exogenous influence on the former (Barry and Slater 2005). Instead, both worlds are aspects of the same activity, although they are produced through framing certain features as relevant to markets and excluding others; the latter come to represent 'externalities' (Callon 1998b). Thus a market is constituted through the actions of its participants, particularly those of economists, and the technologies of economics; i.e. techniques of accounting (Callon 1998a). However, there is an important ambiguity in this formulation; the distinction between the constitutive role of lay and expert 'economising' actors is not clear.

If it is the expert that constitutes the market, then STS approaches – i.e. scientists construct science – have been too crudely transposed into economic sociology – i.e. economists construct economies. Alternatively, if all actors engage in the economy it means that some will benefit from closer relationship to the technologies of economics and, presumably, better calculating abilities. Valorising the activity of economists over everyday market exchange also tends to obscure important elements in the economy highlighted in work in the anthropology of markets (see Carrier 1997; Miller 2002).[2] These elements include a range of different motivations, over and above 'utility maximising' ones, that can be construed in terms reminiscent of Karl Polanyi's (1957, 2001) argument that markets are embedded in institutions (i.e. social norms and values). Despite these cautionary notes, the new approaches in STS help to build upon current research in economic sociology that is questioning the continuing separation of disciplinary fields.

Economic sociology

Greta Krippner (2001) argues that the maintenance of disciplinary boundaries between economics and sociology in the social sciences has stultified research on market causality by, predominantly, separating economy and society. Consequently both economics and sociology share a common theoretical atomism as well as a delineated perception of society whose demarcations presage how the market is understood and thus impact upon political decision-making. This is illustrated in Karl Polanyi's argument that a distinction needs to be drawn between the *substantive* and *formal* meaning of the term 'economic' to clarify our understanding of markets. The former refers to the interchange between 'nature and social environment' which supplies us with the 'means of material want satisfaction', while the latter refers to the logic of means–end relationships where there is a 'definite situation of choice' (Polanyi 1957: 243).

The distinction is important because economic theory rests on the formal meaning concerning questions of 'scarcity' (*contra* 'subsistence') in exchange, entailing the withholding of resources so that a choice can be made, as necessitated by insufficient means, and a market created (Polanyi 1957: 244). Scarcity needs to be a fact, whether 'insufficiency is due to Nature or Law', resulting in the 'introduction of purchasing power as the means of acquisition' which 'converts the process of meeting requirements into an allocation of insufficient means with alternatives uses, namely, money' (Polanyi 1957: 247). Consequently, economic theory concentrates on prices and therefore presents social actions as a causal consequence of 'locational and appropriational movements' (Polanyi 1957). The appropriation of resources enables the withholding of resources from exchange whilst the creation of scarcity stimulates value by avoiding uncertainty. Therefore the market is an effect of simulated scarcity, induced by those who appropriate resources. Thus the need to control access to both inputs (i.e. skills) and outputs (i.e. consumers) is vital, exemplified in biotech with the concern about strong intellectual property rights (IPR).

In this form, market structures are dependent upon inducing the disintegration of social relations and the subsequent integration of disparate individuals divided by an advanced division of labour. Through disintegration actors can offer a scarce resource on the market and achieve a price that will afford them access to other scarce resources. This means that the market is dependent upon a distinction being drawn between those who hold resources and those who have resources withheld from them. Any drive for disintegration is premised upon the theoretical concept that '*social atomization is prerequisite to perfect competition*', where competition arises between resources holders (Granovetter 1985: 484). However, there are two problems with this premise. First, it is presumed to ensue from choices entered into even when such choices 'involve more than one individual' (Granovetter 1985: 487). Second, and more important, economic theory has shown that the success of a firm's behaviour is dependent upon imperfect competition (Schumpeter 1939; Loasby 2000). Thus atomisation does not account for the systemic interaction between actors.

This transforms our understanding of markets, since price competition is both implausible and undesirable because a firm's systemic interactions would quickly collapse in such an environment (Fligstein 1996). Price competition requires the development of a market outside a locality so that choice is enabled under means–end relationships. However, this is always unsuccessful because locality is how scarcity is maintained (Polanyi 1957). Markets are dependent upon locality in two important instances. First, the resources that are withheld are spatially derived; and second, markets depend upon the state (Fligstein 1996; Krippner 2001). Because no firm (or actor) can determine, *a priori*, what activity will maximise profit, firms (or actors) seek to create stable markets through scarcity. Biotech provides a fitting example because firms maximise profit by controlling the 'supply of scientists who have the knowledge about the products' (Fligstein 1996: 666). The ability to exclude others gives them an advantage, but this can easily be lost when a scientist leaves the firm. To alleviate, if not obviate altogether, this situation, biotech firms engage in alliances with both universities and other firms (Fligstein 1996). Thus biotech firms embed themselves in systemic interactions that provide the necessary advantage, but that also threaten ossification and inertia in new circumstances (Uzzi 1997).

The difference between a successful and an unsuccessful firm, then, is the result of systemic interactions – i.e. the ability to influence their context (Carruthers and Uzzi 2000). In his theory of embeddedness Granovetter (1985) provided an initial, yet ultimately limited, means to conceive of such interactions in exchange (see Krippner 2001 for a critique). Granovetter argued that an over-socialised view of markets was incomplete because individuals cannot be assumed to have internalised normative standards of behaviour. Rather, markets are embedded in trust relationships that are meant to deter malfeasance through either force or fraud (Granovetter 1985). The problem with this position is that it implies that there is a core of market activity that exists outside of interactions (Krippner 2001). However, exchange precludes such a perspective because it necessitates transactions between at least two actors since the formal meaning concerns a relationship between people, rather than between a person(s) and resource(s). Thus a *formal* market cannot exist upon the assumption that a person owns a resource; rather, a person has to own the resource at the same time that other people are excluded from ownership. This ultimately requires that political structures be created to enforce a specific set of interactions.

The construction of markets in the United States biotech industry

The discussion in this section concerns a range of legal and political changes instituted in the USA that affect the economic, social and political arrangements shaping the biotech industry. There is a whole raft of changes, designed to boost competitiveness in hi-tech generally and biotech specifically, but this discussion cannot be exhaustive and therefore is limited to a few crucial changes that are indicative of general trends.[3]

Legal changes in the USA

Perhaps the most significant legal change was the 1980 *Diamond v Chakrabarty* (DvC) ruling of the Supreme Court concerning the 'issuance of a patent for a genetically engineered, oil-digesting bacterium': despite its importance, however, it was not a new situation (Iwasaka 2000: 1517). An earlier 1911 ruling in *Parke-David & Co. v H. K. Milford & Co.* had held that a naturally occurring material (adrenaline) could be patented because it had been purified and modified (Ossorio 2002). Therefore what DvC did was uphold a 'broad patent' in biotech specifically (Mowery *et al.* 2001: 103), drawing upon the 1952 Supreme Court judgment that patentable matter could include 'anything under the sun made by man' (quoted in Krimsky 1999). Post-DvC, the US Patent and Trademark Office (USPTO) gradually shifted its position so that by 1987 it allowed multicellular organisms (i.e. animals) to be patented (Dutfield 2003). A year later, in 1988, the USPTO granted Harvard University a patent for its Oncomouse, whilst in 1990 a California Supreme Court ruled that 'a patient did not have a property right to his body tissues after they were used by researchers to develop commercially important cell lines' (Slaughter and Rhoades 1996: 323).

Of secondary significance to *Diamond v Chakrabarty* was the 1982 establishment of the Court of Appeals for the Federal Circuit (CAFC) as the final court of appeals for patent cases (Mowery *et al.* 2001). Post-CAFC, patent infringements were increasingly found in favour of the plaintiff. According to Quillen (1992), two-thirds of pre-1982 litigations were found invalid compared with post-1982 litigations, when two-thirds were found valid. Katz and Ordover (1990) also argued that CAFC upheld patent rights in 80 per cent of post-1982 cases. However, in 2000 CAFC ruled (*Festo Corp. v Shoketsu Kinzoku Kogyo Kabushiki Co., Ltd.*) that a patent claim that had been 'narrowed' was barred from claiming a doctrine of equivalents in relation to the excluded claims (Robertson 2002: 639). This particularly threatened biotech patents because it restricted their scope, an issue that caused the Supreme Court to overturn the decision in 2002. In their ruling they concluded that the bar was impermissible because it 'unfairly diminished the scope of value of existing patents', where scope could cover unforeseen usage (Robertson 2002).

Both CAFC (1982) and DvC (1980) expanded the scope of patent claims because they contradicted the 1966 Supreme Court decision in *Brenner v Manson* that a 'product has no patentable utility if its only use is as an object of further scientific research' (Kaplan and Krimsky 2001: 3). This position was reinforced by the decision in *In re Durden* (1985) which created so-called 'Durden rejections'. Here decisions were based upon the idea that the process leading to a product 'could be found obvious merely because its steps were old and well known' (Maebius 1996: 1). It was not until 1995 in two CAFC rulings that *Durden* was overturned. The first, *In re Ochiai*, ruled that the whole process of invention must be considered (as in *Graham v Deere* 1966), therefore allowing process claims where a patent for the initial and final material exists. The second, *In re Brouwer*, ruled that novelty was based on the final product, therefore allowing process claims as part of the product claim (Maebius 1996). These decisions occurred at the same time that CAFC ruled that *Brenner v Manson* (1966) had reached the wrong conclusion. CAFC ruled that research tools and intermediary processes should be included under patent law section 101 because utility 'necessarily includes the expectation of further research and development' (Kaplan and Krimsky 2001: 4). CAFC highlighted the relevance of this to pharmaceuticals. Following these rulings the USPTO produced new guidelines in January 2001 stating that an invention had to show 'specific and substantial and credible utility', where this could mean a theoretical utility (Laurie 2003: 9). This continued to blur the line between research tool and commercial activity through a circular justification – i.e. research tools are patentable because they are commercial, but research tools can only be sold because they are patentable (Kaplan and Krimsky 2001).

Political changes in the USA

The Cohen and Boyer discovery of recombinant DNA (rDNA) in 1973 led to fears about environmental contamination (see Gottweis 1998). Consequently the 1975 Asilomar Conference sought to establish control over DNA research through, Evans (2002) argues, assuaging public fear rather than legislative action. However, regulatory inactivity was also a result of legislators who 'recognized the value of the technology as a boost for the American economy' (Hughes 2001: 568). Thus in 1981 Senator Al Gore stated that there was a 'widely shared realization of the vast commercial potential of genetic technologies', a potential that could be exploited by foreign firms if the US government restricted industry with regulations (US Congress 1981: 1). In 1984 the Office of Technology Assessment (OTA) produced a report on international competitiveness in biotechnology highlighting the advantages and disadvantages of the various countries under analysis, noting the *beneficial* regulatory environment in the US (Office of Technology Assessment 1984: 11–17). These two instances are indicative of concerns over US competitiveness which political legislation further illustrates.

Early indications of these concerns are evident in the Small Business Investment Companies (SBIC) Act of 1958; the Plant Variety Protection (PVP) Act of 1970; and the Trade Act of 1974. The SBIC Act meant that federal government guaranteed early venture capital (VC) funding, although in the 1970s and 1980s SBIC VC collapsed (Florida and Kenney 1988; Gompers and Lerner 1998). The PVP Act was an equivalent of UPOV (the International Convention for Protection of Plant Varieties) that the USA had not adopted when other countries had done so in 1961 (May 2000; Drahos and Braithewaite 2002). Finally, the Trade Act established the 301 and 301 Special processes whereby the US withdrew favourable trade status from countries not fulfilling US demands on issues like intellectual property protection (Drahos and Braithewaite 2002). These early actions represent a foregrounding to what happened later during the 1980s and 1990s.

Two major developments occurred in 1980. First, the Bayh-Dole Patents and Trademark Amendments Act proved highly significant because it stimulated other legislative acts. It was designed to eliminate restrictions on licensing and allow universities to retain patents from federally funded research.[4] Subsequent legislation, through executive order, in 1983 extended the act to small businesses (Slaughter and Rhoades 1996; Krimsky 1999; Poyago-Theotoky *et al.* 2002).[5] In the same year the Stevenson-Wydler Technology Innovation Act (1980) was introduced which created a technology transfer office at the National Institutes of Health (NIH) as well as a centre to provide information to industry about commercialisable research (Kuhlman 1996; Slaughter and Rhoades 1996). Following on from these, there were at least another fifteen significant changes introduced during the 1980s and at least five during the 1990s.

The following represents a small sample. First, the Small Business Innovation Development Act (1982) required that federal funders allocate 1.25 per cent of their R&D budget to small and medium enterprises (SMEs) (Walsh *et al.* 1995; Slaughter and Rhoades 1996). Audretsch suggests that this had risen to 4 per cent by 2001, representing $411 million in annual grants to biotech SMEs from the NIH alone (Audretsch 2003: 29). Second, the National Cooperative Research Act (NCRA) of 1984 relaxed antitrust laws for joint research ventures (Katz and Ordover 1990).[6] Third, the Federal Technology Transfer Act (1986) encouraged commercialisation of federal R&D and created Cooperative Research and Development Agreements (CRADA) between private and public organisations so that companies could license publicly funded discoveries (Kuhlman 1996; Slaughter and Rhoades 1996; Buctuanon 2001). Fourth, the 1988 Omnibus Trade and Competitiveness Act introduced Special 301s as public law which meant that the US tagged countries that did not concede to multilateral agreements on IP, threatening trade restrictions if they did not comply (Slaughter and Rhoades 1996). It also introduced the Advanced Technology Program (ATP) which was designed to support collaborative research on generic technologies (Poyago-Theotoky *et al.* 2002). Finally, the Biotechnology Process Patent Act of 1995 restricted foreign competitors from using similar processes to create products that were patented in the US (Buctuanon 2001).

Increasingly significant has been the US influence at the international level, where the US negotiation 'about strengthening the IPRs regime was couched in terms of ensuring and maintaining "its competitive advantage"' (Tansey 2002: 580). The most important development was the creation of the World Trade Organisation (WTO) and the Trade Related Aspects of Intellectual Property (TRIPs) measures. In relation to biotech, Article 27 of TRIPs ensured that biotech patents were necessary for all signatories, whilst Article 33 harmonised patent length to a minimum of twenty years. Non-compliance would lead to the withdrawal of General Agreement on Tariffs and Trade (GATT) privileges (May 2000). The WTO also required that patents be available for both product and process patent claims in any technology (OECD 2002).

National competitiveness, biotechnology and social justice

US government intervention in global trade

The changes instigated in the US stem from a general concern with national competitiveness whose origins can be traced back to Nixon's withdrawal from the Bretton Woods agreement in 1973. This boosted the US economy through the dollar's devaluation against the Japanese yen (see Strange 1998) and by ending controls on international capital flows (Harvey 2003). Japan was forced into supporting the dollar to stop further devaluation that would have increased the cost of Japanese imports to US markets. Thus the Japanese Ministry of Finance lowered interest rates to create demand for investment in US currency, industry and government debt (Strange 1998: 46–7). Foreigners now own 48 per cent of US treasury bonds, 24 per cent of US corporate bonds and 20 per cent of all US businesses, totalling $8 trillion (Sharma *et al.* 2004: 54). This debt acquisition is compounded by half of all world exports being denominated in dollars, thus recycling these dollars back into the US economy (Sharma *et al.* 2004). Bruce Carruthers's history of the London financial markets points to an interesting possibility that these US debts tie the creditors to the success of the US economy as they 'acquire a political interest in the survival of the sovereign regime' (Carruthers 1996: 4).

The Bretton Woods withdrawal also has to be considered in light of the 1973 oil crisis. In his 2003 Clarendon Lectures, David Harvey argued that the resultant petrodollars were again recycled by US business, although this time as international loans to developing countries. By 1983 the interest on these debts was three times higher than US banks' profits from foreign direct investment (FDI) (Parenti 1995: 20). Defaults were averted through International Monetary Fund (IMF) bailouts with new loans that essentially subsidised the poor lending policies of US business. The IMF can therefore be seen as an arm of the US economy, functioning as a tribute mechanism enforced by the US military (Harvey 2003). To make it even more profitable for business the IMF funds used to bail out the defaulting countries are primarily derived from the taxes of citizens of developed countries as the result of falling corporate tax take. For example, the share of tax that US corporations pay has decreased dramatically now, representing 7 per cent of federal tax income, compared with 22 per cent in the 1960s and 50 per cent in 1945 (Parenti 1995; Multinational Monitor 2003). Furthermore, according to the US General Accounting Office, in 1993 around 40 per cent of corporations with over $250 million in assets 'either paid no income taxes or paid taxes less than $US100,000' (quoted in McKinley 2001: 9).

Between 1990 and 1996 the pharmaceutical industry had average effective tax rates of 16 per cent against 27 per cent for industry as a whole (Public Citizen 2001). Such hi-tech industries also have higher than usual rates of return on investment, at 20–30 per cent, in comparison with the whole business sector at 10 per cent (Borrus and Stowsky 1997: 2). The differences between tax rates and rates of return crudely mirror each other. Two conclusions can be drawn from this. First, the US public effectively subsidises the high returns received by private investors in the pharmaceutical and biotech industry (through lost tax dollars, public debt and higher drug costs). Second, it could be argued that the US national debt, financed by foreign creditors, funds US industrial policy. The US government encourages such high value-added industries because they are seen as producing benefits across the whole of the economy (CRIC 2000). However, high value-added results from capital-intensive industries for the simple reason that these industries need to charge a higher mark-up than other industries (Krugman 1996). Therefore the level of value-added rises as the prices of products rise – as they have done dramatically in the pharmaceutical sector (see NIHCM 2002 for the difference in average price per prescription, which has increased most in incrementally modified drugs). For example, in the USA between 2001 and 2002 the average price of prescription drugs rose by 9.5 per cent, reaching total retail prescription sales of $166.6 billion (Kaiser Family Foundation).[7]

Whose national competitiveness, whose economic growth?

The concern with hi-tech tends to resolve into specious government concern, according to Krugman (1996), with competitiveness that stretches back to the early 1980s illustrated by Ronald Reagan's 1983 study into US competitiveness (Tyson 1992). Cypher (1987) shows how such concerns cross the military–industry boundary, with competitiveness being cast as vital to national security, resulting in such programmes as MANTECH (Manufacturing Technology). Thus the hi-tech sector can be redefined as a military requirement and therefore reclassified as a national security asset (Stowsky 1992). This creates a desire for 'dual-use' generic technologies, like MANTECH, and ties commercial activity more closely into the security interest.

However, US intervention is not limited to hi-tech as the US intervenes in all commercial sectors (see Chomsky 1997, 1999; Vidal 1999; Chomsky 2000). The preceding cannot be dismissed as conspiratorial musings because even Laura Tyson agrees that the characteristics of national competitiveness in hi-tech sectors 'is less a function of its national factor endowment' and, more importantly, 'a function of *strategic interactions* between its firms and government, and between them and the firms and governments of other nations' (Tyson 1992: 3; my italics). The meaning of 'strategic interactions' can cover both military and economic (i.e. bilateral trade agreements) interventions in foreign countries.

A recent example of a strategic interaction is the shift in global IPR. This appears to be designed to encourage an emphasis on the interpretation of patent applicability as being primarily concerned with 'utility' (i.e. competitiveness) rather than 'novelty' (Padron and Uranga 2001). These changes were driven by economic considerations that some US judges found 'questionable' (Ko 1992: 788). Thus at the international level IP changes appear to benefit the US, in particular, and other developed countries. The TRIPS (Trade Related Intellectual Aspects of Property Rights) agreement – part of the World Trade Organisation (WTO) – requires that foreign and national firms are treated equally by ratifying signatory countries through setting minimum standards of IP protection (Dixon and Greenhalgh 2002). In relation to biotech, there are two issues. First, under Article 34, the burden of proof in patent litigation is shifted to the defendant, who needs to prove their innocence (Polster 2001). Second, wealth transfer benefits the developed nations, as they hold more patents. Thus, according to the World Bank, the USA will benefit from a transfer of $19 billion per year from developing countries in IP rents (Dixon and Greenhalgh 2002: 46). Another study, by the OECD/World Bank, concluded that the developed world would accrue 70 per cent of any additional income from the WTO changes, whilst Africa, for example, would actually lose $2.6 billion by 2002 (in McKinley 2001: 11). As a consequence of these changes, developing countries will not be able to adapt their patent regimes to best suit their national needs. This contrasts with the historical precedent set by developed countries where, for example, between 1790 and 1836 the US limited patent rights to nationals only (Padron and Uranga 2001).

On a simple level, economic growth advocates have to assume technological determinism to justify their theories. An example of this can be found in William Easterley's (2002) discussion of 'creative destruction' and the resistance engendered by technological change. Advocates of economic growth assume that technological progress occurs on a quality continuum, representing successive waves of technical progress – as in Kondratieff Waves. Such growth models, basing themselves on Solow's contention that economic growth results from technology, predict that growth is dependent upon the continual adaptation to new technical developments, rather than that technology is dependent upon continual adaptation to economic developments (for purposes here I equate economic with political and social). A simple illustration of this is current GM technology. Two examples suffice. First, the reason that herbicide-resistant GM crops are the most prevalent (representing 75 per cent of GM crops worldwide in 2002) is because the main agrochemical firms recently lost their pesticide patents (Paul and Steinbrecher 2003: 79 and 187). Second, terminator technology was developed, at public expense in the US, to protect US agricultural technology (Paul and Steinbrecher 2003: 200).[8] These two developments are not technologically inevitable, they are representative of a specific set of socio-economic concerns, now enshrined in global governance organisations; i.e. the TRIPS agreement. TRIPS was essentially written by the US Intellectual Property Rights Committee comprised of thirteen firms including DuPont and Monsanto (Paul and Steinbrecher 2003: 34). Thus the contention that economic growth is predominantly a result of technological change is an inadequate perspective. Rather growth is a result of socio-economic change, how we order ourselves politically, socially and economically.

Conclusion

There is a distinct set of social interactions that have been oriented to benefit the biotech industry, both in the US and – increasingly so – in the UK. Several examples may suffice to illustrate this. First, the biotech industry only employs a tiny proportion of the UK population (around 22,000) yet receives a disproportionate level of funding support and government intervention: i.e. a dedicated research council, R&D tax credits. According to Thompson and Warhurst (2001), the real growth area of the UK economy is in low-level service jobs,[9] which could lead to a concurrent depression in wages as witnessed in the USA. For example, in 2001 a manufacturing worker in the US would have had to work 81 weeks to receive the 1947 median family's annual income (Henwood 2004). The minimum wage in the US has also remained the same since 1997, at $5.15 an hour (Multinational Monitor 2002). Second, what if the loss of comparative advantage is really the result of intra-firm transactions, which do not benefit either the country of manufacture or the import country when tariffs are reduced? Thus in the US transactions between firms and their foreign affiliates/subsidiaries account for '40–50 per cent of total imports and 35–40 per cent of total exports' (McKinley 2001: 7). The countries where production is occurring do not necessarily benefit from the location of industry within their borders, as illustrated by Naomi Klein (2001) in her discussion of Export Processing Zones (EPZ), because of the need to keep wages low (by withdrawing minimum wage laws from the EPZ) and to offer tax breaks to industry.

The construction of a biotech industry appears to have created a reliance on large pharmaceutical firms, both in terms of firm strategy and institutional influence. The current technological regime has made product development particularly costly and uncertain, necessitating access to the wealth and marketing power of large firms. Consequently, biotechnology has primarily been incorporated into a blockbuster drug format that is easily understood (Green 2002: 203), although perhaps now proving both financially and technologically less successful (see Nightingale and Martin 2004). For example, David Rasnick suggests that the biotech industry fails to make money because (a) the products do not work, and/or (b) products are designed for niche markets (i.e. rare diseases) (Rasnick 2003: 356). However, despite some current concerns about the direction of the biotech industry – particularly in relation to pharmaceutical products – the myth of dominance still exists. A spiralling fear of lost national competitiveness has previously lead to the adoption of weak regulatory controls (Wright 1993), which appears to be happening once again as is evident in the UK shift towards an EU-inspired 'outcome-based' regulatory system (DTI 2003: 91). Such a system has been promoted in the European Union, alongside specific changes to the regulation of pharmaceutical products (see Abraham and Lewis 2000), as part of a deliberate strategy to enhance competitiveness across Europe (Abraham and Reed 2003). However, by concentrating on the competitiveness of the multinational pharmaceutical industry, the EU has also had an impact on the regulatory environment of other developed nations (including the USA), especially through the International Conference on Harmonisation (ICH) process, first convened in November 1991 (Abraham and Reed 2002, 2003). Overall, what the changes in both the USA – dealt with only briefly here – and the EU illustrate is that the construction of the biotech industry is an ongoing process that has been instituted across national borders and with an increasingly international structure.

Notes

1 The difference between the USA and Europe can be termed a 'mythic' interpretation because of the factors that policy-makers emphasised when talking about the advantages that the USA had. For example, I will try to show later that the idea that the USA had a much more entrepreneurial business environment is a misrepresentation of deliberate shifts in law and policy.

2 One of Daniel Miller's (2002: 219) critiques is that 'Callon follows the economists in mistaking a representation of economic life for its practice'. For example, Callon (1998a) claims that participants in economies actively shape that which they describe, suggesting that all market participants are performing economics. However, in a particularly interesting article, Ferraro *et al.* (2005) illustrate how economics affects the behaviour and assumptions of economists, making them more self-interested and incentive motivated than non-economists.

3 A fuller list of changes can be obtained from the author.

4 The promotion of licensing was also facilitated with the introduction of a hand's-off policy in relation to IP policing by William Baxter after his appointment as head of the Antitrust Division in 1981 (Drahos and Braithewaite 2002: 166). In 1989 the Antitrust Division also produced guidelines (called *The*

Antitrust Enforcement Policy for International Operations) that relaxed antitrust policy, thereby further promoting licensing (Ordover 1991).

5 As a side note, all agreements under the Bayh-Dole Act are confidential, therefore the public (by whom the research was funded) has no oversight of the decisions made (Kuhlman 1996).

6 The NCRA law was extended in 1989 (through three Acts) to cover joint production, manufacture and marketing of products that result from cooperative R&D (Katz and Ordover 1990).

7 See online at http://www.statehealthfacts.kff.org/cgi-bin/healthfacts.cgi (accessed 7 June 2006).

8 Such genetically modified (GM) crops are also designed to suit intensive farming practices in the developed world and cash-crop production in the developing world (Magdoff 2004).

9 For example, research in the ESRC Future of Work Programme found that the number of people employed as hairdressing managers and proprietors has risen by around 300 per cent during the 1990s, more than any other profession (Thompson 2004).

References

Abraham, J. and Lewis, G. (2000) *Regulating Medicines in Europe*. London: Routledge.

Abraham, J. and Reed, T. (2002) 'Progress, innovation and regulatory science in drug development: the politics of international standard-setting', *Social Studies of Science*, 32(3): 337–69.

—— (2003) 'Globalization of Medicines Control', in J. Abraham and H. Lawton Smith (eds), *Regulation of the Pharmaceutical Industry*. Basingstoke: Palgrave Macmillan.

ACARD (with ABRC and The Royal Society) (1980) 'Biotechnology: report of a Joint Working Party (aka The Spinks Report)'. London: HMSO.

Acharya, R. (1999) *The Emergence and Growth of Biotechnology*. Cheltenham: Edward Elgar.

Acharya, R., Arundel, A. and Orsenigo, L. (1998) 'The evolution of European biotechnology and its future competitiveness', in J. Senker (ed.), *Biotechnology and Competitive Advantage*. Cheltenham: Edward Elgar.

Audretsch, D. (2003) 'The role of small firms in US biotechnology clusters', in G. Fuchs (ed.), *Biotechnology in Comparative Perspective*. London: Routledge.

Barry, A. and Slater, D. (2005) 'Introduction', in A. Barry and D. Slater (eds), *The Technological Economy*. London: Routledge.

van den Belt, H. and Rip, A. (1987) 'The Nelson-Winter-Dosi model and synthetic dye chemistry', in W. Bijker, T. Pinch and T. Hughes (eds), *The Social Construction of Technological Systems*. London: MIT Press.

BIGT (2003) 'Improving national health, improving national wealth'. London: Bioscience Innovation and Growth Team.

Borrus, M. and Stowsky, J. (1997) 'Technology policy and economic growth'. Berkeley, CA: BRIE Working Paper 97.

Buctuanon, E. (2001) 'Globalization of biotechnology: the agglomeration of dispersed knowledge and information and its implications for the political economy of technology in developing countries', *New Genetics and Society*, 20(1): 25–40.

Callon, M. (1998a) 'Introduction', in M. Callon (ed.), *The Laws of the Markets*. Oxford: Blackwell Publishers/The Sociological Review.

—— (1998b) 'An essay on framing and overflowing: economic externalities revisited by sociology', in M. Callon (ed.), *The Laws of the Markets*. Oxford: Blackwell Publishers/The Sociological Review.

—— (1999) 'Actor-network theory – the market test', in J. Law and J. Hassard (eds), *Actor–Network Theory and After*. Oxford: Blackwell Publishers/The Sociological Review.

Carrier, J. (1997) 'Introduction', in J. Carrier (ed.), *Meanings of the Market*. Oxford: Berg.

Carruthers, B. (1996) *City of Capital*. Princeton, NJ: Princeton University Press.

Carruthers, B. and Uzzi, B. (2000) 'Economic sociology in the new millenium', *Contemporary Sociology*, 29(3): 486–94.

Casper, S. and Kettler, H. (2001) 'National institutional frameworks and the hybridization of entrepreneurial business models: the German and UK biotechnology sectors', *Industry and Innovation*, 1: 5–30.

Chomsky, N. (1997) *World Orders, Old and New*. London: Pluto Press.

—— (1999) *Profit over People*. London: Seven Stories Press.

—— (2000) *Rogues States*. London: Pluto Press.

Cohen, J. (2001) *Overdose: the case against the drug companies*. New York: Jeremy P. Tarcher/Putnam.

Cooke, P. (2003) 'Geographic clustering in the UK biotechnology sector', in G. Fuchs (ed.), *Biotechnology in Comparative Perspective*. London: Routledge.

Coombs, R. and Metcalfe, S. (2002) 'Innovation in pharmaceuticals: perspectives on the co-ordination, combination and creation of capabilities', *Technology Analysis and Strategic Management*, 14(3): 261–71.

Coombs, R., Saviotti, P. and Walsh, V. (1987) *Economics and Technological Change*. London: Macmillan.

Coriat, B., Orsi, F. and Weinstein, O. (2003) 'Does biotech reflect a new science-based innovation regime?', *Industry and Innovation*, 10(3): 231–53.

CRIC (2000) 'Biotechnology in the UK: a scenario for success in 2005'. Manchester: University of Manchester DGRC Report.

Cypher, J. (1987) 'Military spending, technical change, and economic growth: a disguised form of industrial policy?', *Journal of Economic Issues*, 21(1): 33–59.

Dixon, P. and Greenhalgh, C. (2002) 'The economics of intellectual property: a review to identify themes for future research'. Oxford: Oxford Intellectual Property Research Centre.

Dosi, G. (1988) 'Sources, procedures, and microeconomic effects of innovation', *Journal of Economic Literature*, 26(3): 1120–71.

Drahos, P. and Braithewaite, J. (2002) *Information Feudalism: who owns the knowledge economy*. London: Earthscan Publications.

DTI (2003) 'Innovation report – competing in the global economy: the innovation challenge'. London: Department of Trade and Industry.

Dutfield, G. (2003) *Intellectual Property Rights and the Life Science Industries*. Aldershot: Ashgate.

Easterley, W. (2002) *The Elusive Quest for Growth*. London: MIT Press.

Ernst & Young (2000) 'Evolution: seventh annual European life sciences report 2000'. London: Ernst & Young International.

—— (2001) 'Integration: eighth annual European life sciences report 2001'. London: Ernst & Young International.

—— (2003) 'Beyond borders: the global biotechnology report 2003'. London: Ernst & Young International.

Evans, J. (2002) *Playing God?* London: University of Chicago Press.

Ferraro, F., Pfeffer, J. and Sutton, R. (2005) 'Economics language and assumptions: how theories become self-fulfilling', *Academy of Management Review*, 30(1): 8–24.

Fligstein, N. (1996) 'Markets as politics: a political-cultural approach to market institutions', *American Sociological Review*, 61: 656–73.

Florida, R. and Kenney, M. (1988) 'Venture capital-financed innovation and technological change in the USA', *Research Policy*, 17: 119–37.

Freeman, C. (1982) *The Economics of Industrial Innovation.* London: Pinter.

Gompers. P. and Lerner, J. (1998) 'What drives venture capital fundraising?', *Brookings Papers on Economic Activity: Microeconomics*: 149–204.

Gottweis, H. (1998) *Governing Molecules.* London: MIT Press.

Granovetter, M. (1985) 'Economic action and social structure: the problem of embeddedness', *American Journal of Sociology*, 91(3): 481–510.

Green, K. (2002) 'Biotechnology, people and markets', *New Genetics and Society*, 21(2): 199–212.

Harvey, D. (2003) 'The New Imperialism', Oxford: University of Oxford, Clarendon Lectures in Geography and Environmental Studies.

Henwood, D. (2004) 'US miracle is based on longer hours for less pay', *The Guardian*, 2 February.

Hughes, S. (2001) 'Making dollars out of DNA: the first major patent in biotechnology and the commercialization of molecular biology, 1974–80', *Isis*, 92: 541–75.

Iwasaka, R. (2000) 'From Chakrabaty to Chimeras: the growing need for evolutionary biology in patent law', *Yale Law Journal*, 109(6): 1505–34.

Kaplan, W. and Krimsky, S. (2001) 'Patentability of biotechnology inventions under the PTO utility guidelines: still uncertain after all these years?', *The Journal of Biolaw and Business*, special supplement.

Katz, M. and Ordover, J. (1990) 'R&D cooperation and competition', *Brookings Papers on Economic Activity: Microeconomics*: 137–203.

Klein, N. (2001) *No Logo.* London: Flamingo.

Ko, Y. (1992) 'An economic analysis of biotechnology patent protection', *Yale Law Journal*, 102(3): 777–804.

Krimsky, S. (1999) 'The profit of scientific discovery and its normative implications', *Chicago-Kent Law Review*, 75(1): 15–39.

Krippner, G. (2001) 'The elusive market: embeddedness and the paradigm of economic sociology', *Theory and Society*, 30(6): 775–810.

Krugman, P. (1996) *Pop Internationalism.* London: MIT Press.

Kuhlman, G. (1996) 'Alliances for the future: cultivating a cooperative environment for biotech success', *Berkeley Technology Law Journal*, 11(2): 331, 314.

Laurie, G. (2003) 'Intellectual property protection of biotechnological inventions and related materials'. Edinburgh: University of Edinburgh, ESRC Innogen Centre.

Loasby, B. (2000) 'Organisations as interpretative systems', paper presented at the DRUID Summer Conference, Rebild, Denmark, 15–17 June.

MacKenzie, D. (2003) 'Long-term capital management and the sociology of arbitrage', *Economy and Society*, 32: 349–80.

MacKenzie, D. and Wajcman, J. (eds) (1999) *The Social Shaping of Technology*. Buckingham: Open University Press.

MacKenzie, M., Keating, P. and Cambrosio, A. (1990) 'Patents and free scientific information in biotechnology: making monoclonal antibodies proprietary', *Science, Technology & Human Values*, 15(1): 65–83.

Maebius, S. (1996) 'The new era of process patentability'. Washington, DC: Foley & Lardner.

Magdoff, F. (2004) 'A precarious existence: the fate of billions?', *Monthly Review*, 55(9): 1–14.

Malinowski, M. (2000) 'Biotechnology in the USA: responsive regulation in the life science industry', *International Journal of Biotechnology*, 2(1/2/3): 16–26

May, C. (2000) *A Global Political Economy of Intellectual Property Rights*. London: Routledge.

McKinley, M. (2001) 'Triage: A survey of the "New Inequality" as combat zone', paper presented at the 42nd Annual Convention of the International Studies Association, Chicago.

Miller, D. (2002) 'Turning Callon the right way up', *Economy and Society*, 31(2): 218–33.

Misa, T. (1997) 'Controversy and closure in technological change: constructing "steel"', in W. Bijker and J. Law (eds), *Shaping Technology/Building Society*. London: MIT Press.

Multinational Monitor (2003) 'Inequality and corporate power', *Multinational Monitor*, May: 8–10.

Mowery, D., Nelson, R., Sampat, B. and Ziedonis, A. (2001) 'The growth of patenting and licensing by US universities: an assessment of the effects of the Bayh-Dole act of 1980', *Research Policy*, 30: 99–119.

Nelson, R. and Winter, S. (1974) 'Neoclassical vs evolutionary theories of economic growth', *Economic Journal*, 84: 886–905.

—— (1982) *An Evolutionary Theory of Economic Change*. London: Belknap Harvard.

Nightingale, P. and Martin, P. (2004) 'The myth of the biotech revolution', *Trends in Biotechnology*, 22(11): 564–9.

NIHCM (2002) *Changing Patterns of Pharmaceutical Innovation*, Washington, DC: National Institute for Health Care Management.

OECD (1996) *The Knowledge-Based Economy*, Paris: Organisation for Economic Cooperation and Development.

—— 2002 *Genetic inventions, intellectual property rights and licensing practice*. Paris: Organisation for Economic Cooperation and Development.

Office of Technology Assessment (1984) *Commercial Biotechnology: an international analysis*. Washington, DC: US Government Printing.

Ordover, J. (1991) 'A patent system for both diffusion and exclusion', *Journal of Economic Perspectives*, 5(1): 43–60.

Ossorio, P. (2002) 'Legal and ethical issues in biotechnology patenting', in J. Burley and J. Harris (eds), *A Companion to Genethics*. Oxford: Basil Blackwell.

Owen, G. (2001) 'Entrepreneurship in UK biotechnology: the role of public policy', Working Paper No. 14: The Diebold Institute.

Padron, M. and Uranga, M. (2001) 'Protection of biotechnological innovations: a burden too heavy for the patent system', *Journal of Economic Issues*, 35(2): 315–22.

Parenti, M. (1995) *Against Empire*. San Francisco, CA: City Lights Books.

Paul, H. and Steinbrecher, R. (2003) *Hungry Corporations: Transnational Biotech Companies Colonise the Food Chain*. London: Zed Books.

Polanyi, K. (1957) 'The economy as instituted process', in K. Polanyi, C. Arensberg and H. Pearson (eds), *Trade and Market in the Early Empires*. Glencoe, IL: Free Press and Falcon's Wing Press.

—— (2001) *The Great Transformation*. Boston, MA: Beacon Press.

Polster, C. (2001) 'How the law works: exploring the implications of emerging intellectual property regimes for knowledge, economy and society', *Current Sociology*, 49(2): 85–100.

Poyago-Theotoky, J., Beath, J. and Siegel, D. (2002) 'Universities and fundamental research: reflections on the growth of university-industry partnerships', *Oxford Review of Economic Policy*, 18(1): 10–21.

Prevezer, M. (2003) 'The development of biotechnology clusters in the USA from the late 1970s to the early 1990s', in G. Fuchs (ed.), *Biotechnology in Comparative Perspective*. London: Routledge.

Public Citizen (2001) *Rx R&D Myths: the Case Against the Drug Industry's R&D 'Scare Card'*. Washington, DC: Public Citizen's Congress Watch.

Quillen, C. (1992) 'Innovation and the United States patent system today'. Paper presented at *Antitrust and Intellectual Property and Policy Issues for the 1990's*, Continuing Legal Education Institute.

Rasnick, D. (2003) 'The biotechnology bubble machine', *Nature Biotechnology*, 21(April): 355–6.

Reenen, J. van (2002) 'Economic issues for the UK biotechnology sector', *New Genetics and Society*, 21(2): 109–30.

Robertson, D. (2002) 'US Supreme Court applies strict limits to patents', *Nature Biotechnology*, 20(July): 639.

Schumpeter, J. (1939) *Business Cycles, Volume 1*. London: McGraw-Hill.

Sharma, S., Tracy, S. and Kumar, S. (2004) 'The invasion of Iraq: Dollar vs Euro', *Z Magazine*, February 2004: 53–6.

Sharp, M. (1996) 'The science of nations: European multinationals and American biotechnology'. Brighton: University of Sussex, STEEP Discussion Paper No. 28.

Slaughter, S. and Rhoades, G. (1996) 'The emergence of a competitiveness research and development policy coalition and the commercialization of academic science and technology', *Science, Technology & Human Values*, 21(3): 303–39.

Stowsky, J. (1992) 'Conversion to competitiveness: making the most of the national labs', *The American Prospect*, 3(11).

Strange, S. (1998) *Mad Money*. Manchester: Manchester University Press.

Tansey, G. (2002) 'Patenting our food future: intellectual property rights and the global food system', *Social Policy and Administration*, 36(6): 575–92.

Thompson, P. (2004) *Skating on Thin Ice: the knowledge economy myth*. Glasgow: Big Thinking.

Thompson, P. and Warhurst, C. (2001) 'Ignorant theory and knowledgeable workers: interrogating the connections between knowledge, skills and services', *Journal of Management Studies*, 38(7): 923–42.

Tyson, L. (1992) *Who's Bashing Whom? Trade Conflict in High-Technology Industries*. Washington, DC: Institute for International Economics.

US Congress House Committee on Science and Technology (1981) 'Commercialization of academic biomedical research, hearings before the Subcommittee on

Investigations and Oversight, June 8–9', 97th Congress, 1st Session Edition. Washington, DC: US Government Printing Office.

Uzzi, B. (1997) 'Social structure and competition in interfirm networks: the paradox of embeddedness', *Administrative Science Quarterly*, 42: 35–67.

Valle, F. della and Gambardella, A. (1993) '"Biological" revolution and strategies for innovation in pharmaceutical companies', *R&D Management*, 23(4): 287–302.

Vidal, G. (1999) *The American Presidency*. Monroe, ME: Odonian Press.

Walsh, V., Niosi, J. and Mustar, P. (1995) 'Small-firm formation in biotechnology: a comparison of France, Britain and Canada', *Technovation*, 15(5): 303–27.

Woiceshyn, J. (1995) 'Lessons in product innovation: a case study of biotechnology firms', *R&D Management*, 25(4): 395–409.

Woolgar, S., Coopmans, C., Neyland, D. and Simakova, E. (2005) 'Does STS mean business too?', Paper prepared for Does STS Mean Business Too? Workshop, University of Oxford, 29 June.

Wright, S. (1993) 'The social warp of science: writing the history of genetic engineering policy', *Science, Technology & Human Values*, 18(1): 79–101.

7 Biopiracy and the bioeconomy

Paul Oldham

Introduction

The concept of the bioeconomy has recently emerged to international prominence through the work of the United Nations Conference on Trade and Development (UNCTAD 2001), the Organisation for Economic Co-operation and Development (OECD 2005) and DG Research within the European Commission. For the OECD the concept of the bioeconomy refers to the increasing convergence of scientific disciplines and technologies directed towards capturing 'the latent value in biological processes and renewable bioresources to produce improved health and sustainable growth and development' (OECD 2005: 5). In contrast, for DG Research at the European Commission, what is variously described as the 'knowledge-based bioeconomy' and the 'bioeconomy' has recently been presented to the public as a progression from the 'Age of Engineering' in the eighteenth and nineteenth centuries, to the 'Age of Chemistry' in the twentieth century, to a 'transition' towards the 'Age of Biotechnology' in the twenty-first century, heralding economic activity and technology that in the words of EuropaBio is 'clean, clever and competitive'.[1]

In practice, the emergence of the concept of the bioeconomy and these tentative and initial steps towards defining and conceptualising this economy reflects a wider process through which the social sciences, international institutions and policy-makers are attempting to grapple with and make sense of the growing convergences between science and technology in the biosciences represented by biotechnology, genomics, proteomics, bioinformatics, bionanotechnology and stem cell research. Among the most heavily contested of the issues surrounding the bioeconomy are those that relate to intellectual property rights and ownership within this emergent economy.

This chapter focuses on the Convention on Biological Diversity as an arena of intense mobilisation and contestation involving multiple actors in relation to intellectual property and its role in the construction of the bioeconomy. In the process this chapter seeks to identify some of the convergences and key fault lines observable in complex negotiations surrounding access to genetic resources and benefit-sharing encompassing an estimated 14 million species worldwide.

The chapter argues that, far from serving as an incentive for the construction of a bioeconomy that is founded on principles of justice and equity, intellectual property protection has become a major obstacle to the pursuit of collaboration between the diverse actors involved in debates surrounding access to genetic resources and benefit-sharing. The chapter concludes that the current situation relating to access to genetic resources and benefit-sharing resembles an anticommons and argues that a wider and more flexible vision is needed in relation to intellectual property if the problems of over-expectation and fear of appropriation that characterise the existing debate in the construction of the bioeconomy are to be overcome.

Negotiating diversity

According to the 2001 *Global Biodiversity Outlook* the concept of biodiversity refers to the diversity of life on this planet, ranging across a spectrum from the genetic diversity of living organisms to the diversity of species and wider ecosystems. As such, biodiversity can be said to constitute the web of life on this planet (SCBD 2001).

The 1992 Convention on Biological Diversity is the primary international legal instrument with responsibility for biodiversity. To date, the Convention has been ratified by 188 governments (or parties) and is directed towards three objectives: the conservation of biodiversity; the sustainable use of biodiversity; and 'the fair and equitable sharing of the benefits arising from the utilisation of genetic resources'. It is this latter objective that will be our focus.

The third objective of the Convention and its detailed provisions are commonly described simply as access and benefit-sharing (ABS) and form part of what Gollin (1993) has described as the 'grand bargain' of the Convention.

The majority of the world's biodiversity is located in developing countries and, under the terms of the 'grand bargain', developing countries as so-called 'providers' of genetic resources agreed to provide access to their genetic resources in return for a share of any benefits arising from the utilisation of those resources by 'users' in developed countries. This bargain is given legal form by a series of provisions set out within the Articles of the Convention. The first of these recognises the principle of state sovereignty over natural resources (Article 15.1). These resources are defined in the following terms: '"Biological resources" includes genetic resources, organisms or parts thereof, populations, or any other biotic component of ecosystems with actual or potential use or value for humanity' (Article 2). 'Genetic resources' are then further defined as 'genetic material of actual or potential value' and 'genetic material' means 'any material of plant, animal, microbial or other origin containing functional units of heredity' (Article 2).

In affirming the principle of state sovereignty over biological resources the Convention also establishes that access to these resources will be subject to the prior informed consent of parties (governments) and that agreements surrounding fair and equitable benefit-sharing will be established on 'mutually agreed terms' (Article 15.5 and 15.4). The benefits that countries providing access to these resources might expect include access to technology and technology transfers on favourable terms with a particular emphasis on biotechnology (Article 16). These arrangements are expected to be combined with research collaborations and information exchange (Article 17).

On the other side of this 'bargain', developed countries, in addition to gaining access to genetic resources in developing countries. also stipulated that: 'In the case of technology subject to patents and other intellectual property rights, such access and transfer shall be provided on terms which recognise and are consistent with the adequate and effective protection of intellectual property rights' (Article 16.3). However, in an important caveat, Article 16.5 goes on to state that 'The Contracting Parties, recognising that patents and other intellectual property rights may have an influence on the implementation of this Convention, shall cooperate in this regard subject to national legislation and international law in order to ensure that such rights are supportive of and do not run counter to its objectives' (Article 16.5).

In considering the terms of the 'grand bargain', it is important to note the scope of these provisions: they refer to the genetic and biological components of an estimated 14 million species worldwide – with the notable exception of humans. When seen from the perspective of the diversity and complexity of life on this planet it is perhaps hardly surprising that the access and benefit-sharing provisions of the Convention have emerged as one of the most intellectually challenging and politically complex areas of its work. In approaching this complexity it is useful to highlight three factors that have shaped the perspectives of developing countries with respect to the grand bargain.

The first of these factors is growing recognition on the part of developing countries of the historical economic importance of biological resources in the context of the emergence of biotechnology. As Calestous Juma, who became the first Executive Secretary of the Convention on Biological Diversity, highlighted in an important 1989 volume *The Gene Hunters: Biotechnology and the Scramble for Seeds*, transfers of valuable biological material such as quinine, rubber, tea and major crop plants from developing countries were central to the success of European empires and emerging agricultural economic powers such as the United States. Demand for new sources of biological and genetic material remains central to international agriculture and an increasing focus of a range of industries from agriculture to pharmaceuticals.

Growing awareness of the historical importance of biological resources in relation to agriculture is also explicitly linked with awareness of the consequences of the loss of control over these resources. Thus, in South America, the collapse of the Amazon rubber boom in the early part of the twentieth century, following the transfer of 70,000 rubber seeds to Kew Gardens and then to plantations in South East Asia, serves as a powerful reminder to countries within the region of the economic consequences of the loss of control over important resources (Juma 1989; Collier 1968). A similar and powerful case can be made for China's loss of the monopoly of tea production to the British (Macfarlane and Macfarlane 2003). In short, biodiversity and control over biological resources affects the fate of nations.

A second factor in understanding developing-country perspectives focuses on expectations surrounding the future potential importance of biodiversity. In particular, debates about access to genetic resources and benefit-sharing have been dominated by high expectations related to the potential of biological diversity to yield income for developing countries through the development of new pharmaceutical products. The origins of these expectations can be traced to the efforts of a number of scientists within the disciplines of ethnobotany and ethnopharmacology to justify the conservation of biodiversity in terms of its economic potential. The work of Norman Farnsworth and colleagues is illustrative in this regard. In a series of important articles Farnsworth sought to draw attention to the dependence of an estimated 64 per cent of the world's population, or around 3.2 billion people, upon plant-based medicines. In relation to the pharmaceutical sector, a global survey of plant life suggested that 119 plant-based chemical compounds are used as drugs or in human healthcare, while an estimated 25 per cent of prescriptions over the twenty-two-year period between 1959 and 1980 contained active principles from plants. The economic value of plant-based medicines was then highlighted by data revealing that, in 1980, US consumers 'paid more than $8 billion (US) for prescriptions containing active principles obtained from higher plants' (Farnsworth 1990: 4).

The economic potential of biodiversity in relation to both agriculture and pharmaceuticals was widely promoted both in the lead-up to the opening of the Convention for signature and in subsequent years. While the promissory nature of these claims is rightly being subjected to increasing scrutiny, and the pharmaceutical sector has sought to dampen expectations with regard to the importance of natural compounds in the era of combinatorial chemistry (see ten Kate and Laird 1999), in practice the fundamental human dependence on biodiversity and the importance of biodiversity-based products in areas such as pharmaceuticals is impossible to deny. Thus, as Newman *et al.* (2003) from the United States National Cancer Institute have recently demonstrated 'yet again', despite expectations surrounding the promise of combinatorial chemistry in the realm of pharmaceuticals during the 1990s, in the period between 1981 and 2002 the percentage of nonsynthetic new chemical entities either of natural origin, based on natural products or mimicking natural products has averaged 62 per cent and rises to 74 per cent in areas such as anticancer drugs. While estimating the overall contribution of biodiversity to the world economy is fraught with difficulty, ten Kate and Laird (1999) suggest that, as a 'ballpark' figure, annual global markets for biodiversity-based products across a spectrum from pharmaceuticals and botanical medicines to agriculture, crop protection, ornamentals, and personal care and cosmetics fall within the region of US$500–US$800 billion per annum.

When seen from a purely economic perspective it is not surprising that developing countries increasingly see biodiversity as a key resource to be protected from exploitation by others until 'fair and equitable' terms surrounding benefit-sharing have been agreed. It is here that the third factor informing developing-country perspectives, in the form of a desire for technology transfer, notably in the realm of biotechnology, constitutes a key strategic aim in the pursuit of development (UNCTAD 2001). However, as we will see, in the 1990s the pursuit of that aim was overtaken by expectations concerning what has come to be described as 'green gold'. and fear of its loss has contributed to a marked 'chilling effect' in relation to biodiversity-related research. Questions surrounding intellectual property and the international patent system lie at the core of these concerns.

Contestations between developing and developed countries about access and benefit-sharing and intellectual property protection under the Convention have also become increasingly bound up with issues of the human rights of indigenous peoples and local communities. The valuation of knowledge, the relationship between knowledge and rights related to biological resources, and intellectual property protection are central to this debate (see also Brown 2003).

The rise of traditional knowledge

> If phytochemists must randomly investigate the constituents of biological effects of 80,000 species of Amazon plants, the task may never be finished. Concentrating first on those species that people have lived and experimented with for millennia offers a short-cut to the discovery of new medically or industrially useful compounds.
>
> (Schultes 1988, cited in Moran *et al.* 2001)

Growing recognition of the economic importance of biodiversity for developing countries is critically associated with a reassessment and *revaluation* of the knowledge of members of societies who have historically been described as 'primitive' or 'backward' and as objects for the exercise of development. This process of reassessment and revaluation is associated with three inter-related developments.

The first of these relates to growing international concern about the situation of the world's indigenous peoples. From the 1960s onwards anthropologists working with what were variously described as 'primitive', 'native' or 'tribal' peoples in areas such as Amazonia increasingly began to focus international attention on the human rights situation of societies who have now reframed themselves as 'indigenous peoples' (Daes 1996). This was reflected in the establishment of specialist human rights organisations dedicated to indigenous peoples, notably the International Working Group for Indigenous Affairs (IWGIA), Cultural Survival, and Survival International, and increasing mobilisations by indigenous human rights activists from the mid-1970s onwards seeking to create wider alliances directed towards securing action in defence of the rights of indigenous peoples within the United Nations system (Barsh 1986).

These mobilisations bore fruit with the formation in 1982 of the United Nations Working Group on Indigenous Populations under the Sub-Commission on the Promotion and Protection of Human Rights. This body has served as a forum for dialogue between a growing number of indigenous organisations and activists with governments, and for international standard-setting with respect to the rights of indigenous peoples. In 1989 mobilisations by indigenous peoples organisations and human rights organisations also witnessed the creation of Convention 169 'concerning Indigenous and Tribal Peoples in Independent Countries' under the International Labour Organisation which has played a critical role in advancing the human rights situation of indigenous peoples in Latin America. The 1990s witnessed further advances with the establishment of a United Nations Decade of the World's Indigenous People (1995–2004), the negotiation of a draft Universal Declaration on the Rights of Indigenous People, and the establishment in 2002 of the United Nations Permanent Forum on Indigenous Issues under the Economic and Social Council (ECOSOC). While work within the human rights arena has primarily focused on issues surrounding recognition of the existence and rights of indigenous peoples, notably in relation to land, issues relating to cultural and intellectual property became increasingly prominent in this arena from the mid-1990s onwards (Daes 1996; Posey 1999; Posey and Dutfield 1996; Cleveland and Murray 1997; Brown 2003; Von Lewinski 2003).

At the same time, developments in the main human rights arenas were accompanied by increasing attention to indigenous and 'peasant' societies in relation to their knowledge of the environment in a context of increasing concern about the failure of development projects and the environmental impacts of standard development models in regions such as Africa and Amazonia (Posey 1999; Ellen *et al.* 2000; Sillitoe *et al.* 2002). These reassessments were marked by an explosion in the scientific literature across a range of disciplines with respect to a subject variously described as 'indigenous knowledge' (IK), 'local knowledge' (LK), or 'traditional ecological knowledge' (TEK), which seeks to explore the nature of these forms of knowledge, their status *vis-à-vis* 'science' and their potential applicability in the pursuit of more effective development and environmental management strategies.

On the policy level the influence of this work is reflected in the outcomes of the 1992 United Nations Conference on Environment and Development (UNCED 'Earth Summit'). Principle 22 of the Rio Declaration on Environment and Development specifies that:

> Indigenous people and their communities and other local communities have a vital role in environmental management and development because of their knowledge and traditional practices. States should recognize and duly support their identity, culture and interests and enable their effective participation in the achievement of sustainable development.

In the case of the Convention on Biological Diversity, growing interest in the subject that is now commonly called 'traditional knowledge' in international policy debates is reflected in Article 8(j) in which each Party to the Convention undertakes,

> [s]ubject to its national legislation, [to] respect, preserve and maintain knowledge, innovations and practices of indigenous and local communities embodying traditional lifestyles relevant for the conservation and sustainable use of biological diversity and promote their wider application with the approval and involvement of the holders of such knowledge, innovations and practices and encourage the equitable sharing of the benefits arising from the utilization of such knowledge, innovations and practices.

However, a close reading of the above quotations reveals that these societies are being revalued in very particular ways. That is, they are being reassessed and revalued in terms of what they know in relation to the environment, conservation and development. In particular, as the opening quote from the ethnobotanist Richard Evans Schultes suggests, in one instrumentalist version of this process they are being revalued in terms of their ability to provide a 'short cut' to the identification of 'new medically or industrially useful compounds'.

Arguments concerning the importance of traditional knowledge are closely linked with scientific recognition of the limitations of existing taxonomic knowledge. Thus, according to the *Global Biodiversity Outlook*, taxonomic knowledge of biodiversity is presently limited to approximately 1.75 million species. This represents approximately 12 per cent of an estimated 14 million species worldwide. When viewed from this perspective the knowledge represented by the estimated 5,000–7,000 language groups worldwide (Maffi 1999) can be seen as an important body of knowledge in relation to taxonomy and as a 'resource' in relation to the identification of the potentially useful properties of biological organisms. This is also linked with wider debates about the relationship between human cultural diversity and biodiversity (Maffi 2001).

These reassessments and revaluations of the knowledge of indigenous peoples and local communities are also linked with highly contested issues regarding the political and legal status of members of these societies and to rights in relation to biological and genetic 'resources'. Thus, Article 8(j) of the Convention establishes that, 'subject to national legislation', parties will promote respect for 'knowledge, innovations and practices' and 'with the approval and involvement of the holders' the sharing of benefits arising from the utilisation of this knowledge. However, this says nothing about what, from the perspective of governments, represents the key issue of rights in relation to the biological and genetic materials to which this knowledge provides a 'short-cut'. In other words, while governments have recognised the rights of 'indigenous and local communities' to their knowledge, rights related to the biological and genetic resources to which their knowledge provides access are generally considered by developing (and a number of developed) countries to fall within the bounds of state sovereignty.

Furthermore, the deliberate ambiguity of the phrase 'indigenous and local communities embodying traditional lifestyles' is linked to underlying tensions between those who describe themselves as indigenous peoples and the states in which they reside in relation to their status under international law. Specifically, indigenous peoples are asserting the right to self-determination enjoyed by all peoples enshrined within the United Nations Charter and the main international human rights instruments (the International Covenants). Thus, common Article One of the International Covenants establishes that: 'All peoples have the right of self-determination. By virtue of that right they freely determine their political status and freely pursue their economic, social and cultural development.' This is directly linked, through United Nations General Assembly Resolution 1803 (1963), to the 'principle of permanent sovereignty over natural resources' as a right enjoyed by all peoples and nations that gives meaning to the right to self-determination (Daes 2003). While the 1990s witnessed growing recognition of the existence and rights of indigenous peoples, notably in Latin America and to a more limited degree in Africa (Hitchcock and Vinding 2004), many governments are reluctant to recognise the rights of indigenous peoples as 'peoples' in the full sense of international law and rights in relation to natural resources remain heavily contested.

As this suggests, mobilisations in the realm of human rights, the status of knowledge and their relationship to biological and genetic resources disguise complex issues and fault lines between indigenous rights organisations, activists, scientists and states. These fault lines and the contestations surrounding them are brought into sharper focus in debates about bioprospecting in the context of the internationalisation of intellectual property protection.

Bioprospecting and biopiracy

In the 1990s the access and benefit-sharing provisions of the Convention provided a spur to a variety of private and public–private initiatives. The emerging transition of economic activity towards the 'bio' is reflected in the way in which these initiatives were re-framed from what Eisner (1989) had called 'chemical prospecting' to 'biological prospecting' or 'bioprospecting', which has been defined as 'the exploration of wild plants and animals for commercially valuable genetic and biochemical resources' (Reid 1993).

Bioprospecting projects take a variety of forms. An early and widely cited example of bioprospecting is provided by the 1991 agreement between the National Biodiversity Institute (InBio) in Costa Rica and the pharmaceutical company Merck. Under the terms of this agreement, and subsequent agreements with other companies, the Institute provided access to biological and genetic material within Costa Rica and exclusive rights over this material in return for payments, equipment and scientific capacity-building (Castree 2003). The InBio example has been seen in international policy circles as an important potential model that other countries might follow.

A second well-known initiative is represented by the International Cooperative Biodiversity Group (ICBG) established by the National Institutes of Health (NIH), the National Science Foundation (NSF) and the United States Agency for International Development (USAID) in 1992 (Rosenthal *et al.* 2000; Hayden 2003; Nigh 2002; Greene 2004). The ICBG pursues a model that focuses on partnerships between developing countries' institutions, universities in the United States and the private sector. ICBG projects focus primarily on drug discovery and include a strong component of local scientific capacity-building. At the time of writing eleven projects have been funded under the ICBG programme, involving ten countries in Africa, Asia and Latin America, including Panama, Madagascar, Vietnam and Laos, Papua New Guinea, Cameroon, Nigeria, Central Asia, Suriname, Mexico, and Peru. Total programme expenditure over the ten years of the ICBG has been reported at US$30 million (Rosenthal *et al.* 2000).

The rise of bioprospecting is also associated with the emergence of start-up companies with an interest in natural product research. Among the best known of these during the 1990s were the now defunct Shaman Pharmaceuticals which specialised in plants research, and the ongoing Diversa Biotechnology which specialises in microorganisms and enzymes (Moran *et al.* 2001). The emergence of the Diversa Biotechnology is associated with trends in bioprospecting towards the unexplored potential of microbial diversity and is increasingly characterised by the use of genomics and bioinformatics techniques. This has coincided with increasing interest in bioprospecting in developing and developed countries in the pursuit of 'thermophiles and extremophiles' (i.e. in Yellowstone National Park) and in areas such as Antarctica and the deep sea bed (ten Kate *et al.* 1998; Lohan and Johnston 2003; Oldham 2004b; Arico and Salpin 2005). These trends have been accompanied by increasing interest in the collection of sea surface marine microbial material and the use of shotgun mapping techniques to identify microbial genomes (see Shreeve 2004). Other bioprospecting-related initiatives focus on the analysis of old herbal medical texts or 'text mining', using electronic scanning to identify leads (Buenz *et al.* 2004).

However, as Hayden (2003) has observed, the recent extension of bioprospecting to northern countries and into areas such as Antarctica, the deep sea bed, and ancient texts can be seen as a reaction to the increasing controversies surrounding bioprospecting. At the heart of this controversy is the concept of biopiracy.

At present there is no internationally agreed definition of biopiracy. However, in the mid-1990s the Rural Advancement Foundation International (RAFI), now the Action Group on Erosion, Technology and Concentration Group (hereafter ETC), defined biopiracy in the following terms:

> Biopiracy refers to the appropriation of the knowledge and genetic resources of farming and indigenous communities by individuals or institutions who seek exclusive monopoly control (patents or intellectual property) over these resources and knowledge. ETC group believes that intellectual property is predatory on the rights and knowledge of farming communities and indigenous peoples.[2]

The history of the evolution of the concept of biopiracy, and its contested meanings, is difficult to trace with precision. RAFI/ETC Group (ETC Group) traces its organisational origins to a 1977 meeting of activists working on issues concerning agriculture, pesticides, the growth of intellectual property protection and trends in corporate control in the seed industry. The early 1990s witnessed an extension of this activism into the wider domain of biological diversity.

This process and its significance can be traced through a 1993 communiqué entitled 'Bio-Piracy: The Story of Natural Coloured Cottons of the Americas' (RAFI 1993). This is the first recorded reference to the concept of biopiracy I have been able to trace. The communiqué focused on plant patents in the United States for plant varieties whose origins could be traced to Central and South America (Kevles 2002). In drawing attention to claims to a monopoly intellectual property right the organisation also sought to highlight the ways in which intellectual property claims over resources and knowledge originating from developing countries are associated with trends in corporate ownership and could be linked to technological trends in relation to genetic engineering.

The concept of biopiracy began to take on an increasingly global dimension the following year. In a paper published in June 1994 entitled 'Microbial

BioPiracy: An Initial Analysis of Microbial Genetic Resources Originating in the South and Held in the North', the organisation sought to highlight the importance of trends in relation to transfers of microbial resources from the South to the North (RAFI 1994). This paper highlighted that a total of 874 deposits of microbial materials from eleven developing countries could be identified within the American Type Culture Collection (ATCC). Of these an estimated 89 samples were the subject of patent protection and a further sixteen were the subject of patent claims (RAFI 1994). The report clearly sought to articulate biopiracy as a 'north' vs. 'south' issue.

By 1995, and with the hallmark tongue-in-cheek style that characterises the organisation's work, biopiracy had become a 'global pandemic' with cases of the patenting of resources originating in developing countries ranging from patents belonging to the University of Wisconsin over a plant protein from *Pentadiplanddra brazzeana* in Gabon, to an Oxford University researcher's patents over a fish poison (*barbasco*) used by the Wapishana of Guayana and indigenous peoples throughout Amazonia, and to concerns surrounding Pfizer's interest in surveying areas of biodiversity in Ecuador and the initiation of an ICBG project among the Huambisa in the Peruvian Amazon (RAFI 1995; Greene 2004). Other cases addressed during the mid-to-late 1990s included patents and other intellectual property claims in relation to neem (India), turmeric (India), *Banisteriopsis caapi* or *ayahuasca* (Amazonia), the enola bean (Mexico), golden rice (GM), and on a wider level patents in relation to human genetic material. These cases form part of a mounting succession of reports during the 1990s that sought to draw attention to research and patenting by individual researchers, universities and companies, across a spectrum from plant material, to human DNA and tissues, GM and genetic restriction or 'terminator' technologies, to genomics and most recently to nanotechnology (i.e. ETC Group 2005). In the process, ETC Group has become the leading international non-governmental organisation and 'clearing-house' dealing with trends in research, technology and intellectual property and bringing the implications of these trends to a wider international audience.

In particular, as the list above suggests, the concept of biopiracy has drawn attention to the centrality of science in the extraction, commodification and commercialisation of knowledge and resources from indigenous peoples, farmers and local communities in many parts of the world. This has had real consequences, as in the heavily contested circumstances surrounding the cancellation of the US$2.5 million ICBG-Maya project whose bruising fallout continues to echo in the literature (RAFI 1999; Dalton 2001; Rosenthal *et al.* 2002; Nigh 2002; Berlin and Berlin 2003). As we will see, growing awareness of the commercial dimensions of scientific research in relation to bioprospecting has also contributed to increasing 'chilling effects' upon biodiversity-related research in developing countries.

However, in seeking to understand increasing contestations concerning the role of science in bioprospecting, and declarations such as 'all bioprospecting is biopiracy', it is important to recognise the wider context and target of discourses surrounding biopiracy. Specifically, the work of ETC Group forms part of a wider critical questioning of the implications of intellectual property claims in relation to biological and genetic materials by a wide range of NGOs and civil society networks, notably GRAIN, Greenpeace, Friends of the Earth, Third World Network (among many others) and activist-scholars such as Shiva (1998). In particular, the concept of biopiracy has served as a powerful banner for counter-mobilisations directed towards the 1994 Agreement on Trade-Related Aspects of Intellectual Property Rights (TRIPS) that emerged from the Uruguay round of GATT negotiations under what is now the World Trade Organisation (WTO).

Biodiversity and the trouble with TRIPS

The Trade-Related Aspects of Intellectual Property Rights (TRIPS) agreement introduces a requirement for the now 148 member states of the World Trade Organisation (WTO) to extend patent protection as a form of industrial intellectual property to all areas of invention irrespective of the subject matter. This requirement is embodied in Article 27.1 which establishes that 'patents shall be available for any inventions, whether products or processes, in all fields of technology, provided that they are new, involve an inventive step and are capable of industrial application'.

This provision represents a major departure in international law by introducing high minimum standards for patent protection across all areas of invention (Dutfield 2000). In practice this constitutes a requirement for member states to provide patent protection as a form of exclusive temporary protection that includes the right to exclude others from 'making, using, offering for sale, or selling' or 'importing' the protected invention into a jurisdiction where the patent protection is in force, or to charge others for any uses or purposes involving the protected invention within such jurisdictions (i.e. through licensing) (TRIPS Article 28).

The extension of this form of intellectual property protection is qualified in a variety of ways such that countries can exclude 'inventions' from patentability on the grounds of *ordre public* or morality necessary to protect human, animal or plant life or health (Article 27.2). However, these exclusions cannot be established by legislative fiat and must be justified (Article 27.2). In connection with biological and genetic material Article 27.3(b) goes on to specify that:

> [Members may also exclude from patentability:] (b) plants and animals other than micro-organisms, and essentially biological processes for the production of plants or animals other than non-biological and microbiological processes. However, Members shall provide for the protection of plant varieties either by patents or by an effective *sui generis* system or by any combination thereof. The provisions of this

subparagraph shall be reviewed four years after the date of entry into force of the WTO Agreement.

While framed in the language of exclusions, it is important to note two points in relation to Article 27.3(b). The first of these is that intellectual property protection must be provided for plants, either through patents or *sui generis* (of its own kind) protection such as Plant Variety Protection certificates under the Union for the Protection of New Varieties of Plants (UPOV), or combinations of the two (Dutfield 2000). In the context of the rise of biotechnology and genomics Article 27.3 (b) is also significant because patent protection must be provided for microorganisms and microbiological processes (Adcock and Llewelyn 2000; Oldham 2004b).

The provisions of the TRIPS agreement have generated a vast literature across a range of disciplines and this chapter does not seek to review that literature (see UNCTAD-ICTSD 2005; Maskus and Reichman 2005). However, the following points stand out. The first of these is that the TRIPS agreement enshrines and gives legal force to underlying trends towards the extension of patent protection as a form of temporary monopoly to biological organisms and their components which have their origins in a 1980 United States Supreme Court Decision *Diamond v Chakrabarty* (see Kevles 2002). This decision held that a microorganism that had been modified by the hand of man could not be considered to be a product of nature and was therefore eligible for patent protection. In reaching this decision the court overturned an earlier doctrine that organisms and their components were ineligible for patent protection on the grounds that they are products of nature (Kevles 2002). At the same time, the TRIPS agreement – by requiring either patent protection or *sui generis* forms of protection for plants – reflects trends within developed countries towards classifying plants and their components as eligible for intellectual property protection either at the genetic level (i.e. in the case of GM crops) or at the level of varieties (in the case of plant patents and plant variety protection) or both. TRIPS thus enshrines trends towards the reclassification of nature as a form of industrial property on the international level.

In formal terms the introduction of these requirements and their extension into the realm of biological organisms and their components has been justified in terms of the promotion of enhanced trade in goods and services, Foreign Direct Investment (FDI) and technology transfer (World Bank 2001). In particular, patent protection has been presented in terms of promoting security for companies interested in investing in foreign countries and is related to arguments supporting the promotion of technology transfer as a key goal for developing countries (World Bank 2001). However, while the internationalisation of patent protection in these areas has not been a focus of detailed empirical research, evidence for such positive effects in other areas of patent protection in developing countries is presently limited and mixed (see Fink and Maskus 2005).

A more convincing argument than what often appear to be *post facto* justifications is provided by the negotiating history of the TRIPS agreement. Drahos and Braithwaite (2002) reveal that the TRIPS provisions can be traced to concerns among a number of large companies and trade associations about intellectual property 'piracy' in developing countries. More specifically, the history of TRIPS can be traced to private-sector mobilisations marked by the establishment of an Intellectual Property Committee (IPC) by Pfizer and IBM, and their subsequent success in creating a group of countries known as the 'Friends of Intellectual Property' within the GATT process which tabled versions of what became the TRIPS agreement. In short, as Maskus and Reichmann (2005a) have argued, the TRIPS agreement and wider internationalisation of intellectual property protection is perhaps best regarded as a result of 'policy capture' within developed countries. When viewed in light of economic theory the TRIPS agreement constitutes an outcome of a form of rent-seeking behaviour on the part of private-sector interests that is global in its scope (see Tullock 1993; Krueger 1974).

The impacts of TRIPS, and related agreements such as the Patent Cooperation Treaty (PCT), in establishing an intellectual property regime that is global in scope should not be underestimated. Between 1990 and 2000 the total number of estimated patents granted worldwide rose to 7.6 million and accelerated to a provisional 8.3 million in the year 2001.[3] It is not readily possible to generate data on the precise number of patent applications and patent grants worldwide in relation to biodiversity. However, my research on trends in patent publications (consisting of applications and grants) as a measure of demand using the European Patent Office worldwide database of publications from over 70 countries for the period 1990–2000 and provisional data for 2001 to mid-2005 reveals escalating demand. In the case of traditional medicines a total of approximately 51,765 patent publications are recorded in the database for the period between 2000 and August 2005 and a further 37,227 patent publications are related to new plants and processes for producing them in the realm of agriculture. In the case of organic chemistry 1,054,000 patent publications are recorded between 1990 and August 2005 with 107,737 publications relating to 'sugars, nucleosides, nucleotides and nucleic acids'. In the case of biotechnology, the significance of the inclusion of microorganisms as a requirement for patentability under the TRIPS agreement is revealed by the dominance of the category 'microorganisms or enzymes' within the data for biotechnology, with 340,219 patent publications, of which 229,204 refer to recombinant genetic engineering. Furthermore, the inclusion of a wide range of human, plant and animal material in the category of 'microorganisms or enzymes', such as undifferentiated human, animal and plant cells and tissues, suggests that 'microorganisms' have become the eye of the needle through which biotechnology patents in relation to humans, animals and plants are being threaded (Oldham 2004b).

This ongoing research, involving over 500 categories of patent claims across a spectrum ranging from medicinal plants to bionanotechnology, reveals that existing accounts focusing on patent activity in the major patent offices (notably the United States) seriously underestimate international demand for patent protection (Oldham 2004a). In particular, existing accounts fail to appreciate that the key vehicle for the operationalisation of the TRIPS agreement in the realm of biology is the 1980 Patent Cooperation Treaty (amended 2001). The Patent Cooperation Treaty allows patent applicants to submit a single patent application for possible patent grants in up to 133 countries and introduces a major multiplier effect into the international patent system. In the year 2000 developed countries accounted for 89.5 per cent of all patent applications submitted under the Patent Cooperation Treaty and in January 2005 the Patent Cooperation Treaty celebrated receipt of one million applications (WIPO 2003; WIPO website 14 January 2005).

As this suggests, one problem confronting researchers, governments and civil society organisations is the international scale of demand for patent protection in the realm of biodiversity. This in turn raises serious questions surrounding the implications of such claims from ethical, human rights, social, environmental, economic and legal perspectives. The data provided above suggest that, despite the very significant work conducted in this area, there is still a long way to go in understanding the full scope and longer-term implications of permitting this form of monopoly protection for human welfare in the context of the rise of the bioeconomy. However, growing concern surrounding 'biopiracy' in an era of global intellectual property protection is manifest in highly defensive responses on the part of both indigenous peoples and developing countries. This is producing marked chilling effects in relation to biodiversity research.

Towards an anticommons?

> Mr. Chairman, until such time that the Parties recognize the existence and rights of Indigenous Peoples, our peoples will not be in a position to consider providing our free, prior and informed consent to the commercial exploitation of such knowledge and resources. We have suffered discrimination, exploitation and marginalization for generations. The constant insistence that we commodify our knowledge and resources must stop. Indigenous peoples cannot be forced to share our knowledge and resources.
> (Opening statement, International Indigenous Forum on Biodiversity,
> 22–26 October 2001, Bonn, Germany)

In the year 2001 the Convention on Biological Diversity convened an Ad-hoc Working Group on Access and Benefit-Sharing to consider the development of a set of international guidelines to regulate access to genetic resources and benefit-sharing. In preparation for this meeting, the International Indigenous Forum on Biodiversity (IIFB) met during the previous week to consider the proposals. The IIFB was established by indigenous delegates in 1996 to serve as an open forum for delegates from indigenous peoples' organisations and indigenous activists to discuss the issues raised by the Convention and to seek to develop common positions to present to parties. In the year 2000, the Forum was recognised as an advisory body to the Conference of the Parties and has served as a platform for increasing participation by indigenous peoples' organisations throughout the work of the Convention.

In discussing the proposed guidelines with indigenous delegates it became clear that the question of access to genetic resources and benefit-sharing presents acute dilemmas from the perspective of indigenous peoples' organisations. On the one hand, participating in discussions could be seen as legitimating the commodification of life and of culture, and could thus undermine the repeated emphasis that indigenous organisations and activists have placed on the cultural and spiritual values of biodiversity (Posey 1999). At the same time, delegates were aware that participation in these debates could open the floodgates to the exploitation of indigenous peoples around the world. On the other hand, the failure of indigenous delegates and activists to participate in these discussions would be welcomed by a significant number of governments who were keen to exploit the possibilities of 'green gold' without reference to the indigenous people involved in making such exploitation possible. This dilemma was resolved in Bonn by a decision to 'do no harm' that focused on defending the rights of indigenous peoples to decide for themselves with a particular focus on prior informed consent and the right to say no to bioprospecting.

On a wider level, mounting concern among indigenous peoples' organisations and activists is also linked to broader issues of how the rights of such peoples and societies might be protected in the context of the internationalisation of intellectual property instruments. This is reflected in increasing proposals about prior informed consent and research ethics and, in a stronger form, in recent proposals for the potential development of a new international instrument on the protection of indigenous peoples' cultural heritage, a category that extends from traditional knowledge and biodiversity to wider cultural property in the form of art, symbols and designs (Yokota and Saami Council 2005). This latter proposal is an elaboration of earlier draft guidelines for the protection of the heritage of indigenous peoples which has been characterised by Brown (2003) as promoting 'Total Heritage Protection', i.e. a desire to maintain control over every aspect of culture as 'property' that extends to biological material. While it is unclear to the author what alternatives might realistically exist for indigenous peoples' organisations when confronted with unscrupulous governments and unscrupulous scientists seeking to pursue 'green gold', what is clear is that responses from indigenous peoples' organisations and activists are highly defensive in nature.

In the case of developing countries this defensive reaction is even more marked. Thus, the Working Group Meeting in Bonn and the Sixth Conference of the Parties in 2002 (COP6) were dominated by the tortured negotiation of what became known as the Bonn Guidelines on Access to Genetic Resources and Benefit Sharing. These consist of a set of voluntary guidelines relating to almost every aspect of access and benefit-sharing which is twenty-five pages long and which the author, who participated in the process, now finds difficult to understand (see Parry 2004). At the closure of COP6, even as delegates were congratulating themselves on this achievement, delegates from Africa began to demand legally binding guidelines. Shortly afterwards, as developing country negotiators and ministries began to digest the contents of the guidelines, a view emerged that the balance of responsibilities under the guidelines was falling on the 'providers' (developing countries) rather than the 'users' or developed countries to whom the guidelines were in theory mainly directed.

In 2002, in the lead-up to the World Summit on Sustainable Development (WSSD), the newly formed 'Group of Like-Minded Megadiverse Countries', consisting of fifteen developing countries classified as 'megadiverse', began to demand a new and legally binding international instrument concerned with benefit-sharing related to genetic resources.[4] In the Plan of Implementation that emerged from the WSSD this demand found form in a recommendation to begin the negotiation of a new 'international regime' concerning benefit-sharing and genetic resources. The recommendation was subsequently endorsed by the United Nations General Assembly which invited the Convention to begin the negotiations. These negotiations were initiated in 2003 and, during the Seventh Conference of the Parties (held in Kuala Lumpur in February 2004), they resulted in a negotiating mandate for a new international regime on access to genetic resources and benefit-sharing under the Convention. The framework for these negotiations is set out in COP7 decision VII/19, which consists of an extensive list of international instruments that need to be considered in developing a regime that is likely to consist of one (or more) instrument(s) that may or may not be legally binding. In short, much remains in play and it seems unlikely that the complex issues involved in the negotiations will be resolved in the near future.

In the meantime, in the face of these uncertainties and concerns regarding biopiracy, developing countries have increasingly introduced restrictions on biodiversity-related research and the granting of research permits, or have introduced or are introducing new regulations (i.e. Brazil, the Philippines, Costa Rica and Venezuela). This can be characterised as a process of closing down biodiversity-related research in response to concerns about biopiracy. It affects researchers from developed and developing countries and has provoked increasing protests from members of the scientific community. These protests are reflected in headlines in the scientific press such as: 'Biologists Sought a Treaty; Now They Fault It' (Revkin

2002); 'Biodiversity Treaty called "disastrous"' (Agres 2003); 'Brazil's bio-piracy laws "are stifling research"' (Massarani 2003); and the rather more pedestrian 'Biodiversity law has had some unintended effects' (Pethiyagoda 2004). As the curator of Amazon botany at the New York Botanical Gardens describes it, concerns relating to biopiracy have led to 'bioparanoia' in developing countries, while another researcher from Colombia University suggested that this concern has led to the 'criminalization of the biological researcher', and a situation in which 'everyone is suspect' (Agres 2003).

As I have suggested in relation to the responses of indigenous peoples, in practice it is difficult to see what alternatives are available to developing countries in countering concerns about biopiracy and the internationalisation of intellectual property protection. In particular, it is difficult to see how else developing countries might respond in a context of ongoing difficulties in securing reform of the TRIPS agreement (Helfer 2004). Although scientists have been among the first to criticise the Convention for stifling research, it is also clear that, while science has become a casualty in this process, it is far from being an innocent casualty. Specifically – as developing-country governments and indigenous peoples' organisations are well aware – the commodification of knowledge and the biological components of organisms through intellectual property is fundamentally dependent on science and occurs in the wider context of the commercialisation of science (e.g. Rai and Eisenberg 2003; OECD 2003). In short, science and scientific institutions have been deeply complicit in the creation of the phenomenon that is now derided as 'bioparanoia' (see also McNeely 1999).

However, while science has been complicit in these processes it has not been the main driver of these developments. As I have argued, following Drahos and Braithwaite (2002), in practice these processes have their origins in the interests behind the promotion of the internationalisation of strong forms of international property protection. I have also argued that, when viewed from the perspective of economic theory, the origins of TRIPS are best understood as a form of rent-seeking behaviour. When considered in this light, the primarily defensive postures of indigenous peoples, mobilisations concerning biopiracy, and counter-rent-seeking by developing countries, can be said to constitute externalities or costs generated by the original rent-seeking that motivated the TRIPS agreement. In practical terms, one consequence of that original rent-seeking is the increasing emergence of 'anticommons' effects that are global in scope (Heller and Eisenberg 1998). In their study of the case of intellectual property claims in the realm of upstream biomedical research, Heller and Eisenberg argue that an anticommons constitutes a situation in which people overestimate the likelihood in which an individual contribution will result in a particular outcome (e.g. a blockbuster drug) and at the same time 'systematically overvalue their assets' in these arenas (1998: 701). At the same time, an anticommons is also characterised by a situation in which the holders of resources or property fear appropriation of those resources by others, and fear the consequences of infringing the rights of others. The practical out-comes are a climate of mistrust, chilling effects and ultimately losses to welfare. The language of the 'anticommons' and anticommons effects aptly describes the tortured landscape of international policy debates concerning access to genetic resources and benefit-sharing as the situation presently stands (see also Safrin 2004; Brush 2002; Oldham 2004a).

Conclusions

In this chapter I have been concerned with mapping out the contestations and fault-lines involved in debates on access to genetic resources and benefit-sharing under the Convention on Biological Diversity in the context of the rise of the bioeconomy. In the course of this brief analysis of large-scale, complex and shifting configurations I have sought to highlight the role of intellectual property as the central point of articulation around which these contestations are being played out.

In seeking to map out this landscape in relation to the emergent bio-economy it has become clear that the emergence of a bioeconomy cannot be conceived in simple instrumentalist terms as the extraction of 'latent value' or an apparently innocent progression from one 'Age' to another. Rather, in seeking to understand the complex mobilisations and configura-tions surrounding intellectual property and the bioeconomy it is also necessary to grasp the historical, ethical, social, economic and related dimensions of this emergent economy from multiple perspectives.

In the case of debates around access to genetic resources and benefit-sharing under the Convention on Biological Diversity, I have argued that the expectations concerning the value of knowledge and biodiversity, config-ured in multiple ways, and anxieties about its appropriation and loss in the context of internationalised intellectual property is generating an anti-commons and anticommons effects on multiple levels. While it is tempting to dismiss these anxieties and their effects as 'bioparanoia' produced by the provisions of the Convention, it is important to bear in mind that those provisions were negotiated in a context of historical awareness of the impor-tance of biodiversity and emerging trends relating to the internationalisa-tion of intellectual property protection. At the core of these concerns are the provisions of what is now the TRIPS agreement. Seen from the per-spective of rent-seeking, the present difficulties attributed to the Convention and to developing countries and indigenous peoples constitute externalities attributable to the rent-seeking behaviour embodied in the TRIPS agree-ment. These externalities are to be measured not only in the tortured nature of these debates, but also in terms of what does not happen, and indeed cannot happen – in relation to the pursuit of the wider objectives of the Convention and in advancing human welfare – until these wider ques-tions concerning international equity are addressed.

In highlighting the importance of the concept of biopiracy in relation to the emerging construction of a bioeconomy, my purpose has been to emphasise the need for a wider view that simultaneously recognises the historical and related dimensions of the emergent bioeconomy and also recognises that we do not yet know how a 'new' bioeconomy is best organised or whose interests it will serve. However, the concept serves as a useful reminder that the vast majority of humanity has always lived in a bioeconomy that has been configured in multiple ways. The arrival of the 'new' bioeconomy provides a rare and important opportunity to consider how this emergent economy might best be constructed to serve human welfare.

Notes

1 See http://www.bio-economy.net (accessed 15 November 2006).
2 ETC Group, Keyword Definitions. Available online at http://www.etcgroup.org/key_defs.asp (accessed 13 March 2006).
3 WIPO Summary – Patent Statistics 1990–2001. World Intellectual Property Organisation website. Available online at http://www.wipo.int/ipstats/en/statistics/patents/index.html (accessed 13 March 2006).
4 The fifteen members of the 'Like-Minded Megadiverse Countries' are signatories to the Cancun Declaration of 18 February 2002. The countries are: Bolivia, Brazil, China, Costa Rica, Colombia, Ecuador, India, Indonesia, Kenya, Mexico, Malaysia, Peru, Philippines, South Africa and Venezuela.

References

Adcock, M. and Llewelyn, M. (2000) 'Micro-organisms, definitions and options under TRIPS and micro-organisms, definitions and options under TRIPS supplementary thoughts', Quaker United Nations Office Programme on The TRIPS Process: Negotiating Challenges and Opportunities, Occasional Paper 2.

Agres, T. (2003) 'Biodiversity treaty called disastrous', *The Scientist*, 10 September 2003. Available online at http://www.plant-talk.org/stories/17genes.html (accessed 25 October 2005).

Arico, S. and Salpin, C. (2005) *Bioprospecting of Genetic Resources in the Deep Seabed: scientific, legal and policy aspects.* UNU-IAS Institute of Advanced Studies. Tokyo: United Nations University.

Barsh, R. (1986) 'Indigenous peoples: an emerging object of international law', *The American Journal of International Law*, 80(2): 369–85.

Berlin, B. and Berlin, E. (2003) 'NGOs and the process of prior informed consent in bioprospecting research: the Maya ICBG project in Chiapas, Mexico', *International Social Science Journal*, 179: 629–38.

Boyle, J. (2003) 'The second enclosure movement and the construction of the public domain', *Law and Contemporary Problems*, 66: 33–74.

Brown, M. (2003) *Who Owns Native Culture.* Cambridge, MA: Harvard University Press.

Brush, S. (2002) *Farmer's bounty: the future of crop diversity in the modern world.* New Haven, CT: Yale University Press.

Buenz, E., Schnepple, D., Bauer, B., Elkin, P., Riddle, J. and Motley, T. (2004) 'Techniques: bioprospecting historical herbal texts by hunting for new leads in old tomes', *Trends in Pharmacological Sciences*, 25(9): 494–8.

Castree, N. (2003) 'Bioprospecting: from theory to practice (and back again)', *Transactions of the Institute of British Geography*, NS 82: 35–55.

CIPR (2002) *Integrating Intellectual Property Rights and Development Policy*. Report of the Commission on Intellectual Property Rights. London, September.

Cleveland, D. and Murray, S. (1997) 'The world's crop genetic resources and the rights of indigenous farmers', *Current Anthropology*, 18: 477–515.

Collier, R. (1968) *The River that God Forgot*. London: Collins.

Crosby, A. (1972) *The Columbian Exchange: biological and cultural consequences of 1492*. Westport, CT: Greenwood Press.

—— (1991) 'The biological consequences of 1492', *North American Congress on Latin America XXV* 2: 5–13, September.

Daes, E.-I. (1996) *Working Paper by the Chairperson-Rapporteur, Mrs Erica-Irene A. Daes, on the Concept of 'Indigenous People'*. Sub-Commission on Prevention of Discrimination and Protection of Minorities, Working Group on Indigenous Populations, Fourteenth Session, 29 July–2 August. Document: E/CN.4/Sub.2/AC.4/1996/2.

—— (2003) *Indigenous Peoples; Permanent Sovereignty over Natural Resources. Preliminary Report of the Special Rapporteur, Erica-Irene A. Daes*. Submitted in accordance with Sub-Commission resolution 2002/ 15. Sub-Commission on the Promotion and Protection of Human Rights, Fifty-fifth session, 21 July. Document: E/CN.4/Sub.2/2003/20.

Dalton, R. (2001) 'The curtain falls', *Nature,* 414, 685.

—— (2002) 'Bioprospectors turn their gaze to Canada', *Nature,* 419: 768.

Doremus, H. (1999) 'Nature, knowledge and profit: the Yellowstone bioprospecting controversy and the core purposes of America's national parks', *Ecology Law Quarterly,* 26 401–88.

Drahos, P. and Braithwaite, J. (2002) *Information Feudalism: who owns the knowledge economy?* London: Earthscan.

Dutfield, G. (2000) *Intellectual Property Rights, Trade and Biodiversity*. London: Earthscan.

Eisner, T. (1989) 'Prospecting for nature's chemical riches', *Issues in Science and Technology*, 89–90: 31–4.

Ellen, R., Parkes, P. and Bicker, A. (eds) (2000) *Indigenous Environmental Knowledge and its Transformations: critical anthropological perspectives*. London: Routledge.

ETC Group (2005) 'Special report – NanoGeoPolitics: ETC group surveys the political landscape', *Communiqué*, July 28.

Farnsworth, N. (1990) 'The role of ethnopharmacology in drug development', in D. Chadwick and J. Marsh (eds), *Bioactive Compounds from Plants*. Ciba Foundation Symposium 154. Wiley and Sons: Chichester.

Fink, C. and Maskus, K. (2005) *Intellectual Property and Development: lessons from recent economic research*. Oxford: Oxford University Press.

Gollin, M. (1993) 'An intellectual property rights framework for biodiversity prospecting', in W. Reid, *Biodiversity Prospecting: using genetic resources for sustainable development*. Washington, DC: World Resources Institute.

Greene, S. (2004) 'Indigenous peoples incorporated? Culture as politics, culture as property in pharmaceutical bioprospecting', *Current Anthropology*, 45(2): 211–37.

Hayden, C. (2003) 'From market to market: bioprospecting's idiom of inclusion', *American Ethnologist*, 30(3): 359–71.

Helfer, L. (2004) 'Regime shifting: the TRIPs agreement and new dynamics of international intellectual property lawmaking', *The Yale Journal of International Law*, 29(1): 2–81.

Heller, M. and Eisenberg, R. (1998) 'Can patents deter innovation? The anticommons in biomedical research', *Science*, 280: 698–701.

Hitchcock, R. and Vinding, D. (2004) *Indigenous Peoples' Rights in Southern Africa*. Copenhagen: International Working Group for Indigenous Affairs.

Juma, C. (1989) *The Gene Hunters: biotechnology and the scramble for seeds*. London: Zed Books.

Kevles, D. (2002) *A History of Patenting Life in the United States with Comparative Attention to Europe and Canada*, European Group on Ethics in Science and New Technologies to the European Commission. 12 January. Luxembourg: Office for Official Publications of the European Communities.

Krueger, A. (1974) 'The political economy of the rent-seeking society', *American Economic Review*, LXIV, 291–303.

Lohan, D. and Johnston, S. (2003) *The International Regime for Bioprospecting: existing policies and emerging issues for Antarctica*. UNU/IAS Report. UNU-IAS Institute of Advanced Studies. Tokyo: United Nations University.

Macfarlane, A. and Macfarlane, I. (2003) *Green Gold*. London: Ebury Press

Maffi, L. (1999) 'Language and the environment', in D. Posey (ed.), *Cultural and Spiritual Values of Biodiversity*, United Nations Environment Programme. London: Intermediate Technology Publications.

—— (ed.) (2001) *On Biocultural Diversity: linking language, knowledge and the environment*. Washington, DC: Smithsonian Books.

Maskus, K. and Reichman, J. (2005a) 'The globalization of private knowledge goods and the privatization of global public goods', in K. Maskus and J. Reichman (eds), *International Public Goods and Transfer of Technology under a Globalized Intellectual Property Regime*. Cambridge: Cambridge University Press.

—— (eds) (2005b) *International Public Goods and Transfer of Technology under a Globalized Intellectual Property Regime*. Cambridge: Cambridge University Press.

Massarani, L. (2003) 'Brazil's biopiracy laws "are stifling research"', News, SciDev.Net, 21 July 2003. Available online at http://www.scidev.net/News/index.cfm?fuseaction=readnews&itemid=920&language=1 (accessed 25 October 2005).

McNeely, J. (1999) 'Hands off our Genes', *Plant Talk*, 17, April. Available online at http://www.plant-talk.org/stories/17genes.html (accessed 26 October 2005).

Moran, K., King, S. and Carlson, T. (2001) 'Biodiversity Prospecting: lessons and prospects', *Annual Review of Anthropology*, 30: 505–26.

Newman, D., Cragg, G. and Snader, K. (2003) 'Natural products as sources of new drugs over the period 1981–2002', *Journal of Natural Products* 66 (7): 1022–37.

Nigh, R. (2002) 'Maya medicine in the biological gaze: bioprospecting research as herbal fetishism', *Current Anthropology*, 43(3): 451–77.

OECD (2003) *Turning Science into Business: patenting and licensing at public research organisations*. Paris: Organisation for Economic Co-operation and Development.

—— (2005) *Proposal for a Major Project on the Bioeconomy in 2030*. Organisation for economic co-operation and development. Paris: Organisation for Economic Co-operation and Development.

Oldham, P. (2004a) 'Global status and trends in intellectual property claims: geno-mics, proteomics and biotechnology', *Global Status and Trends in Intellectual Property Claims*. No. 1.

—— (2004b) 'Global status and trends in intellectual property claims: micro-organisms', *Global Status and Trends in Intellectual Property Claims*, No. 2.

Parry, B. (2004) *Trading the Genome: investigating the commodification of bio-information*. New York: Columbia University Press.

Posey, D. (ed.) (1999) *Cultural and Spiritual Values of Biodiversity*, United Nations Environment Programme. London: Intermediate Technology Publications.

Posey, D. and Dutfield, G. (1996) *Beyond intellectual property: toward traditional resource rights for indigenous peoples and local communities*. Ottawa: Interna-tional Development Research Centre.

RAFI (1993) 'BIO-PIRACY: the story of natural coloured cottons of the americas', *Communiqué*, 30 November 1994.

—— (1994) 'Microbial BioPiracy: an initial analysis of microbial genetic resources originating in the south and held in the north', *Occasional Paper Series*,1(2), June.

—— (1995) 'Biopiracy Update: a global pandemic', *Communiqué*, 30 September.

—— (1999) 'Messages from the Chiapas "bioprospecting" dispute', *Genotype*, 22 December.

Rai, A. and Eisenberg, R. (2003) 'Bayh-Dole reform and the progress of biomedi-cine', *Law and Contemporary Social Problems*, 66: 289–314.

Reid, W. (ed.) (1993) *biodiversity prospecting: Using Genetic Resources for Sus-tainable Development*. Washington, DC: World Resources Institute.

Revkin, A. (2002) 'Biologists sought a treaty; now they fault it', *New York Times*, Science section, 7 May.

Pethiyagoda, R. (2004) 'Biodiversity law has had some unintended effects', corre-spondence, *Nature*, 429, 129.

Rosenthal, J., Beck, D., Bhat, A., Biswas, J., Brady, L. *et al.* (2000) 'Combining high risk science with ambitious social and economic goals', *Pharmaceutical Biology*, 37 Supplement: 6–21.

—— (2002) 'Curtain has fallen on hopes of legal bioprospecting', *Nature*, 416: 15.

—— (2004) Untitled presentation to the International Expert Workshop on Access to Genetic Resources and Benefit-Sharing, Cuernavaca, Mexico, 24–27 October. Workshop report available online at http://www.canmexworkshop.com/record.cfm (accessed 25 October 2005).

Safrin, S. (2004) 'Hyperownership in a time of biotechnological promise: the inter-national conflict to control the building blocks of life', *The American Journal of International Law*, 98(4): 641–85.

SCBD (2001) *Global Biodiversity Outlook*. Montreal: Secretariat of the Conven-tion on Biological Diversity.

Shiva, V. (1998) *Biopiracy: the plunder of knowledge and nature*, Totnes: Green Books/Gaia Foundation.

Shreeve, J. (2004) 'Craig Venter's epic voyage to redefine the origin of species', *Wired Magazine*, August.

Sillitoe, P., Bicker, A. and Pottier, J. (eds) (2002) *Participating in Development: approaches to indigenous knowledge*. London: Routledge.

Stork, N. (1997) 'Measuring global biodiversity and its decline', in M. Reaka-Kudla, E. Wilson and D. Wilson, *Biodiversity II*. Washington, DC: Joseph Henry.

ten Kate, K. and Laird, S. (1999) *The Commercial Use of Biodiversity.* London: Earthscan.

ten Kate, K., Touche, L. and Collis, A. (1998) *Benefit-Sharing Case Study: Yellowstone National Park and the Diversa Corporation,* Submission to the Executive Secretary of the Convention on Biological Diversity by the Royal Botanic Gardens, Kew, 22 April.

Tullock, G. (1993) *Rent Seeking,* The Shaftesbury Papers, 2. Aldershot: Edward Elgar.

UNCTAD (2001) 'The new bioeconomy: industrial and environmental biotechnology in developing countries', Ad Hoc Expert Group Meeting, Palais de Nations, Geneva 15–16 November.

UNCTAD-ICTSD (2005) *Resource Book on TRIPS and Development.* Cambridge: Cambridge University Press.

Von Lewinski, S. (ed.) (2003) *Indigenous Heritage and Intellectual Property: genetic resources, traditional knowledge and folklore.* Dordrecht: Kluwer Law International.

WIPO (2003) *Yearly Review of the PCT: 2002.* Geneva: World Intellectual Property Organisation.

World Bank (2001) *Global Economic Prospects and the Developing Countries 2002.* Washington, DC: The International Bank for Reconstruction and Development/ The World Bank.

Yokota, Y. and Saami Council (2005) 'Expanded working paper submitted by Yozo Yokota and the Saami Council on the substantive proposals on the draft principles and guidelines for the protection of the heritage of indigenous peoples', United Nations Working Group on Indigenous Populations, Document E/CN.4/ Sub.2/AC.4/2005/3.

8 Identifying John Moore

Narratives of persona in patent law relating to inventions of human origin

Hyo Yoon Kang

Introduction

Intellectual property law is commonly understood to be a regime that is concerned with the management of intangible goods, such as information, artistic creativity and scientific ingenuity. The rationale of intellectual property law is taken to serve an economic function. On the one hand, it is to provide incentives and reward inventors for their work, and on the other hand, to promote public interest by the dissemination of the ideas embedded in an invention by their publication in patent documents. Therefore patent law is understood to entail a trade-off between the assignment of a temporary monopoly right in the tangible invention to the patentee and making freely available intangible information about the patented invention to the public (Eisenberg 2002). Consequently, most intellectual property scholarship has been commonly based on utilitarian considerations, with the ultimate aim of identifying the ideal equilibrium of the bargain, whilst taking into account the different political and economic interests arising in specific situations occasioned by new technologies.

Seen from such a functionalist, regulatory viewpoint, patent law's operation does not seem to affect the meaning of human personhood or of human constitution in any way. However, with continuing controversies surrounding the practice of patenting human genetic materials and information, and also in the context of the question of patenting of stem cells of human origin (Laurie 2004), there have been calls for a 'larger debate over gene patents' (Boyle 2003). The difficult process of adoption and implementation of the 1998 Biotechnology Directive (98/44/EC) has been a poignant case, which depicted the political and moral contentiousness of patent law. Not only did the Directive result in several legal disputes between a number of member states and the Commission at the European Court of Justice,[1] but France banned the patenting of inventions related to human genetic sequences outright, whereas Germany limited the scope for patents on human and primate gene sequences (Report from the Commission to the Council and the European Parliament 2005). Various national bioethics committees have recognised the moral unease about the practice of patenting human genetic material and have especially raised concerns that patenting practices may be contrary to human dignity (e.g. Nuffield Council on Bioethics 2002; German National Ethics Council 2004).

It is by no means clear, however, what the value of human dignity entails in relation to the question of property and especially that of gene patents. The meaning of humanness itself appears to be subject to many shifts by technoscientific practices in the area of genomics, which makes it difficult to pinpoint what one should understand as the essence of humanity (Pottage 1998; Strathern 1999). At first sight, the problem seems to relate to the fear of commodifying the human person and treating it like an object. Nevertheless, this recognition does not suffice to explain exactly how inventions relating to the human genetic material or information have come to be associated with the essence of human personhood. In what ways does patent law implicate questions of personhood? Can a patented invention relating to human genetic material be understood as a human element, despite the fact that it has acquired an existence apart from the human body? What kind of picture of the human person, in other words what kind of human persona, does patent law portray?

The aim of this chapter is to examine the way in which patent law brings about a certain image or persona of the human subject by reference to the orthodox property law division between persons and things. Precisely because the legal property doctrine holds that persons cannot be objects of property relations, it seems pertinent to examine what kind of status and association inventions acquire that are in some sense held to be 'human'. Rather than approaching the property distinction as an object of criticism or evaluating its adequacy in the context of human genetic patents, I take the property distinction as a starting point into the analysis of the legal understanding of personhood and seek to infer meanings of human persona which are engendered by the legal application of the property distinction. This chapter does not mean to evaluate the internal or moral adequacy of the present patent law framework with respect to its application to genetic material and information, nor for that matter, to assess the desirability of application of an (intellectual) property regime on these matters. More precisely, what seems important at this point is to trace the narrative of human personhood that is implicitly contained within patent law practice in order to get a better understanding of the ways in which patent law delineates and shapes these ideas. I refer back to the controversial case of *Moore v Regents of University of California* (henceforth *Moore 1990*)[2] in order to explore these concerns. The case has been enormously well documented and discussed, especially from the perspective of ethical dilemmas that the reality of commodification of human bodily materials has raised (e.g. Andrews 1986; Gold 1995; Kahn 2000; Harrison 2002). It is not the intention of this chapter to add another voice to those concerns, as valid and important as they may be, but here the focus of analysis lies in Moore's understanding of the distinct material which his body produced and which subsequently gave rise to the US patent no. 4,438,032 on the so-called 'Mo cell line' that was assigned to the Regents of the University of California and named Dr David Golde and Shirley Quan as its inventors. The *Moore* case represents a curious concoction in which questions of intellectual property law, human material body, technoscientific practice and human subject/object positions have become entangled, and thus offers an ideal context from which to explore the status of human subjects and the processes of objectification in relation to these questions.

The chapter is organised as follows. The first part examines Moore's relationship to himself and the patented Mo cell line. The discussion suggests that, contrary to the common belief that the 'living body is part of the person, it is not an object' (Moufang 1994), Moore seems to relate to himself already as an object and a potential commodity. The commodification of genetic material and information in the form of a patent seems to extend an already existing logic of self-ownership, which had long been the basis of the political rationality of citizenship, and particularly of legal subjecthood, but which had not been recognised in the legal discourse of property (Collier *et al.*; Davies and Naffine 2001). What seems novel, however, is the visual form in which property claims have to be embedded as they will have to fit the patent law's definitional requirements of a patentable invention. The second part offers some theoretical observations on the legal property distinction between persons and artefacts and on the relations which this distinction engenders. Despite patent law's rhetoric of neutral utility optimisation, patent law practices in the area of human genetic technologies seem to quietly and significantly contest, as well as uphold, a peculiar vision of the human persona. It appears that patent law reflects the human person's fragmented relationship to herself as an object, while at the same time reconstructing and affirming the wholeness of human personhood by self-reference to the property law doctrine of the division between persons and things.

Identifying John Moore

> What the doctors had done was to claim my humanity, my genetic essence, as their invention and their property. They viewed me as a mine from which to extract biological material. I was harvested ... I believe that all genetic material extracted from human beings should belong to society as a whole, and not be patentable.
>
> (Interview with John Moore, *Guardian* 12 November 1994)

In the above quote, John Moore, the famous plaintiff of the case *Moore v. Regents of the University of California*, asserts that one should have the right to exert control over one's body. However, he seems to be riddled by a muddled conception of who should own genetic material and genetic 'essence': should it belong to oneself or 'to society as a whole'? Who is meant to be 'society as a whole'? Moore's confusion about 'who ought to own what' conveys many of the problems that are raised and remain unresolved in his case.

After briefly recounting the facts of the case, I turn to Moore's own account of the invention at contention. I explore Moore's representation of his cells and his association of the invention of the Mo cell line with 'himself'. This analysis serves to illustrate the confounding tales of the subject and/or object John Moore himself thinks he is, as opposed to what the judges hold him to be. An examination of the construction of object/ subject positions within the case helps to clarify what or who patent law objectifies through its practice. Clarifying this point is crucial before undertaking an analysis of the process of commodification in patent law in order to gain a better understanding of *what it is* that is being commodified.

Facts of the case

John Moore received treatment for hairy-cell leukaemia, a rare form of cancer, between 1976 and 1983 at the Medical Center of the University of California at Los Angeles under the guidance of the physician Dr David Golde. Golde realised that Moore's blood products and blood components promised a significant commercial potential as Moore's body overproduced t-lymphocytes, a type of white blood cell, that were deemed scarce. They represented valuable sources for ascertaining the mRNA that were under-stood to code for lymphokines, and thus opened up the possibility of pro-ducing larger quantities of lymphokines with the use of recombinant DNA technique for further research purposes. For purposes of cancer treatment, as well as with the awareness that 'access to a patient whose blood con-tained these substances would provide "competitive, commercial, and sci-entific advantages"' (Moore 1990: 126), Golde advised Moore to undergo an operation which would remove his spleen. Golde justified this recom-mendation on the basis of 'fear for his life'. Before the operation Golde arranged with Shirley Quan, a researcher, to 'obtain portions of [Moore's] spleen following its removal' for further research. He did not inform Moore about the research activity, which was wholly unrelated to Moore's medical treatment, and he never did so for the entire duration of their doctor–patient relationship. Golde performed the operation in 1976, and then proceeded to perform unnecessary additional post-operational tests in order to remove further additional samples of 'blood, blood serum, bone marrow aspirate, and sperm' from Moore's body during several visits between November 1976 and September 1983. These visits were paid for by Moore himself upon Golde's recommendations, who argued that they were necessary for Moore's health and well-being. During these visits, Moore enquired about any commercially valuable traits in his blood, which Golde denied and stated that 'they had discovered nothing of any commercial or financial value in his Blood or Bodily substances, and in fact actively discouraged such inquiries'.

After successfully engineering a cell-line from Moore's t-lymphocytes with the recombinant DNA technique in 1979, the Regents of the University of California applied for a patent on the cell line, called the Mo cell line, in 1981. The patent was granted as US patent no 4,438,032 on 20 March 1984. The patent identified Golde and Quan as the inventors, and the claims covered the molecular compound of the cell line, its derived products, as well as various methods of lymphokine production which would be based on the cell line. With the help of the Regents, Golde subsequently concluded lucrative contracts with Genetics Institute Inc. and later with Sandoz Pharmaceutical Corporation, in which he negotiated a large share of equity in Genetics Institute, as well as securing a substantial amount of monetary incentives for himself. Moore indicated the potential market value of the lymphokine-producing cell line and its application to be an estimated three billion US dollars by 1994.

John Moore's relationship with himself

John Moore's causes of action against the Regents of the University of California represent an interesting mixture of ownership claims, ranging from rights of use of the physical bodily materials derived from his body (especially: conversion, fraud and deceit, unjust enrichment), to claims of economic rights in the ensuing potential financial gains resulting from the patent in the Mo cell line (intentional interference with prospective advantageous economic relationships), as well as claiming compensation for harm suffered to his sense of personhood (claim of intentional infliction of emotional distress, negligent misrepresentation, slander of title, declaratory relief). But the most controversial has been the claim of conversion which essentially entails a claim to property.[3]

What exactly is the object of his claim of conversion? Moore's pleaded allegations do not seem to claim that the patent on the Mo cell line is not valid, nor does he expressly argue that the patent is wrongly attributed to the Regents and should be rather assigned to him. Therefore, Moore does not contest the commodification of 'his cells' through the patent on the Mo cell line. By the claim of conversion, Moore rather seems to claim proprietary interest in his body and thereby a 'proprietary interest in each of the products that any of the defendants might ever create from his cells or the patented cell line' (*Moore 1990*: 135).

The cells' semantic and symbolic transformation from, at first, a pathological symptom of leukaemia threatening Moore's life to a source of valuable genetic information, up to their transposition into the form of a financially valuable patent over which competing property interests are asserted, is astonishing. Moore claims, for instance, that 'genetic sequences ... are his tangible property' (*Moore 1990*: 135), but how would such a proprietary right be conceived in practice? In what sense are genetic sequences tangible, and how would such a notion of tangibility be experienced by Moore? Does it make sense, at all, to speak of ownership in genetic information? As Justice Panelli in the case remarks, such a notion of ownership in genetic information does not make sense in light of the fact that 'the genetic code for lymphokines does not vary from individual to individual'. But by a similar nonsensical token, property right has still been claimed by and granted to the Regents. Would Moore's property claim also extend to the informational quality of the cells? Moore does not address such questions at all, but is primarily interested in the economic aspect of the conversion claim.

Despite Moore's claim of conversion and the thereby implied notion of his innate ownership right over his body, it can be doubted whether Moore's proprietary interests would have arisen *prior* to the knowledge of his cells' potential financial value. Moore's claims of conversion and proprietary interest in the financial gains made from all products derived from the cells of his body do not seem to relate to the patented Mo cell line, because the cells' financial value materialised only after the replication and manipulation of Moore's 'original' cells had brought about the knowledge of the valuable genetic information for the production of certain lymphokines. Therefore it was arguably the biotechnological practice and the ensuing patent that enabled him to realise the value of the cells which his leukaemia-suffering body unfortunately (or with hindsight, fortunately) overproduced.

Although Moore's wish to obtain a fair share of the financial gains made by Golde and other parties from the unauthorised removal of his bodily materials is only understandable from an economic point of view, as well as from considerations of equity and fairness, Moore's claim effectively results in an acceptance of the treatment of his body as a potential commodity through an implicit approval of the patentability of the Mo cell line. This belief is also underlined by his demand that he should be a benefactor of the financial proceeds stemming from the Mo cell line patent. Hence, what Moore's claim seems to imply in effect is that his body has already and all along been an object under his control, particularly with the potentiality of turning that object into a commodity by his will. However, Golde and others had interfered with his exclusive rights of use, which would have been properly dealt with by granting him a share of the patent income, as well as by properly adhering to the principle of informed consent during his medical treatment.

In the judges' majority opinion, Moore is portrayed as understanding his body as an object of property by not putting into doubt the validity of the Mo cell line patent as a whole, but by wishing to partake in the profit proceeding from the invention. Indeed, Moore's argument amounts to an admission of a doubly objectified body: the body is an object at the disposal of the person who occupies it, and moreover it can be commodified through the techniques of informed consent and patent. The processes of objectification and commodification, in other words the transformation of a bodily substance into a commercial 'thing', were precisely the mechanisms by which Moore related to his particular cells as embodying value. And such a relationship between Moore and his body, between the human subject and the body as an object, seems to have been constituted by the practice of patent law, which served as a conceptual linkage between the individual person, technological practice and economic rationality. Moore seems to relate to himself as a reified agent by accepting the process of double objectification and by wishing to partake in the income which the patented invention derived from his body might yield. What patent law has effected was the transformation of Moore's self-regarding relationship to his body from one of objectification into one of commodification.

Moore's cells and the Mo cell line patent

In what precise ways did patent law establish the link between the Mo cell line and Moore or Moore's 'primary cells' found in his body? Moore's 'primary cells' have only become known as being valuable after the research process into the Mo cell line had been initiated. This is to say that patent law created the linkage which associated Moore's cells with the patented Mo cell line by reference to the commercial potentiality, i.e. the utility, of the patented cell line. Despite the fact that both entities have been embedded in different temporal and material circumstances, they can only be identified as specific particulars by reference to the existence of the other. Such an epistemological identification of both entities as two faces of a legally created association or relationship does not amount to a concession to the California Supreme Court's majority opinion, which argued that it was only through Golde's scientific labour that the cells within Moore's body had started to 'exist'. Rather the existence and knowledge of both embodiments of the 'Moore cell' are predicated on each other: the Mo cell line would have not come into existence without the cells within Moore's body, and Moore's cells would not have been recognised and become known as having a body-external, economic value without the research that led to the patenting of the Mo cell line. The 'primary cells' embodied negative value as they were expressions of leukaemia, but later on came to represent a source of positive economic value to Moore.

I draw two conclusions from the discussion: first, the external object of property in this case, the patented invention of the Mo cell line, cannot be clearly delineated from the original cells found in Moore's body after the patent has been granted. The effect of patent law was the diffusion of temporal and material differences that existed before the patent was issued. Second, it also appears that the application of patent law together with biotechnological practice effected a retrospective change of the nature of the commodified objects, i.e. of Moore's cells and the Mo cell line. This is because the existence of the patented object, the Mo cell line, overlapped temporally with the disappearance of the initial 'unique' character of the cells found in Moore's body. By the time the Mo cell line existed, Moore's body stopped over-producing t-lymphocytes as he was cured of the disease. But instead the Mo cell line could be multiplied indefinitely. Therefore, the patent did not only signal a change of representation of Moore's cells, but the 'properties' of the cells were transformed as well. These observations point to larger conceptual questions about patent law's temporality and its social transformative power over its narrowly defined subject (invention) and not so clearly defined, yet somehow related objects (human agents, information, bodily materials).

This section has attempted to address some of the problems that the practice of patent law has posed to the traditional dichotomies between persons and things, and subject and object by reference to the *Moore* case. What is curious is the impression that these dichotomies have not all turned hybrid, in the sense that they have merged and become materially and temporally indistinguishable. On the contrary, certain definitions of things and persons seem to have become reinforced through the practice of patent law, such as by the insistence on the difference between the Mo cell line (a thing) and Moore's primary cells (a part of a person). Moreover, the process of commodification has both been affirmed and at the same time also rejected by John Moore and the courts. And although the subject of the patent, the Mo cell line, could not be clearly differentiated from its source, Moore's cells, the official legal discourse made a palpable distinction between what it regarded to be its legitimate subject matter, i.e. the invention, as opposed to others and thereby, in effect, 'produced' the invention itself. And strangely, what this legal differentiation seems to have brought about is a conflation, rather than a distinction, of the present and past 'properties' of the object.

Property and the black box of the human

It is not unknown that the legal property discourse of persons and things is riddled with a central paradox. The human person is not held to be an object of property relations. Davies and Naffine (2001) convey succinctly that 'a person cannot be property and so cannot be a thing which can be owned, for it is impossible to be a person and thing, the proprietor and the property'. However, the human body seems to hold an unclear ownership and property status (Gold 1996; Rabinow 1999). Although there exists a widespread belief that one's body belongs to oneself, such a sense of ownership can be attributed to a subjective sense of ownership rather than to a legally recognised property right. In law, with the exception of corpses which are regarded as quasi-properties of the immediate family of the deceased, the human body is expressly not seen as an object over which property claims can be asserted. Both observations may be traced back to the concern about an almost universal rejection of slavery from the viewpoint of the humanistic norms embedded in liberal Western legal systems. To recount a simplistic analogy, it is argued that a human body should not occupy the same object status as a car or a house. Radin's theory of fungible and infungible property (1982) has developed this consideration further as an interpretation of Hegel's theory of property.

However, there is undoubtedly an intimate link between personhood and property. The notion of property is closely, perhaps covertly, linked to the concept of legal personhood, most notably to the liberal notion of citizenship. The Kantian interpretation that a person ought not to be treated as a thing has been taken up by the liberal jurisprudence of property; but the same line of jurisprudence bases the notion of non-property status of the human person on an understanding of one's relation to oneself as self-owning. Indeed, the concept of personhood which is broadly understood to denote a specific identity appears to be closely connected to the idea of property: '[T]he idea of the person is in fact deeply imbued with the idea of property. To be a person is to be a proprietor and also to be property – the property of oneself' (Davies and Naffine 2001). In legal anthropology, a historical link between social contract theory and the rise of the capitalist economic organisation has been suggested by Collier *et al.* (1995). They argue that the mutual inter-relationship between market, property and the individual seems to have been reversely interpreted as being indicative of human nature itself. In their account, human nature had been taken to mean an essential trait in line with John Stuart Mill's assertion that human beings have a 'natural' capacity to participate in the market, which, in turn, had led to an understanding of the market as 'natural' and self-regulating. Collier and colleagues contend that the outcome of this double process of 'social naturalisation' has been the tendency to essentialise human differences and inequalities as stemming from a 'natural', pre-social, and thus extra-legal, realm. These myths were maintained by resort to a fictional private/public distinction, then the concept of 'race', and more recently with reference to a monolithic understanding of 'culture'. According to Collier and colleagues, the liberal double-sided coin of natural equality and difference occasioned the rise of identity politics defined by these cleavages. This critique of the Lockean narrative of individuation and property, in which the status of a legal subject is closely linked to the 'ownership' of labour and the conception of natural rights and which entails a specific vision of citizenship, raises many questions. The conceptualisation of ownership of rights as quasi-property would be an interesting point for further exploration: do human rights represent commodified rights? Or, put differently, are human rights quasi-commodities? How could one understand the relationship between inalienable human rights and the *de facto* alienable human body?

The property paradox may serve as a useful primary explanation, but it does not suffice for understanding the details of the disquiet that the link between the human person, the body and property (or patent) has given rise to. For if property is commonly understood to denote a relationship between persons in relation to things, then the pertinent question to ask is how such a relationship might change if it is conducted in relation to things that are, to various inconsistent degrees, seen to be 'human'. The mediating thing, i.e. the object of property relations in the form of a patented invention related to a human gene, may be 'human' in either the material or informational, intangible senses (or both). It is by no means clear what the distinctive humanness of a patented invention entails. What this black box of humanness implies is that the frequently voiced concern about degrading the human person into an object is inadequate for explaining the underlying legal conception of the relationship, if any, between the human body and human personhood. Thus, it is quite pro-blematic to stipulate certain rights as human rights in the human genome in light of the fundamental uncertainty and soul-searching for the meaning of humanness. A part cannot be recognised as being a part of something without a conception of an underlying unity or a reference to a whole. Is one variable missing in order to solve the equation? For example, if a property object possessed some 'human' characteristics, would it in turn effect a changed perception of the human subject in law? If the object of a property relation changed its character as a 'thing', would it result in changes to the character of (human) subject in a property relationship? It is in this context of the legal technique of boundary-drawing between what is a legally permissible object of property or not, that questions of patenting inventions related to human genetic material have to be situated.

John Moore's relation to himself: some theoretical observation and scenarios

One of the points made in the preceding discussion was the observation that John Moore seemed to understand himself as an already objectified agent. He was engaged in an object relationship to himself by regarding his body as an object with the potential of transforming it into a commodity by his will. There are two ways of interpreting this point. On the one hand, Moore detached 'himself' from his body, believing that the essence of his personhood remained unharmed by the partial commodification of his genetic material and information, which the Mo cell line patent might have signified to him. On the other hand, one could conceive of Moore's self-pertaining object relation to his persona as not only comprising his body as the object of property, but also his subjectivity or his sense of self-reflection.

Interestingly, what both interpretations have in common is the point of view from which Moore relates to both his body and his subjective con-sciousness from a transposed, outside point of observation. The difference between the two would, however, consist in the nature of the relationship between body and subjectivity, and this difference would also result in different understandings of the human agent. At first glance, the first interpretation would, relatively speaking, be less extensive, as the body would be seen as something external to subjectivity, so that it would constitute an object to be controlled by the mind. In this case, the notion of property in oneself could be understood as the ownership of one's body. The implication of the division between body and subjectivity would be to regard oneself as a Cartesian dualistic entity, a person split between body and mind, the latter of which may broadly connote the capacity to self-consciousness. The material body might be objectified and perhaps even represent a potential commodity, but the locus of legal subjecthood would remain with human consciousness and therefore outside of relations of commodification.

What remains to be identified in such a scenario is *what it is* outside the body that constitutes the subject of property, i.e. what or who the legal subject of property relations would be. The subject or the owner of the body would be the mind, and the mind would then be understood to constitute the essence of a person. But how would the law be able to identify and recognise the mind? The legal subject can only be identified in the form of an embodied person rather than an abstract mind. An exclusively psychic legal subject seems practically inconceivable. The human legal subject requires an external form and visualisation, as well as physical materiality (Foucault 1977, 1978).

The second interpretation of John Moore's relation to himself is to include as well the very idea of personhood as an object of property relations. By personhood, one would not only denote a human subject position or agential capacity, but also the biographical process of internalising and making sense of events that one experiences. In this sense, both body and consciousness would constitute objects of self-ownership, but it is not clear what such a notion of ownership would entail, as its meaning would differ in both the legal practical sense – can one alienate one's consciousness or personhood? – and in a subjective sense of being engaged in some form of relationship to oneself.

This interpretation of property in oneself poses some considerable problems. From a legal technical perspective, it is difficult to make any sense of personhood as an object to be owned, at all. Personhood is not easily identifiable in that it is not materially expressed in an external form, as would be necessary for a legal recognition of a property relationship. Even the concept of intellectual property – as granting proprietary rights in either the creative 'essence' or ingenuity – is only applicable to the production of expressions and to external materialisations of intangible ideas, i.e. in the form of works in the context of copyright or to inventions in patent law (Bently and Sherman 1999). If indeed 'ownership conventions are coupled to a particular conception of production as the means by which potentialities are made actual' (Pottage 2004), then personhood is clearly not legitimately 'produced' in a legally recognisable and recognised manner of production, not least because the legal technique of property takes personhood as the presupposition for the existence of things that can be owned rather than as the result of its practice of producing distinctions. However, if personhood constituted an object that could be owned, then the outcome of the legal technique would be inverted as its foundation. What has been the effect of property technique would be taken to be its condition.

For the sake of exploration, if one was to accept that both the body and the consciousness are objects of property relations, it is important to note that in such a constellation, subject positions would still need to be identified. They would not be eliminated, but it seems that the subject positions would have to be identified externally from human subjectivity. In order for the human person to relate to herself as an object of ownership, there has to be a subject position to which the relationship can be attributed, as otherwise one could not conceive of such a relation in the first place. Two questions arise: what or who is the subject, and what position would the human agent occupy in such a constellation? The body or body parts as objects of property and self-ownership cannot be perceived without corresponding subject positions.

Therefore, where or what could the subject position in self-regarding property relations be? It has to be located externally to the human person because the legal orthodoxy holds that a person cannot be a thing at the same time. In the context of self-regarding relations, the subject position seems to be assumed by the legal constitution of property relation, i.e. it appears to be produced by the legal practice itself, for example by patenting practices. And where does the human agent come into the picture? In such a constellation, the human agent would recognise and reflect on herself by reference to the relationship between her body/body parts and their legal reproduction through property practices. Precisely because social relations, such as patenting practices that 'must' occur outside the human person, occupy positions of subjectivity by which the human agent relates to herself as an object, the objectifying legal practices seem to engender specific forms of personhood or persona.

The common legal convention holds that the human person is not a property object of property relations. It is unclear whether the human person has the capacity to operate independently of the prevailing legal ownership norm, but as seen in *Moore 1990*, it seems that the legal narrative of property as not being applicable to persons seems to be increasingly contested by human agents themselves who already understand themselves as objects and potential commodities. At this point, one would need to try to make sense of the practice of mutual reference and co-constitution that the legal narrative and human persons engage in through the mediating technique of property. Taking this interaction as a starting point of analysis and not treating it as an anomaly might yield more interesting observations.

The process of identification of any external subject positions would require the delineation of the object position that the human agent would occupy. Such a recognition would mandate a particular kind of sub- jectivity, that of being engaged in practices of ownership and property in relation to oneself, to others, and in relation to other objects of property. Property could then be understood as a way of seeing and making an object of oneself, i.e. as the establishment of a reflective position which would take oneself as one's object of reflection amidst social relations. The effect of objectification as a result of subjectivity corresponds to Foucault's ana- lysis (1982) of the double meaning of the word 'subject'. Moreover, the possibility for self-perception can only be realised by differentiation and distinction. The way in which objectification of oneself and other persons is articulated and recognised is by differentiating techniques which, in turn, underlie a certain convention or moral economy (Kopytoff 1986). Strathern expresses the web of relations more elegantly:

> The objectifications are realised in persons. This, in turn, constrains the manner in which difference, and thus the basis for connection, can be registered. Connections between social relations, like connec- tions between persons, have to be experienced in terms of their changing effects upon one another. Relations and persons become in effect homologous, the capabilities of persons revealing the social relations of which they are composed, and social relations revealing the persons they produce. But revelation also depends upon the technique by which one entity – a person, a relationship – is seen to be differentiated from and to have an impact upon another.
>
> Strathern 1988: 173)

To return to the Moore example once again, the relevant starting point of the analysis would be Moore's recognition that his sense of self and his body are property objects of either himself, or of someone or something else. What remains to be identified is what these are. If the human body was no longer the privileged locus of a possible (human) subject position because it was conceived as an object of property (for example, by the associative commodification through patent law), then John Moore would no longer be able to perceive his personhood as being contained within the corporeal bounds of his body. The localisation of personhood would have to be extended and related to a body-external sphere. This would be the result of thinking about personhood through the legal technique of prop- erty, which must decide that it was not Moore who was commodified in the Mo cell line patent because Moore as a person could not be the object of property relations. The essence of Moore, or what Moore thinks of as himself, has to be located externally, but not within his body or in the patented Mo cell line. Accordingly, one is left with the task of having to locate Moore, the person, somewhere, but not by reference to the body, because the body would be an object of property. The resulting persona of Moore is of a person who understands some aspects of himself as legitimate objects of property, but also as someone else, who can be identified independently from those elements that have already been commodified. This suggests that John Moore as a person relates to himself as John Moore as a human object, by reference to the articulation of the positions that have resulted from patenting or commodifying practices. Who is John Moore? The question is rather, which Moore one seeks to identify.

One consequence of understanding the human person as a reflective process of taking an object position or as situating herself in the external world is that such process of subjectivity would still have to be made visible in order to be identifiable. And one location among the possible multiple locations of human self-reflection is the act of legal boundary-drawing between property objects and subjects. Developing the observation further, this suggests, in reverse, that if the law has adopted the 'human person' as part of its own operation, it has to cope with the different, internal practices of a self-regarding relation, both in relation to the body and with reference to one's consciousness. Legal operation itself is only sustained by a constant process of selective, social engagement in the autopoietic, operative sense of the word 'engagement' by structural coupling (Teubner 1992). This process depicts a highly fluctuating model of constantly shifting meanings of property and the human, social self. It portrays a rather complex, circular constitution of seeing oneself as being owned, through the loop of external reflection and dispersion of oneself within in the legal process, which is in turn drawn back and related to oneself as the object of such practice. Such a mutual constitution of both subject and object positions of the human agent by the legal property discourse may help to explain the seemingly paradoxical facets of John Moore's self-understanding, who understood himself as an object and potential commodity, while at the same time claiming that his 'humanity, [his] genetic essence' had been unjustly taken from him.

Conclusion

Patent law continues to conflate notions of the legal subject, human subject, the human person and the physical body into one unified whole. By this logic, an object of property cannot be a representation or part of a human person because persons cannot be 'de-totalised' by (legal) definition: persons cannot be parts or things. This suggests that personhood is constituted by an internal reflection of the legal application of the property division itself, for example within the patent law division drawn between discovery and invention. The outcome of these lines of thought is that both persons and things can be either objects or subjects. Therefore, the term 'property', and for that matter, also patents, can entail both subject and object positions: property can be acted upon, as well as bring about action.

Furthermore, the notion of agency, which is implicitly contained or denied in subject or object positions, is independent and disconnected from the property division of persons and things.

This leads to some unexpected places to look for social articulations and reflections of human personhood. Rather oddly, the human person may be more readily identified by the externalisation of her characteristics, for example by digitised genetic information or through a patent related to genetic characteristics, precisely because it cannot be owned by or form an exclusive link to the human person herself. A person 'possesses', 'contains' and embodies human genetic information, yet she cannot own it as her property, because only the process of externalisation, or the process of giving form with the help of the legal technique of making an invention, allows for the visible transformation of genetic information into a proprietary object.

It is in this legal technical and aesthetic sense that one might argue that patent rights in human genetic information and material produce new self-regarding objects, which in turn bring about new subjectivising relations that had not existed before the invention. And these new legal subjects are not located 'within' the human body, but rather in the process of an agent's relation to the legal and technoscientific practices that take the human body as their object of knowledge. Thus, the legal subject seems neither to be wholly fabricated by law, nor can it be equated with the human body. The legal subject is the relationship that occurs between these two entities, i.e. it resides in the third space or in the event of relations between law and the human agent.

If one accepts that the legal technique of making property and patents has important implications that are closely interlinked with the under-standing of the self, then the next task consists of a more differentiated delineation of the processes of objectification and commodification of the human agent. The analysis would need to shift its focus from taking the human as the primary field of the study to the human agent's relation to oneself as being mediated by social systems or networks within which she is situated. The environment or the external sphere is turned into oneself or enfolded as an internal technique of relating to oneself. In the context of this chapter, it is then patent law's technology of making an invention which produces new subject positions.

Notes

1 In *Netherlands v European Parliament*, Case C-377/98 [2002] *OJEPO* 231, the Netherlands and Italy tried to challenge the validity of the Biotechnology Directive (98/44/EC) at the ECJ, which was rejected. Implementation of the Biotechnology Directive in member states had been significantly delayed: by the 30 July 2000 deadline, only six member states had implemented the Directive in national legislation (Denmark, Finland, Greece, Spain, Ireland, the UK). The Commission referred Germany, France, Italy, the Netherlands, Sweden, Austria, Belgium and

Luxembourg to the ECJ for non-implementation. Germany only implemented the directive on 21 January 2005, but with a more limited scope for patents on human and primate gene sequences. Legislation available available online at 217.160.60.235/BGBL/bgbl1f/bgbl105s0146.pdf (accessed 9 June 2006). France has banned the patenting of inventions related to human genetic sequences. Italy, Luxembourg, Latvia and Lithuania have not yet implemented the Directive to date.

2   California Supreme Court decision at 51 Cal. 3d (1990), superceding the California Court of Appeals' decision at 202 Cal. App. 3d 1230 (1988).

3   Conversion is 'a tort that protects against interference with possessory and ownership interests in personal property' 51 3d. Cal. 3d (1990), 134. This means that the conversion claim could only be upheld if it was deemed that Moore was indeed holding a proprietary interest in his body and bodily materials before the removal of the cells.

References

Andrews, L. (1986) 'My body, my property', *Hastings Center Report*: 28, 16(5): 28–38.

Bently, L. and Sherman, B. (1999) *The Making of Modern Intellectual Property Law: the British experience, 1760–1911*. Cambridge: Cambridge University Press.

Boyle, J. (2003) 'Enclosing the genome: what the squabbles over genetic patents could teach us', Paper available online at http://www.law.duke/edu/ip/pdf/enclosing.pdf (accessed on 14 April 2004).

Collier, J., Maurer, B. and Suarez-Navaz, L. (1995) 'Sanctioned identities: legal constructions of modern personhood', *Identities*, 2: 1–27.

Davies, M. and Naffine, N. (2001) *Are Persons Property? Legal Debates about Property and Personality*. Aldershot: Ashgate.

Eisenberg, R. (2002) 'Molecules vs information: should patents protect both?' *Boston University Journal of Science and Technology Law (Symposium on Bioinformatics and Intellectual Property Law, 27 April 2001)*, 8: 190–202.

Foucault, M. (1977) *Discipline and Punish*. London: Penguin.

—— (1978) *History of Sexuality Vol. 1*. London: Penguin.

—— (1982) 'The subject and power', Afterword in H. Dreyfus and P. Rabinow, *Michel Foucault: beyond structuralism and hermeneutics*. Hemel Hempstead: Harvester Wheatsheaf.

German National Ethics Council (2004) 'Zur Patentierung biotechnologischer Erfindungen unter Verwendung biologischen Materials menschlichen Ursprungs. Stellungnahme', available online at http://www.nationalerethikrat.de/stellungnahmen/pdf/Stellungnahme_Biopatentierung.pdf (accessed on 9 July 2006).

Gold, E. R. (1995) 'Owning our bodies: an examination of property law and biotechnology', *San Diego Law Review*, 32: 1167–247.

—— (1996) *Body Parts: property rights and the ownership of human biological materials*. Washington, DC: Georgetown University Press.

Guardian, The (1994) 'Feature: lambs to the gene market'. 12 November.

Harrison, C. (2002) 'Neither Moore nor the market: alternative models of compensating contributors of human tissue', *American Journal of Law and Medicine*, 28: 77–105.

Kahn, J. (2000) 'Biotechnology and the constitution of the self: managing identity in science, the market and society', *Hastings Law Journal*, 51: 909–52.

Kopytoff, I. (1986) 'The cultural biography of things: commoditization as process', in A. Appadurai (ed.), *The Social Life of Things: commodities in cultural perspective*. Cambridge: Cambridge University Press.

Laurie, G. (2004) 'Patenting stem cells of human origin', *European Intellectual Property Review*, 26: 59–66.

Moufang, R. (1994) 'Patenting of human genes, cells and parts of the human body? – The ethical dimensions of patent law', *IIC*, 25: 487–515.

Nuffield Council on Bioethics (2002) 'The ethics of patenting DNA', available online at http://www.nuffieldbioethics.org/go/ourwork/patentingdna/introduction [(accessed on 9 July 2006).

Pottage, A. (1998) 'The inscription of life in law: genes, patents and biopolitics', *Modern Law Review*, 61: 740–65.

—— (2004) 'Introduction', in M. Mundy and A Pottage (eds), *Law, Anthropology and the Constitution of the Social*. Cambridge: University of Cambridge Press.

Rabinow, P. (1999) *French DNA*. Chicago, IL: University of Chicago Press.

Radin, M. (1982) 'Property and personhood', *Stanford Law Review*, 34: 957.

Report from the Commission to the Council and the European Parliament (2005) 'Developments and implications of patent law in the field of biotechnology and genetic engineering' (SEC(2005) 943), available online at http://www.europa.eu.int/comm/internal_market/en/indprop/invent/com_2005_312final_en.pdf (accessed on 20 August 2005).

Strathern, M. (1988) *The Gender of the Gift*. Berkeley, CA: University of California Press.

—— (1999) *Property, Substance, and Effect: anthropological essays on persons and things*. London: Athlone Press.

Teubner, G. (1992) *Law as an Autopoietic System*. Oxford: Blackwell.

9 Sampling policies of isolates of historical interest

The social and historical formation of research populations in the People's Republic of China and the Republic of China

Margaret Sleeboom-Faulkner

Introduction

Genetic sampling and gene-banking serves various areas of research. As part of the Human Genome Project (HGP), which was initiated in 1991, it served the completion of the human genome map. As part of the Human Genome Diversity Project (HGDP) samples from indigenous populations served to avoid the irreversible loss of genetic information. Referring to indigenous populations as isolates of historic interest (IHIs), the HGDP planned to immortalise the DNA of disappearing populations for future study. One aim of the gene banks for scientists is to reconstruct the history of the world's populations by studying genetic variation to determine patterns of human migration. As will become clear in this chapter, these new scientific findings concerning our origins can be used in disputes on aboriginal rights to territory, resources and self-determination. The initial conceptualisation of the HGDP has been widely criticised for its consideration of indigenous peoples as mere research subjects, with little regard for their continued livelihood. For this reason, according to the American National Institute of Health (NIH), the HGDP has been substituted by the haplotype effort. As part of the global initiative to create a haplotype map (participants in the initiative are USA, UK, Japan, China and Canada) and commercial haplotyping, genetic sampling is supposed to be exclusively used for the advancement of medical knowledge that emphasises the differences between the genetic make-up of different individuals.[1] This claim, however, is open to dispute, as patterns of variation between individuals are still generalised over entire populations, which are likely to be stigmatised in situations of conflicting political and socio-economic interests (cf. Lee 2003).

Case-studies of genetic sampling in China and Taiwan are interesting in this context for political, economic and cultural reasons. First, the different socio-political and economic circumstances in Mainland China and in Taiwan affect the ways in which genomics research takes place. The various criteria for group delineation include examples of their effect on genetic sampling and the formation of genetic knowledge. Second, in China and Taiwan research target groups, i.e. IHIs, are defined through different cultural and political perceptions. Taiwan itself has become an IHI in a political sense, as it has been almost completely ostracised diplomatically from the international community under the weight of the PRC, while the PRC has become recognised as a full political and sovereign player on the international diplomatic platform. Their respective histories and cultural and national identities have brought with them different forms of state organisation and different perceptions of marginal economic, cultural and ethnic groups. These perceptions always involve ideas about the extent to which we define groups as static or dynamic. Relatively static criteria are associated with perceptions of nature, such as genetics, race, stable biological features, and language (*sic*), while relatively dynamic criteria are associated with a changing environment, such as economic development, social improvement, cultural change and other forms of potential advancement. But in the field of genomics, the main problem is precisely how to distinguish populations from one another on the basis of genetic criteria. In other words, before genetic sampling of populations begins, estimates are made about the genetic nature of these populations. The question in this chapter is, therefore, how in the PRC and the ROC the criteria used for population group definitions and the delineation of bioethical interest groups influence practices of genetic mapping, and vice versa.

It seems safe to assume that the suitability of the criteria researchers use for group delineation depends on their perceived relevance to the research problem at hand. The problem is, however, that the definition of such groupings is intimately related to cultural perception and political outlook. I illustrate this by discussing criteria used for the delineation of groups in Mainland Chinese and Taiwanese contexts.

Problems in defining relevant sample populations

There are a number of factors relevant to delineating sampling populations. In this chapter I mainly discuss socio-political factors, but it is also important to draw attention to the linguistic and environmental criteria used to define sampling groups, assuming a correspondence between linguistic and genetic development or environmental isolation and genetic development. Some population geneticists make use of linguistics,[2] while others use genealogical records. These criteria lead to familiar research problems in history and culture, which do not assume languages to be static: languages migrate and change not necessarily parallel with the biological make-up of genetic groups; and genealogical records are often faulty, or are manipulated for religious, personal and political ends. As the 'family' is not just a biological concept, but also a phenomenon understood through changing cultural meanings, it is difficult to define a stable research population. This is also true for what we regard as communities, including linguistic communities. Thus, in many cases the horizontal transfer of language is paralleled by genetic transfer; that is, languages spread through migration. However, a parallel genetic development does not automatically follow horizontal transfer. For instance, Melanesian islanders represent substratum speech communities of a different origin from Austronesian, which is related to conquest by other peoples and not to genetic admixture. The phenomenon in Melanesia is referred to as the 'pidginisation hypothesis' (Arthur Capell), and would be a good explanation for the absence of a Melanesian-specific haplotype among the Polynesians (Li 2001: 237–9). Conversely, genetic interchange between groups may take place over linguistic boundaries. Thus, the genetic profiles of twenty-eight populations sampled in China supported the distinction between southern and northern populations. Linguistic boundaries are often transgressed across language families studied, reflecting substantial gene flow between populations (Chu 1998).

The difficulty with defining a research population on the basis of genetic particularity seems obvious: we don't know its suitability unless we first sample the population. And even if we find a shared genetic trait, we still don't know if groups elsewhere share the same trait until we have sampled all people. Neither do we know whether the trait is relevant as a criterion of distinction in the case concerned. The research experience of Harvard epidemiologist Xu Xiping shows that this problem lies at the basis of his research choice (Sleeboom 2005). Xu had to change the focus of his research from asthma to blood pressure, as he had overestimated the genetic homogeneity of his research population in Anqing and underestimated its complexities *vis-à-vis* the environment, which of course is of central concern to genetic epidemiology. Another example shows that decisions on genetic sampling are made on the basis of confusing scientific cultures. A programme for liver cancer and lung examination in the Central Region of China (*Zhongbu*) looked into polymorphisms related to coronary atherosclerosis (Lin *et al.* 1999). The authors compared the distribution of the ACE genotypes and allele frequency in 195 Han and 195 Bunun individuals categorised by age and sex. The frequencies for homozygous deletion genotype and deletion allele were greater in Bunun than in Han groups (Lin *et al.* 1999), and indicated the relevance of genetic differences in the epidemiology. Lin Juei-Jueng and colleagues, however, argue that there is no significant relation between the ACE gene and the genesis of myocardial infarction (Lin *et al.* 1999). Such different conclusions are not uncommon, and are attributable not only to different definitions of research units and samples, but also to the different methods of analysis applied to the collected serum.[3]

The socio-political construction of IHIs

Certain types of communal lifestyle are used as an indication for the existence of an IHI. Thus, the lifestyles of mountain peoples, slash-and-burning communities, and isolated sedentary communities form an indication for their candidacy as IHI. A difficulty when defining such groups is that the groups themselves take part in the ways in which they are defined, and therefore, perceived. Some have no historical records and others do their best to define themselves strategically as different from other peoples. In some cases this is a form of self-preservation against what is regarded as the 'modern' world (Barth 1969), and in others, against the threat of being assimilated into a greater power.

In the case of Mainland China (PRC), the official creation of national minorities in the 1950s (of which there are now 55) was largely a political decision.[4] In the 1950s, China began to adhere to the Stalinist definition of a nation, according to which a nation is a historically formed stable community of people arising on the basis of common language, common territory, common economic life and a typical cast of mind manifested in a common culture (Stalin 1913). Paradoxically, in the 1950s after 1954 and in the 1960s the concept of nation meant the rejection of the idea that minorities and their territories were distinctive. Discriminatory appellations for minorities (for example, Luoluo with the pictograph 'dog' or 'pig' for the Yi) were permanently abolished by legislative decree in 1951.[5] Ultra-leftist policies demanded that national minorities were to be treated the same as the majority Han, and all special privileges were to be eliminated in a class-society that recognised only one lifestyle. After the late-1970s, however, the special nature of the national minorities was gradually recognised again. But different lifestyles were mainly attributed to different historical and cultural traditions. In the 1970s, the view that lifestyle and biological features go together came to be advocated more widely. Policies to improve the quality of the population were directly linked with this view and had consequences for social policies towards marginal groups such as criminals, the handicapped and illiterates. Developments in the life sciences and the post-1978 reforms stimulated the explanation of social and cultural features on the basis of static biological principles. In turn, this gave impetus to the 1990s initiative of genetically classifying national minorities and storing them in ethnic databanks.

In the same period, the world's largest gene bank for ethnic minorities was set up in Kunming, Yunnan Province of Southwest China. It aims to study the diversification of inheritance and inherited diseases for the ethnic minorities. It has stored the DNA from at least twenty-five ethnic groups in the province. As they live in remote areas, they are thought to have high purity and separation degrees.[6] Thus, China is interesting to geneticists because its rural population and the so-called national minorities are thought to have remained static for centuries, making each region different in its pattern of genes and diseases. This would make it easier to trace hereditary diseases back to a specific defective gene, which may be unusually abundant where the disease is prevalent. But as many historical cultural and political factors play a role in their definition, geneticists can easily be steered in the wrong direction.

For example, Harvard researcher Xu Xiping, in his grant application to the National Institute of Health, used the following arguments to persuade the board of the special genetic nature of his research population in Anqing:

- The population of Anqing has been stable over the last 2,000 years.
- The type of asthma prevalent is the same type as in the West.
- Anhui's chronic sufferers of asthma, compared to Western ones, hardly ever receive medicine, enabling researchers to trace distortions in the medical syndrome.
- Divorce is rare in China, households are stable and villages are usually closed off (facilitating the collection of harmonious data).
- The Chinese household is bigger in scale compared with the Western home.

But, as we have seen, Xu had to change the direction of his research as his assumptions had been wrong: the population was less homogeneous and stood in a more complex relation to the environment than he had expected.

It can be seen therefore that the choice of IHI is largely influenced by government policies and prevailing images of ethnic communities in ancient history as isolated and static groups in the remote mountains of China. In Mainland China, the trend of regarding human groups as socio-politically malleable increasingly converged with perspectives that root human difference in their genetic make-up. This trend, together with an increased emphasis on the (genetic) quality of the population, has continued until the present. Moreover, in governmental and scientific circles the interest in competing internationally in the field of genomics has encouraged this trend of seeking solutions to social, economic and medical problems in the knowledge (and possibly the alteration) of the human genome.

In Taiwan, by contrast, the discovery of IHIs happened later. It was only five days after the announcement of the decoding of the genome on 4 July 2000 that members of the prestigious Academia Sinica urged the government to establish a Taiwanese genetic database of Taiwan's ethnic groups. The database would serve the purposes of medical research and further study of the diversity of human populations and languages. Some researchers expressed reservations about the plan, saying the database should be used only for disease research, not for ethnic studies, lest these give rise to political dispute (Liu 2000). Furthermore, the purpose of defining a 'Taiwanese genome' did not seem feasible, as the Republic has only been in existence since 1911, and the information to be entered into the gene bank was not thought to be capable of shedding much light on a 'Taiwanese genome'. Nevertheless, in the 1990s, a political trend towards the recognition of aboriginal culture made it politically desirable to overcome such an argument. Substantial evidence for the recognition of Taiwan's cultural roots became crucial to the formation of a new Taiwanese identity.

The definition of IHIs in Taiwan has come about as part of a particular constellation of interest groups on the island. General agreement exists, however, about Taiwan's main historical events: that Portuguese sailors named Taiwan *Ilha Formosa* in 1544, and that the island had been occupied for a long time previously. After the Portuguese, the Dutch and the Spanish briefly occupied a pocket of land in southern and northern Taiwan separately, before the Ming Dynasty loyalist Koxinga expelled the Dutch in 1662, and recruited Han settlers, the so-called Hoklo, to cultivate the land. The Chinese Qing Dynasty in 1683 annexed Taiwan and, nearly two centuries later, in 1895, the Qing ceded Taiwan to Japan. After World War Two, Taiwan was given up to the Republic of China, whose nationalist government, having been defeated by the Communist Chinese, took refuge in Taiwan in 1949 (Brown 2004).

In order to understand the discussion on ethnic identity relations between various groupings in Taiwan, it is necessary to look further into its history. In the 1920s, Sun Yat-sen, the founder of the Kuomintang (KMT), the nationalist party, had recognised China merely as a 'republic of five nationalities'. After World War Two, the government under Chang Kai-shek regarded the five as merely branches of the Han. After 1947, the past conquests, pacification, and at times segregation and containment gave rise to political co-optation, economic domination, forced cultural assimilation and social prejudice. But, while the KMT strictly denied the existence of any ethnically, culturally or socially unequal treatment, members of the second and third Mainlander generation had privileged access to professions in the military, state, and the educational sector until far into the 1980s (Chang 1994).

Partly thanks to the rise of social movements in the 1980s, the fate of the aborigines began to draw the special attention of the media. At the time, the popular perception of the aborigines was mainly one of social pathology. They were seen either as a needy class of losers or as a breed of inferior humans. It was not until the mid-1990s, when the ruling KMT failed to win a stable majority in the Third Legislative Election (1995), that the government took the appeals of the indigenous movement seriously. These appeals had been carried forward by the Democratic Progressive Party (DPP). To demonstrate their willingness to compromise, the central government created a Commission of Indigenous Peoples in 1996.

In the mid-1990s, the cultural and ethnic rights of aboriginal groups were recognised by a majority of the population, and their legal status became an important issue of public contention. The tribal image of the 'mountain compatriots' (*shandi tongbao*) as they were called,[7] disappeared from the media, after the lifting of martial law in 1987. It gave way to an image in which they were increasingly recognised as the forebears of the Taiwanese (*yuanminzu*) and as essential to the formation of a new Taiwanese identity.[8] In other words, the DPP, and to a lesser extent the KMT, mobilised the indigenous aspect of aboriginal identity to form a buffer of Taiwanese national identity against the perceived increase in Chinese nationalism in the 1990s.

In addition, among Han Chinese in Taiwan, the idea of a pure Han breed weakened. This was an acknowledgement of the possibility of interbreeding amongst their forefathers.[9] Similarly to the DPP, the KMT government has experienced a profound transformation since the early 1990s. It assigned symbolic functions to aborigines in two directions. Internally, the particular acknowledgement of their existence and their cultural achievements supported the development of a new Taiwanese and the construction of a new cultural centre. This also included the perception of their communities as being vested with a strong feeling of solidarity among its members – a condition that had to be protected and that could serve as a model to Taiwan's Han society. Externally, however, the protection and the fostering of this ethnic minority served to signify not only the government's democratic and multicultural attitude, but also its new cultural and political orientation.

It is of basic interest when conducting any research among ethnic, social or cultural groups, therefore, that before defining these IHIs we need to know why and how we use the criteria by which they are defined. Political, economic and cultural motives might have shaped the formation of groups that might seem suitable candidates as IHIs. In formulating a research project, it also seems (bio)ethical to check whose interests will be served by the research, its outcomes and its future applications. And as the state in both countries has played a crucial role in the definition and administration of groups that are seen as potential IHIs, it is important to understand the historical background of the relationship between these groups and the state.

The politics of group characteristics and IHIs

The different political histories of Mainland China and Taiwan have resulted in the use of different political, social and cultural criteria for the delineation of so-called ethnic groups. In both countries, the state has played a major role in choosing the criteria for the delineation of ethnic communities, and determining their status and treatment.

In China, relatively rigid political and administrative criteria are used to define national minority groups and the difference between peoples in rural and urban areas. Officially recognised national minorities have special administrative status, and city and countryside have different forms of public healthcare and administration. In the 1950s, basic rights were established for minorities by edict, concerning the practical implementation of regional autonomy, protection of the rights of dispersed minorities, financial rights, and equal rights for all nationalities. Since 1954, however, the constitution has not provided sufficient constitutional guarantee for these rights. This situation, together with the Cultural Revolution, caused much unrest. Additionally, the 1978 Constitution failed to grant minorities the rights that they had previously enjoyed under the 1954 constitution. It was only in the 1980s that regional autonomy provided self-administration within the contours of PRC law (Heberer 1989: 41; Heberer 2001).

Autonomy, in this context, does not mean that they may secede from the sovereign territory of the PRC but that, under the direction of higher authorities, they enjoy certain special rights over other administrative units, such as drawing up their own productivity plans.

Other, semi-biological categories used in public healthcare policy-making result from the application of eugenic criteria to groups of mentally handicapped people. In the official press, links are also made between genetic make-up, having a criminal record, and being a country bumpkin.[10] The introduction of eugenic legislation in 1995 supports the systematic 'implementation of premarital medical check-ups' for hereditary, venereal or reproductive disorders as well as mental disorders, so as to prevent 'inferior births' (PRC's Maternal and Infant Health Law 1994: 3–8). It is a practice that labels a large part of the peasant population and criminals as retarded, so forging a link between biological make-up and socio-cultural conditions of health, education and marital customs. In the early 1990s, great concerns over the differential birth rate between the urban and the rural sectors of the population were justified in biological terms of difference in genetic make-up. For instance, the journal '*Population and Eugenics*' (*renkou yu yousheng*) advocated a eugenic policy similar to that enacted in Singapore in the 1980s (Dikötter 1998b); that is, genetically fitter elements should be encouraged to have more than one child while massive disincentives would contribute to checking dysgenic trends in the countryside.

In the political life of Taiwan since the 1990s, it is generally agreed that the population of Taiwan is made up of four relevant major ethnic groups: the nine indigenous groups (*Yuanzhu Minzu*) (1.7 per cent), Mainlanders (*Waishengren*) (11 per cent), Hakkas (*Kejiaren*) (9 per cent), and Hoklos (*Helaoren*) (77 per cent). While the aborigines are thought to be of Malayo-Polynesian stock, the latter three are descendants from Han refugee-migrant-settlers of Mongolian origin that sailed from China 400 years ago. Although friction occurs between many ethnic and social groups, the main ethnic cleavages are to be found between the indigenous peoples and Hans (*Hanren*) (including Mainlanders, Hakkas and Hoklos); Mainlanders and. Natives (*Benshengren*) (including indigenous peoples, Hakkas and Hoklos), and Hakkas and Hoklos.

If we look at Taiwanese national identity formation, it becomes clear that Taiwanese Han elites have tried to turn the memory of the Mainlanders (who had ruled Taiwan for over forty years after the withdrawal of the Japanese in 1948) into collective memory. In the 1990s, other memories that had been suppressed for more than 90 years were retrieved, recreated and expressed in a new discourse. Among the most dedicated cultural designers were members of the opposition party, the DPP. The DPP had been founded in 1986, but was not fully legalised until 1989. Most of its membership was of Hoklo origin (at 77 per cent the biggest Han Chinese group on the island). They contended that the Taiwanese had a 400-year-long history on Taiwan. This history included the common experiences of a creative pioneer settler people from South China that developed a particular language and culture after their exodus from China in the seventeenth century and that had endured domination by several foreign powers, each of these using force to subjugate Taiwan's population. Incidents that were still remembered by the people, such as the incident of 28 February 1947 and the Formosa incident in 1979, were experiences that constituted the culture of the Taiwanese. The DPP contended that in Taiwan's schools it was these experiences and history that should be mediated, not Mandarin, the Yangtse River, Peking opera, the Great Wall and the Anti-Japanese Resistance War.

A group that showed some discontent with the unilateral request formulated by the Hoklo was Taiwan's Hakka, another Han-Chinese group in Taiwan whose members constituted about 9 per cent of Taiwan's total population and who had always had difficulty in asserting themselves against the Hoklo. In a well-organised 'Return our mother-language' (*huan wo muyu*) movement in 1988, representatives of this group argued that they spoke their own Chinese language and that their ancestors had been living on Taiwan at least as early as Taiwan's Hoklo. They also claimed that their culture was closer to the core of the 5,000-year-old Chinese culture (*zhongyuan*) than that of most other Han. But by the end of the 1980s, this 'closeness to the *zhongyuan*' had lost much of its former appeal and their movement did not receive significant support from the Hoklo.

Taiwan's Aborigines, the fourth ethnic group on the island, constituted 1.7 per cent of the population. The social movement representing this group had reached prominence at the end of the 1980s due to the new political freedom people suddenly enjoyed in Taiwan. Nevertheless, the focus of their movement at that time was not yet directed at the attainment of cultural rights, but against cultural discrimination and social marginalisation. In protests against the 'myth of Wu Feng'[11] in Taiwan's schoolbooks and in demonstrations for the 'return of land' seized by Han Chinese in the course of the last few centuries, the KMT government, and the Han in general, were petitioned to treat the *Yuanzhumin* in accordance with internationally recognised indigenous peoples' rights.

Only when plans for a constitutional reform became a distinct possibility after the official vote of Li Denghui as president in 1990 did intellectuals and political representatives of this group also begin to concentrate on the question of the status of aborigines in Taiwan's society.[12] They emphasised the value of indigenous languages and cultures and the necessity of safeguarding aborigines' physical and cultural survival as a people by implementation of special administrative and educational organs at the central government level, and by giving them parliamentary status. Contrary to the movement of the Hakka, the supporters of a sovereign Taiwan – especially the Hoklo elites – welcomed the aboriginal movement, as their demands did not obstruct the supporters' nationalist aims. Due to these political and social changes in perceptions of Taiwanese and Mainland Chinese society, the meanings of IHIs in genetic research were influenced on a fundamental level. In Taiwan, the emphasis on biological differences between Taiwanese insider groups seemed to diminish, while the ascription and recognition of biological differences between the Taiwanese and 'foreign' populations increasingly attracted attention in the media and public opinion. At the same time, the study of the genetic roots of the Taiwanese and the redefinition of Taiwanese genetic identity were used to augment the unity of Taiwanese cultural identity, whose essence now increasingly found its expression in the public image of the aborigines.

On the Mainland, however, the cultural pluralism of the Chinese (Zhonghua) was increasingly recognised in the 1990s, but was accompanied by an emphasis on a diversity of genetic roots. This diversity, according to some authors, has even contributed to the superior unity of the Chinese genetic gene pool (Chen 1992: 39–46). This perceived advantage, however, did not prevent the media, policy-makers and researchers from using the genetic argument to further their educational, academic and (inter-)national policies.

Conclusion

In the media and in academic discussions in China and Taiwan there are different ways of defining IHI for socio-political and historical reasons. If only issues of language group and environment were involved, the problem of defining a genetic sampling group would be difficult enough. Using linguistic clues for finding genetic research units may be a common-sense strategy as most people have both genes and a language, but language changes have different causes from genetic changes and may be the result of different processes. Furthermore, in the process of globalisation, the formation of nation-state borders and the standardisation of language in writing, the relationship between genetic and linguistic change may have weakened. Moreover, both linguistic and environmental factors are intimately linked with socio-economic and political processes that geneticists wish to both isolate from and connect with evolutionary genetics in a controlled manner. The problem is, however, that human groups have their own views on what IHIs consist of, and the people that make up an IHI have bioethical qualms, political currency and cultural value. In this chapter I have focused on the social construction of IHI by examining the way it is related to ideas of community, political and cultural identity, and development. It is clear that the choice of IHI is largely influenced by government policies and prevailing perceptions of the ancient history of isolated and static communities in the remote mountains of China. In Mainland China, the trend of regarding human groups as socio-politically malleable increasingly converged with perspectives that root human difference in their genetic make-up. This trend, together with an increased emphasis on the (genetic) quality of the population, is growing stronger.

In Taiwan, a reverse trend took place from the 1990s onward. The former tribal image and perception of biological inferiority attached to aborigines altered into a perception of their cultural particularity as the forefathers of the Taiwanese (*yuanminzu*). The human particularity now warrants cultural, legal and ethnic rights for the people who are increasingly considered as essential to the formation of a new Taiwanese identity. Furthermore, the DDP, and to a lesser extent the KMT, mobilised indigenous identity in support of a Taiwanese national identity to act as a buffer against the perceived increase in Chinese nationalism in the 1990s. The different political and cultural histories of Mainland China and Taiwan have led to very different discussions about and motives for defining IHI. In Mainland China, national minorities occupied a special position in the political administration, and after many upheavals they have increasingly gained the possibility of self-administration within the political and legal framework of the state. Other politically defined groups, including criminals, the handicapped, and inhabitants of the countryside have been defined genetically, though they have not been systematically examined separately from the national minorities. Nevertheless, genetic analysis is still used in political arguments supporting the policy of a united China. In Taiwan, however, the aborigines have not had a special status in the government administration and only came to be recognised as ethnic groups in the 1990s. In this capacity, they also have become important as both carriers of a new Taiwanese identity and targets of genetic sampling used to give Taiwan a more prominent cultural and political place in Austronesia and Southeast Asia. This trend partly came about as a result of internal conflict and as a reaction to increasing nationalism in Mainland China. These differences indicate that the perception of what are populations cannot be imagined or understood out of their historical, political and social contexts.

Notes

1 Sites in the DNA sequence where individuals differ at a single DNA base are called Single Nucleotide Polymorphisms (SNPs). A haplotype is a pattern of SNPs on a block of inherited sets of nearby SNPs on the same chromosome.

2 The diagnosis and discussion of such correlation can perhaps be traced, at least in terms of influence on the semi-popular scientific literature, to the parallel linguistic – genetic tree of Cavalli-Sforza (1998).

3 Different analyses of serum samples include Y-chromosome analysis, the analysis of red blood cells, blood types, HLA, microsatellites, mtDNA and others.

4 In the 1950s, 400 ethnic groups responded to an initial call for registration of national minorities. By 1957, 54 ethnic groups were officially recognised (the Jinuo in 1979 made it 55). In the late 1970s, re-petitioning for recognition occurred again: 80 groups (totalling over 900,000 persons) petitioned in the Province of Guizhou alone. The population census of 1982 included numerous new groups in other regions (Heberer 1989: 35–8).

5 Nevertheless, the Han Chinese majority fixed most of their current names. Thus, the Yi constitute many groups that have other names, but are called Yi in communication with Han Chinese (Heberer 2001).

6 Professor Xiao Chunjie from Yunnan University explains that all the samples are taken from males (because they offer both X and Y chromosomes) selected

from fifty men from each ethnic group. The sample donors have no blood or spouse relations with other ethnic peoples, and every man has the same ethnic origin for at least three generations in succession. According to the *People's Daily* (28 November 2000), the gene bank is so far the largest of its kind in the world that has adopted the standard of the international human genome programme.

7 The origins of different groups were inscribed in people's identity cards (from one of thirty-five mainland provinces existing prior to 1945, from Taiwan province, or from the Mountain area of Taiwan province).

8 Although the first public articulation of essential cultural differences between the Taiwanese group and the Mainlanders group occurred in 1983/4, in the course of the dispute on 'Taiwan consciousness' and 'China consciousness', public and political debate only gained momentum in the 1990s.

9 The following passage is taken from a speech by Hakka Luo Rongguang – a Hakka church minister speaking for Taiwan's independence – at a DPP conference on the problem of 'name correction in Taiwan' in 1994: 'I admit that I'm a Han, my ancestors come from Canton. Hence, I can say that I'm a Han and that I belong to the Han nation/people. However, my ancestors here in Taiwan may very well have a blood relationship with the Pingpu Aborigines. Perhaps I am not a pure Han anymore; I might very well be a new Han who has melted together with the Aborigines . . . just a new Han.' (*Minzhong shibao*, 12.12.1994, reprint of the record of the conference in the Legislative Yuan on 18.10.1994).

10 Educated people allegedly are encouraged to have more babies, while peasants are discouraged from doing so in order to improve the nation's quality stock (Dikötter 1998a: 1–13; 1998b: ch. 7).

11 Wu Feng is the name of a seventeenth-century Han merchant who is said to have dedicated himself to educating Aborigines and who in exchange was cruelly killed by the former headhunters. From 1923, the Aborigine Management Section in the Japanese Governor's Office produced films on the subject of prohibiting facial tattoos and the establishment of schools for aborigine children. The film 'Hero Wu Fong' produced by Taiwan Provincial Film Production Studio in 1932 was a drama about the elimination of the customs of beheading to support the government policy of taming the aboriginal tribes.

12 The concept of 'Taiwan's fate community' (*Taiwan mingyun gongtongti*) was used to unite the 'four ethnic groups'. Shortly after its creation by the DPP in 1990, President Li Denghui picked up on the term and altered it to 'Taiwan's life community' (*Taiwan shengming gongtongti*). Now even officials appealed to Taiwan's inhabitants to form an autonomous national community with an autonomous national identity, emphasising the necessity of the 'Management of Great Taiwan and the Construction of a New Centre of Chinese Culture' (*jingying da Taiwan, jianli xin zhongyuan*).

References

Anonymous (2002), *Minshengbao*, 'Lin Mali rang dalu hen "ganmao"'(Lin Mali gives the Mainland a severe 'cold'), 18 September.

Barth, F. (1969) 'Introduction', in F. Barth (ed.), *Ethnic Groups and Boundaries: the social organization of cultural difference*. London: Allen and Unwin.

Brown, M. J. (2004) *Is Taiwan Chinese?* Berkeley, CA: University of California Press.

Cavalli-Sforza, L. (1998) 'The Chinese human genome diversity project', *Journal of the Proceedings of the National Academy of Sciences, USA*, September 29; 95(20): 11501–3.

Chang Mao-kuei (Zhang Maogui) (1994) 'Towards an understanding of the Sheng-chi Wen-ti in Taiwan – focusing on changes after liberalization', in Chen Chung-min, Chuang Ying-chang and Huang Shu-min (eds), *Ethnicity in Taiwan – Social, Historical and Cultural Perspectives*. Taiwan: Institute of Ethnology Academia Sinica.

Chen Liankai (1992) '*Zhonghua Minzu jie*' (Interpreting Zhonghua Minzu), *Zhongnan Minzu Xueyuan Xuebao* 5: 39–46.

Chen Shuzhuo (2000) '*Yuanzhumin renti jiyin yanjiu zhi lunli zhengyi yu lifa baohu*' (The ethical debate and legal protection of human genetic research of aborigines), *Shengwu Keji yu Falue Yanjiu Tongxun*, 6 October.

Chu Jiayou *et al.* (1998) 'Genetic relationship of populations in China', *Journal of the Proceedings of the National Academy of Sciences, USA,* September 29; 95 (20): 11763–8.

Dikötter, Frank (1998a) 'Reading the body: genetic knowledge and social marginalization in the People's Republic of China', *China Information,* 13(2–3): 1–13.

—— (1998b) *Imperfect Conceptions: medical knowledge, birth defects and eugenics in China.* London: C. Hurst & Co.

Du Ruofu and Xiao Chunjie (1997) '*Cong yichuanxue tantao Zhonghua Minzu de yuan yu liu*' (The origin and development of the Chinese nation from a genetic perspective), *Zhongguo Shehui Kexue,* 4: 139–49.

Heberer, T. (1989) *China and Its National Minorities, Autonomy or Assimilation?* Armonk, NY: M. E. Sharpe.

—— (2001) 'Nationalities, conflict and ethnicity in the PRC, with special reference to the Yi in the Liangshan Yi Autonomous Prefecture', in S. Harrell (ed.), *Perspectives on the Yi of Southwest China.* Berkeley, CA and London: University of California Press.

Lee, S. (2003) 'Racial profiling of DNA samples: will it affect scientific knowledge about human genetic variation?', in M. Knoppers, *Populations and Genetics: legal and socio-ethical perspectives.* Leiden/Boston, MA: Martinus Nijhoff.

Li, Paul Jen-Kuei (2001) 'Some remarks on the DNA study on Austronesian origins', *Language and Linguistics,* 2,(1): 237–9.

Lin Juei-Jueng *et al.* (1999) 'Angiotensin-I converting enzyme gene deletion/insertion polymorphism and myocardial infarction in Taiwan Chinese', *Show Chwan Med. J.* 1(4): 201–7.

Lin Mali (2001) 'Cong DNA de yanjiu kan Taiwan yuanzhumin de laiyuan' (The Origin of the Taiwanese seen from DNA research), *Language and Linguistics* 2 (1): 241–6.

Liu Shao-hua (2000) 'Genes, ethics and Aborigines', *Taipei Times,* 28 August.

PRC's Maternal and Infant Health Law [Zhonghua renmin gongheguo muying baojian fa] (1994) *Zhonghua renmin gongheguo quanguo renmin daibiao dahui changwu weiyuanhui gongbao,* No. 7, 3–8.

Slatkin, M. (1995) 'A measure of population subdivision based on microsatellite allele frequencies', *Genetics,* 139: 1463.

Sleeboom, M. (2005) 'The Harvard case of Xu Xiping: exploitation of the people, scientific advance or genetic theft?', *New Genetics and Society,* 20, April.

Stalin, J. S. (1913) *Marxism and the national question.* Available online at http://www.marxists.org/reference/archive/stalin/works/1913/03.htm#s1 [accessed 18/09/2005].

Su Bing, J. H. Xiao, J. P. Underhill, R. Deka, W. L. Zhang, J. Akey and W. Huang (1999) 'Y-chromosome evidence for a northward migration of modern humans into East Asia during the Last Ice Age', *American Journal of Human Genetics,* 65: 1718–24.

Wang Xiaohui (1996) 'Yanzheng Taiwan Yuanzhumin xueyuan zhi yuan' (Testification of the genealogical origin of Taiwan's *Yuanzhumin*), *Renmin ribao* (Overseas edition), 16 February 1995, 5.

10 The making of scientific knowledge in the anthropological perspective
Case studies from the French scientific community

Angela Procoli

The way to STS

In the USA and UK, during the 1970s, a new stream of studies developed which, whilst it does not amount to a specific branch, could better be described as an interdisciplinary approach. These are the 'laboratory studies' or 'science and technology studies' (STS) which incorporate the work of sociologists, anthropologists and indeed, in certain cases, historians (Latour and Woolgar 1986; Kevles 1987; Traweek 1988; Knorr-Cetina 1999), who cross the threshold of scientific laboratories, and who assess the way scientific facts emerge within so called 'branches of science' (the invention of radioastronomy, the detection of gravitational waves, the wave theory of light, Mendelian genetics, the development of statistics, etc.). In the STS approach, science is considered as one of society's activities amongst many others, and thus can be subjected to the form of sociological, historical or ethnographic investigation usually carried out within social groups, institutions or traditional societies. Michel Callon and Bruno Latour, who introduced the new 'sociology of sciences' in France in the 1980s, carried their approach as far as it could go. This approach rested on a demand for symmetry by virtue of which the success and failures of science, the so called rational activities and attitudes formerly cast out as irrational, are dealt with in similar fashion and where the 'great divide' between the 'pre-modern' and the 'modern' world is removed from the stage (Chateauraynaud 1991).

As against the epistemological approach, the new sociology of the sciences considers that 'the products of science themselves have come to be seen as cultural entities rather than as natural givens "discovered by science"' (Knorr-Cetina 1995). The new sociology of the sciences attempts to reveal the practices and values of scientific communities harbouring knowledge and using languages which may seem *prima facie* impenetrable. The assumption that the world of science is at a remove from society at large may well give rise to methodological problems where ethnographic surveys are concerned: should the social science researcher be well grounded in hard sciences (possibly a special course in the area he will be surveying), so as to carry out his investigation?[1] The difficulties possibly encountered by the ethnographic method, were it to be applied inside our own societies, points to a fundamental question: does science remain isolated from other factors at work in society?

In fact the more recent STS agree in representing the scientific world as one whose borders are permeable, by no means a citadel or a 'stronghold of science' according to Emily Martin's metaphor (1998). The development of STS has waged an all-out attack on the notion of borders set out to mark the 'sacredness' of science. Science is constructed according to cultural categories (Gieryn 1995) which apply well beyond the area of science. It attempts to understand how science becomes integrated in the social field, the way it moves beyond the walls of the laboratories within which it has been confined for years. Such an analysis may follow different lines of thought. Surveying the mode of scientific production as it is evolving in Western society at the present juncture, Gibbons *et al.* evoke a new mode of knowledge production (Mode 2), where science also emerges outside the area it traditionally occupied, and which is coming to light alongside an older and more familiar pattern (Mode 1) where problems are analysed in an academic context governed by the interests of a community (Gibbons *et al.* 1994). This division between Mode 1 and Mode 2, however, has been challenged by French historian Dominique Pestre, who argues that the world of science has always maintained close links with the world outside. As early as the eighteenth century, science was closely related to the market (Pestre 1997).

Whether the present mode of scientific production is taken to be a break with the past or not, the development of STS has pointed to the necessity of interpreting science in the community, of making it clear to the public and having it regulated by political management. The recent work of Michel Callon and Bruno Latour has insisted on the necessity for science to be encompassed in the world of democracy. Scientific activity should be redefined so as to become part of the day-to-day workings of society and brought closer to political activity, involving a fresh look at the role of experts in public debate (Latour 2004).[2] This point has been hotly debated in France in the scientific community (mainly human and social sciences). Faced with numerous instances in which governments have attempted to conceal or cover up the failures of applied science and technology,[3] the scientific community has underlined the necessity for governments to make use of experts to assist in the management of technoscience. A new form of democracy must be founded on a dialogue between scientists, politicians and actors in the public sphere.

As Michel Callon has rightly stated, in view of the degree of uncertainty brought about by these crises, the controversies must be brought into the open and resolved in 'hybrid fora' where political personnel, technicians and outsiders[4] may establish a dialogue with experts and scientists. These groups discuss options which affect society in a number of areas (environment, health, ethics, economy, physiology, atomic physics, etc.). Hybrid fora under-line the citizen's right to confront the powers-that-be and to challenge the technocratic aspects of particular political decisions (Callon 1991).

Michel Callon has also pointed to the fact that the dialogue between scientists and outsiders greatly enriches the whole process of knowledge production. Walled-in research (cut off from the world outside), as it has often been represented, is nothing but a stage in the scientific process which is also built up through a network of permanent exchanges between spe-cialists and the world around them (Callon *et al.* 2001). Scientific knowl-edge circulates and is built up in exchange between the laboratory and the world outside. The latter can provide data (i.e. information) which are analysed, simplified and systematised by the scientists, who then re-direct the results to the 'world outside'.[5]

With more than an eye on anthropology, Paul Rabinow's *French DNA* (1999) also comments on the relationship between the scientific world and the world outside. Starting from an ethnographic investigation carried out in the Research Centre on Human Polymorphism in Paris (in the 1980s and 1990s it was committed to the implementation of the Genopole project), this work shows clearly how the process of scientific production is by no means an isolated phenomenon but is subject to the influence of the same social, economic and cultural factors that are at work in the world outside and which even play a role in the way the world of science is organised.

In this chapter I shall be developing the fundamental question of the STS: how is scientific knowledge built up by exchanges with the outside world, and arranged around 'thought categories' whose matrix is not necessarily scientific? I shall present three case studies taken from an anthropological survey of research in France which I undertook between 2002 and 2003. First, I look at the area of 'surface sciences' at the cross-roads of various branches (or sub-branches) such as physics, physical chemistry and chemistry. I shall attempt to show that scientists call upon structural oppositions which are by no means specific to their milieu in order to establish internal hierarchies.

I then present two case studies of different aspects of animal genetics – its relation with practitioners, i.e. breeders, and the breaking down of the scientific area into two opposing sub-branches within the genetics depart-ment of an important organisation working on applied research. The second case study will show how, at the present juncture, the know-how of a cate-gory of practitioners (the cattle breeders) may be influenced by the knowl-edge of scientists (researchers in animal genetics), but also that knowledge emerges within a pattern of exchanges organised between practitioners and scientists. The third case study will look further into the world of animal genetics and will show, as the actor's discourse is analysed, that they represent their sphere of work (their laboratory) as tightly closed, with sharply delimited borders which must be defended. This defensive attitude becomes even more marked when scientific policies are at work to weaken the local context of science. This case indicates a substantial area that has not been adequately dealt with within STS and which therefore, above all, has failed to yield a methodological principle. The analysis of the actor's discourse – a feature of laboratory studies since their early stages – must be placed into perspective by the ethnographer. Actors provide their vision of the world to the ethnographer, but their 'version' may well be in contradiction with the way the ethnographer reconstructs, in his own terms, the way that world operates. On this particular point anthropologists may, at the present juncture, provide an interesting new opening to STS.

Pure and impure in 'surface sciences'

My first example concerns the so-called 'surface sciences', in other words the study of the specific physical and chemical properties of surfaces. Mental categories and behaviours that are not 'scientific' in origin structure the organisation of the milieu and the way research is carried out, in a hierarchical fashion, with obvious consequences for scientific progress. In this case, the 'pure' and 'impure' categories have been fully studied in anthropological literature pointing to a wide field of application from religion (see *Purity and Danger* by Mary Douglas 1966) to that of nanoscience (Johansson 2003).

The surface science community involves both chemists and physicists concerned with the surface structure of crystals as distinguishable from bulk, which leads to important applications in catalysis and material science, respectively. Specifically, surface physicists have concentrated on the surface of semiconductor crystals. These are materials of high purity. Over the last forty years they have observed the way atoms are rearranged over the surfaces, in an order dissimilar to the one prevailing in the volume (see Figure 10.1).

This phenomenon had to be understood so as to manufacture, at a later stage, new structures and new materials, through a 'controlled' deposition of metal and semiconductor atoms likely to be workable in terms of electronics. Before they are studied, the surfaces must be prepared in an ultra-high vacuum (less than one ten thousand billionth of atmospheric pressure). Several procedures are applied to clean away the 'dirt' (oxides, hydrocarbons) accumulated on their surface before introduction into the vacuum. This cleaning operation is long and difficult. Further, these 'clean' surfaces become clouded as they react to the molecules of the residual gas, carbon oxide, water, hydrocarbons rejected by the pumping system. The 'impure' atoms (carbon or oxygen – known as 'contaminants') are the researcher's archenemy. When contaminants are present on the surface they modify it, make observation impossible, and ruin any hope of controlled modification (see Figure 10.2).

Under the circumstances, who would ever think of depositing a hydrocarbon molecule on the surface of a semiconductor? Dirt on a clean surface!

In 1987, it occurred to Jun Yoshinobu, a Japanese physical chemist, to deliberately drop a hydrocarbon (ethylene C_2H_4) onto a silicon surface. No

Top surface

volume

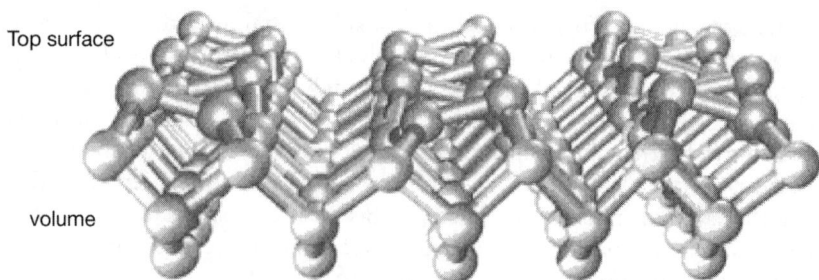

Figure 10.1 'Ball (atoms) and stick (bonds)' image of the clean silicon surface (silicon is the main material used in the microelectronic industry).

Notes:
Note that the coordination (three first neighbours) of the top atoms is different from that of the bulk atom (four first neighbours). Hence the surface chemistry is different from that of the volume. Due to its high reactivity to contaminant species, the surface must be prepared and kept under ultra-high vacuum. Its understanding is a prerequisite

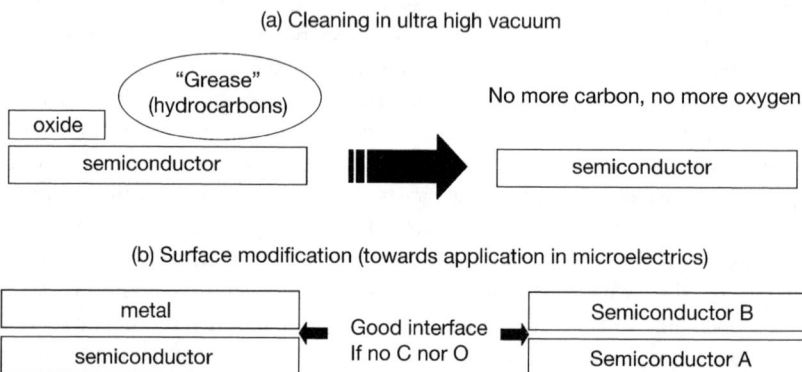

(a) Cleaning in ultra high vacuum

(b) Surface modification (towards application in microelectrics)

Figure 10.2 The chemical reaction between an organic molecule and an inorganic substrate.

Notes:
For nearly forty years, surface physicists dealing with semiconductor surfaces, (a) have cleaned them and (b) have modified them by depositing metal or semiconductor overlayers in highly controlled conditions (ultra-high vacuum, etc.) to fabricate interesting metal/semiconductor or semiconductor/semiconductor heterostructures. In these processes carbon (C) is the archenemy.

one but a chemist would consider dropping such molecules on to metal surfaces, because such studies are of catalytic interest. Yoshinobu was the first to discover an interesting reaction which led to the molecule being grafted to the silicium surface (see Figure 10.3a), thereby opening up the same possibility for many other molecules of organic chemistry.

This type of reaction paves the way for an inorganic semiconductor such as silicon being functionalised by organic molecules, with many possible applications in electronics, nanoelectronics, optics, sensors, etc. (see Figure 10.3b). Yoshinobu's work, however, had only a slight impact until the year 2000. The community of surface physicists, representing the bulk of the workforce in the area of semiconductor surfaces, largely ignored his work. One of my informants, who had been trained in solid-state physics, took a course studying the chemical reactivity of surfaces. He had been working on the same topic as Yoshinobu since 1994. He still remembers his colleagues scoffing at the very idea of the 'dirt' he intended to smear on the surfaces.

Things have changed sharply since the year 2000. Organic 'dirt' (in particular the conjugated organic molecules) appeared to physicists as more worthy of attention, and the tendency to consider them as 'impure' is on the wane. Between 2000 and 2005, fifty articles on the subject, both experimental and theoretical, have been published in the American journal *Physical Review B*. This journal enjoys a considerable reputation in the field of pure physics. How then can one account for the fact that physicists have taken so much time to appear in this new field, nearly twenty years after Yoshinobu's seminal paper? The value of molecular electronics (using the rectifying properties of conjugated molecules) had already been underlined in 1974 by Aviram and Ratner (1974). Such papers had been available for quite some time. The fact that they were withheld from view may be due to the conception of the inorganic being associated with purity and the organic (carbon) with 'dirt'.[6] On inorganic matter, nothing other than inorganic matter could be deposited (a semi-conductor or a metal atom) – this is what physicists have been doing for the last forty years. Because materials are represented in terms of pure vs. impure, and because of the fact that sciences follow a given hierarchical order (top-to-bottom) physics vs. chemistry, the field of research has been patterned accordingly, and activity in this field has been consequently hampered for the last twenty years.

It is also possible that this change of mind by the community of surface physicists is due to the growth of nanoscience and nanotechnologies, as they attract ever increasing financial resources. Molecules are interesting nano-objects *par excellence*, as they are able (i) to transform a light excitation into a modification of their electronic structure, eventually associated with a change in their geometrical conformation; or (ii) to recognise other molecular species, for instance by making hydrogen bonding with them (this possibility is suggested by Figure 10.3b). On the other hand, inorganic semiconductors have demonstrated their capability of amplifying and processing an electric signal (the transistor effect), something that purely molecular devices cannot do yet with a high level of integration. Hence the attempt to produce hybrid structures combining the properties of both systems is easily understood. However, nanotechnology funding is not the only driving force. The attraction exerted by life sciences on material sciences, and on physics in general, has increased over the last decade. Typically the ability of 'light harvesting' and molecular recognition displayed by molecular nano-objects is shared by key molecules of life, such as retinol

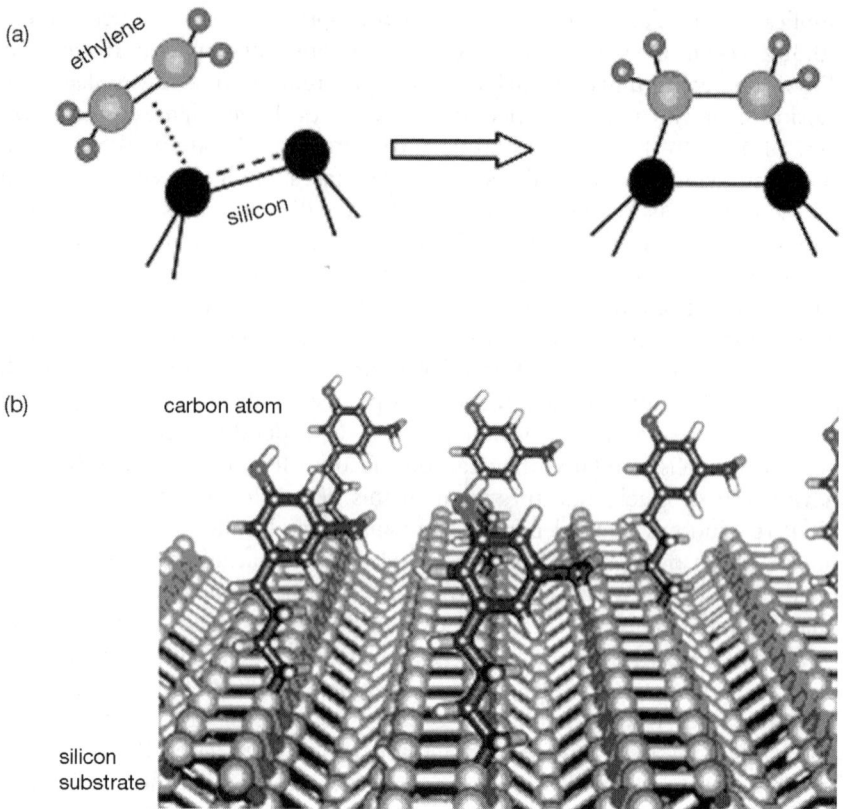

(a) ethylene

silicon

(b) carbon atom

silicon
substrate

Figure 10.3 The chemical reaction between an organic molecule and an inorganic
substrate.

Source: Image reprinted from *Surface Science* 500: 879, Stacey F. Bent, 'Organic
functionalization of group IV semiconductor surfaces: principles, examples, applications, and prospects' (2002), with permission from Elsevier.

Notes:
(a) In 1987 Jun Yoshinobu discovered a new class of chemical reaction between an
    organic molecule (here ethylene) and an inorganic substrate (silicon).
(b) Prospective view of an organic/inorganic (semiconductor) hybrid structure with
    potential application in molecular electronics, sensors, etc. The organic
    molecule – an alkane chain (carbon atoms) terminated by a conjugated carbon
    ring – is grafted on to the silicon substrate. In this illustration the hybrid nature
    of the device is emphasised by the opposition between the infinite periodic
    representation of the silicon crystal and nanometre size of the attached mole-
    cule. The molecule end protruding into the vacuum is 'decorated' with amine
    (NH2) and alcohol (OH) groups, suggesting that this 'device' could be used for
    molecular recognition via hydrogen bonding, emphasising the analogy with
    biomolecular systems.

(the molecule of vision) and DNA. Therefore the 'organic' would extend its meaning beyond that given by chemistry, encompassing bio-mimetic systems. Despite the initial reluctance in the surface physicists' milieu, slowly but surely the 'organic' paradigm invades the field of surface science research.

The construction of a networking knowledge

The second case study illustrates the notion that knowledge is not confined within the walls of laboratories, but may be nurtured by the experience born of working practices applied outside the world of science.

Here I draw on the ethnographic study which I carried out in 2003 in the French National Institute of Agronomic Research (INRA). Since it was founded in 1946, INRA has worked on the improvement of animal and plant species used in agriculture (Cranney 1996). Selection is the time-honoured practice for species improvement in both modern and traditional societies. In the West, a degree of breeding rationalisation took place in the middle of the eighteenth century in the historical context of the English Industrial Revolution, as race selection produces animals with better productivity performances in terms of quantity and quality (Minvielle 1998). In France, after World War Two, new reproduction technologies such as artificial insemination appeared. Artificial insemination is carried out in the world of agriculture on almost all animal species and has enabled rapid progress in the field of genetics. On cattle farms, dairy cows are inseminated with the semen of pre-selected 'race-begetter' bulls.

During the 1950s, a genuine 'selection science' appeared, along with the development of quantitative genetics, bringing about a larger variety of selected species. Above all, these developments created the ability to 'manufacture' animals with desirable characteristics such as (i) the Holstein cow, whose milk yield is three times that of their local counterparts; (ii) the 'cardiac pig', sensitive to stress, all muscle and hardly any fat; and finally (iii) dwarf hens needing less food than standard-sized hens and occupying less space, but able to lay eggs containing standard-sized chicks.

In France, quantitative genetics is developed in INRA where I carried out my ethnographic study, having previously worked with breeders/selectors of cattle and pigs in Brittany. Since this area of France is at the forefront of food and agricultural production, it provides a clear demonstration that, within an industrial agricultural economy, 'genetics' – the work of a category of breeders concentrating on selection – are instrumental in maintaining the animals' 'economic worth'. It is precisely through genetics that the two worlds of breeding and research become closely interlocked. This is underlined by the way in which selection is networked.

Animal selection is conceived as an all-embracing process within which zootechnical information must be collected and processed, along with genealogical data and performance control, all of which are necessary for the selection of animals most suitable to ensure an unbroken line. Figure 10.4

shows a network organisation linking cattle breeders, as the first stage in selection, to the animal genetics laboratory of INRA.

This connection is established at regional level, by bodies responsible for animal identification, status and performance control. The data are then made available to the national centre for the processing of genetic data within INRA. The data are filed, and then passed on to the animal genetics laboratory, again in INRA. In particular, the quantitative genetics laboratory computes an index of the animal's 'genetic worth', passes it on to the national centre for genetic data processing, downwards to the local centres and further to individual breeders. Furthermore, INRA passes on the data to breeder institutes who provide permanent technical back-up to the National Unions for Race Development (these are associations where breeders and industrialists together handle the genetic improvement of a given race). Breeders' institutes in turn pass the INRA lab data to the 'semen production centres' (these are commercial establishments), who monitor the programmes for testing and selecting males which are carried out by Centres for Artificial Insemination. These latter then sell the semen of 'begetter bulls' to breeders.

In spite of this networking, INRA geneticists and breeder/selectors work on different scales and, more importantly, the perception they have of their role in the selection process tends to differ. INRA geneticists, especially researchers in quantitative genetics working with animals, believe that in order to improve a race, a strain, or the livestock of a region (or a country)

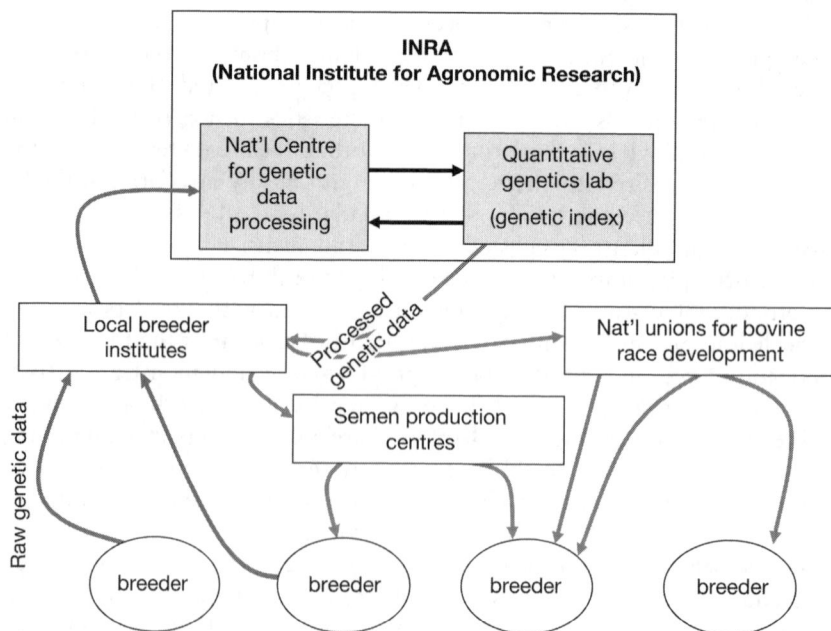

Figure 10.4 Scheme of the elaboration of genetic information and its diffusion.

as a whole, isolated selection is ineffective, and any plans to be carried out in breeding centres must be part of an overall programme. As far as cattle breeders are concerned, this collective attitude to selection stands in opposition to their more individualistic approach. Differences in methods of measurement account for the contrast. Geneticists hold that selection should be applied to a 'population' – described as a closed community of individual animals of the same species – which by sexual reproduction exclusively within the species reproduce the pattern. In this conception the notion of the animal's 'race' may no longer be taken into account. However, from the breeders' standpoint, selection takes place within the flock which still bears the animal's racial hallmark.

While knowledge and know-how do indeed circulate between the 'popular' and the 'scientific' worlds, it remains true nevertheless that the outlook on selection varies from the one to the other. What, then, are the similarities and discrepancies between the two approaches?

When raising questions about the embeddedness of science in society, it is worth considering how the scientific knowledge of researchers working in an institute of applied science such as INRA may draw on popular knowledge emerging from the network's other extremity. Conversely, the question may well arise as to how scientific knowledge may impose a pattern on the way work is organised and may influence the representations of people working in the field. I have focused on the relationship and exchange of data (obliquely in most cases since they are channelled by professional bodies and industrialists in food and agriculture) between the quantitative geneticists in the department of animal biology of INRA and the cattle breeders responsible for selection. The gap between these two professional categories is more apparent than real: their patterns of thought are close enough. On the contrary, any connection between the visions and the practices of a breeder and those of a specialist in molecular biology would be far more difficult to establish. The cultural gap is far too wide, even though molecular biology is conveyed to the public by the media.[7] In fact, quantitative biologists and breeders are brought together by their work on animal lines and on the phenotype.

Quantitative genetics, as an applied version of population genetics, developed in a straightforward Mendelian line. It studies genetic polymorphism, i.e. phenotype variations. Any change in the phenotype cannot always be associated with a given gene. A phenotype may derive from several genes and/or environmental factors, such as age, food, ecology and climate, etc. These multifactor problems are the quantitative geneticist's chief concern and she deals with them with sophisticated statistical methods – under no circumstances will the animal be reduced to its DNA. As far as cattle selection is concerned, a quantitative geneticist attempts to improve characteristics such as animal morphology, rate of growth, calving performance, milk yield, etc. Such characteristics are measurable, expressed in number arrays, using the tools of biometrics. 'Quantitative' genetics are thus defined as a sub-division by the very fact that these characteristics are measured. Apart from measurement as an important aspect of their work, quantitative geneticists focus on antecedents and decedents. To compute the genetic value of a male begetter, geneticists take into account the performances of the applicant's parents and that of its offspring. Data on the parental line are collected in indexes. To compute the offspring's genetic value, a sample of females is artificially inseminated. If the genetic evaluation of the descending line is satisfactory, the bull is accepted as a 'race improver' and its semen is marketed. The quantitative geneticist builds a model of the animal-to-be. She computes the mean value of the begetter's offspring taking into account, on the one hand, the mean impact of the genes received from its parents (father and mother of bulls) and, on the other hand, the interaction taking place, on the descending line, between paternal and maternal genes. INRA geneticists have informed me that their research places them ahead of the breeders because they are able to predict the future genetic value of the herd based on the semen which they have selected.

The picture I have drawn of the techniques and purposes of quantitative genetics points to strong analogies with the traditional practices of animal selection. In both cases it is a matter of observing the animal, of keeping records[8] on animal lines, and of cross-breeding to improve performances.

The working practices of breeder/selectors in this techno-scientific context now come into the picture. Because they create 'begetter' male animals, selectors constitute a small elite, clearly distinguishable from their farming colleagues who produce milk or meat. As advancing scientific knowledge has established an irrefutable link between the notion of selection and genetics, breeder/selectors are described by producers as *'those who do genetics'*. Their work consists in selecting the right semen with which to improve their animals' performance. This choice depends above all on their capacity to look at an animal and evaluate it. It is a know-how they have acquired over the years in their day-to-day contact with the animal. Breeders describe this cognitive process as 'having a keen eye'. It cannot be translated in terms of methodology; it is a form of uncoded knowledge. It consists in being able to size up the animal at a glance, its girth, back, udder, gait, mood (Grasseni 2003). This qualitative approach evaluates the same information that biometrics make formal, quantified and therefore amenable to statistics. Locally, owners keep track of their flocks, keeping books, records and photographs of the animals that have passed through their farms. Here again the analogy with the more formal methods of quantitative genetics is clear: the important point is that the family relationship is given prominence. Where data collection is concerned, the quantitative genetics approach is very close to the practices of those at work in the field. The extent to which animal biometrics has taken on board the practices and know-how of agricultural workers remains to be seen. Thus the role of vets and zootechnicians, who as part of their training spend a long period of time in the field, deserves further investigation.

While biometrics calls, to a certain degree, on breeders' know-how, conversely, the concepts of genetics shape the breeders' work practices. When the breeder identifies a problem in his herd (a problem concerning an animal's morphology), he knows it can be corrected by selecting semen from the catalogue of race-begetter bulls. He then examines the index of inherited characteristics which he will use to improve the morphology of his cows, in the descending line, through systematic artificial insemination.

When visualising animals, breeders are necessarily influenced by the 'ideal animal' model which is produced in a scientific laboratory, broadcast by the media for their audience in agriculture, and in the rings at agricultural shows. It is worth noting that while the quantitative geneticist is at a remove from the 'real' animal, he has his own particular view of the ideal animal. The quantitative geneticist's view is that of an engineer for whom productivity considerations are built into an aesthetic system, in the same way that optimum penetration through the air determines the ideal shape of a car for an engineer in the car industry. However, breeders have *not* adopted the geneticists' standpoint unreservedly. Thus, for example, one breeder/selector denied any validity to the pronouncements of an INRA laboratory that his calf was too inbred to become a begetter and that it should go straight to the butcher. As he said to me: 'my calf cannot be assessed on the basis of a blood sample', meaning that the long and patient selection work over several generations, and his own closeness to the animal, could not be brushed aside by a laboratory test. For a breeder, an animal will never be just 'an array of figures'.

The complexity of scientific knowledge

The third and last example will work back to the holistic–reductionist opposition. Life science researchers, when they belong to competing subgroups (for example, quantitative and molecular biology) underline their differences and call upon the same structural opposition (holistic–reductionist) which divides practitioners from researchers. In doing so, they stress their disagreement with the image that life sciences are attempting to broadcast at the present juncture: that of all-embracing (or integrative) science.

The INRA is a cluster of laboratories organised into departments under a directorate. The directorate actively promotes a policy favouring interdisciplinarity (in the sense of an interplay between the different branches of biology) and developing what has come to be described as 'all-embracing biology'. This new organisation of research in biology (first adopted in the USA) is essential to organise the great mass of available data concerning the genome of living beings.

All-embracing biology is an attempt to organise a theoretical framework for the data provided experimentally by molecular biology. In doing so it calls upon other branches of knowledge such as mathematics/statistics and computer science which provide models and tools involved in studying living organisms. So the new biology encompasses new branches, hitherto foreign to biology, and alters the balance between the various branches of biology. In my observation of INRA, this point has appeared of particular relevance. The directorate has reorganised its departments so as to give greater prominence to those branches of science likely to play a role in elucidating living organisms in all their complexities. Such is the case with animal physiology. Today, animal physiology is recovering the status it lost during the 1960s and 1970s when molecular biology (with its reductionist vision) had swept it aside. Because it raises questions about the interplay between the animal main functions (reproduction, food and health), animal physiology takes a global approach, calling upon other branches of know-ledge, including molecular technology, which it absorbs particularly well. It makes the study of the complexities of living organisms possible, starting from the level of genes upwards to the complex functions of the body as it interacts with its social and physical environment. Networking is an inte-gral part of INRA's research strategy, to ensure it remains competitive with the new international trends prevailing in the life sciences in which the notion of 'complexity' is given prominence.

At the time of my ethnographic study, in which I was focusing on researchers' discourse, it soon became apparent that, at the laboratory level, this new 'all-embracing' model coexisted with 'older' methods of work in which the area of science was still divided into watertight departments. The 'all-embracing' model was more of an aspiration than a reality. This is not always the case, however. For example, knowledge flow and cooperation are organised between molecular biologists and physiologists, although they do not belong to the same department. However, the divide appears clearly between molecular and quantitative geneticists. They all belong to the same department but the latter operate in line with the traditional, Men-delian and evolutionist school, where the phenotype is taken to be influ-enced by hereditary transmission and the environment. Their methods are also patterned to the objectives of their respective research, and these are not the same. Molecular geneticists develop medical research: animal dis-eases in herds and diseases affecting humans (studied through animals); whilst quantitative geneticists develop agronomic research and the selec-tion of farm animals.

The opposition between these two branches of genetics (schematised in Figure 10.5) materialises as two different ways of conceiving the animal come to light. As we have seen, in quantitative genetics the concept of an animal as a living being tends to be weakening (especially when compared with the attitude of breeders). Today, however, when in opposition to their 'molecular' colleagues, the quantitative school of thought reclaims the notion of the animal as a living being. To illustrate the point, one of the senior researchers in quantitative genetics stated that in interdepartmental meet-ings of the molecular and physiological departments, he had protested against the use of the word 'molecule' to signify an animal, and had suggested

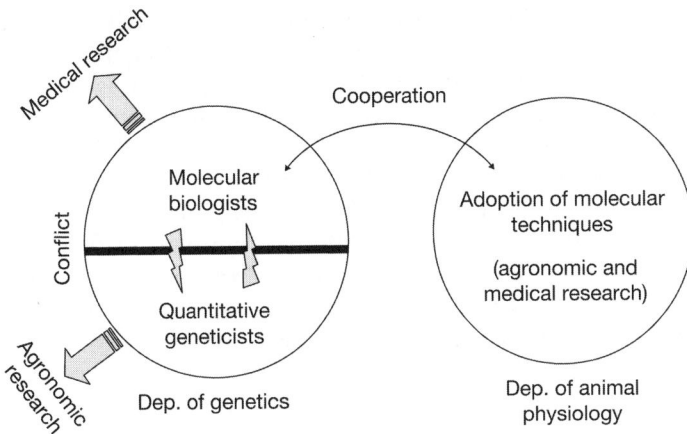

Figure 10.5 Cooperation and conflicts within and beyond the department of genetics in INRA.

instead *'functional biological unit'*: the word 'animal', as we may note, had not recovered its status.

The quantitative school criticises its molecular colleagues for reducing an animal to its DNA. As the controversy proceeds, the very concept of 'gene' loses part of its centrality for the quantitative school, which considers the gene to be an abstract entity. The very object of its research, the phenotype, whose heredity may be complex, inasmuch as it involves several concurrent genes (some of which may be unknown), compels it to move away from the very notion of 'gene', which it takes to be an *'assumption'* or a *'black box'*. The gene is what cannot be seen, as opposed to the phenotype, which is visible and obvious: the gene is an abstract entity, a 'figment of scientific imagination'. This is in direct opposition to the conception of the molecular school which views the gene to be a concrete entity. Since the primary objective of the molecular school is gene identification – to bring to the surface what was hitherto concealed – molecular geneticists have a mechanistic and material view of the gene and consider the position of quantitative genetics as 'outmoded'.

The way research in genetics is thus subdivided is, as already indicated, a pattern characteristic of this particular area and is not necessarily duplicated elsewhere. It results in the quantitative school's refusal to adopt molecular technology even though INRA's directorate is attempting to inject more molecular research into animal selection so as to improve cost effectiveness in this area.

This tug of war between the molecular and the quantitative school is typical of a research institute such as INRA, which works against the backdrop of a European policy which is inimical to the application of biotechnologies to animals in the world of agriculture. European policy is seeking a new identity by promoting quality-oriented production, sustainable development, and animal health. Thus geneticists in agronomic research have to camouflage their activities as 'medical research' when attempting to secure national or European financing. At INRA, this 'medical research' is carried out by both laboratories working on molecular biology and laboratories working on animal physiology – mixed teams are engaged in the drafting of projects. Specifically, both schools of thought cooperate in the chromosome mapping of farm animals (so far, pigs), and these maps are the first step towards the sequencing of animal genomes.[9] But the purposes of both schools of thought are not the same. For the animal physiologist, it is a matter of identifying the genes which might be of interest to agriculture – they have been able to identify, for example, the RN gene responsible for the fact that pork is waterlogged. For the molecular biologists, mapping and sequencing aimed at detecting genes responsible for diseases is the primary focus. The molecular school has for some time been developing research on the immune system of the pig, which has made it possible to cut the amount of antibiotics in its diet. In the past, they have cooperated at some length with a laboratory run by Professor J. Dausset, a Nobel Prize winner in medicine working on the human immune system. The interesting point is that man and animal are brought much closer together as the comparative mapping of pig and man point to homologies between human and porcine chromosomes. As the genetic studies of human and porcine immunity have been compared, advances have been made possible in the sequencing of the genome of pigs, which Professor Dausset coordinates worldwide. An important result of this work is that significant headway has been made in the fields of xenografts (pigs' organs being grafted onto human beings), and of melanoma.

This type of medical research, in which human conditions are studied using animal subjects, lends weight to the critical standpoint of quantitative geneticists, who charge their molecular colleagues with reductionism. Whilst their own research also aims at developing some advantages for human beings (for instance, by providing quality meat) their discourse excludes any references to animals and humans being on the same level.

The way the molecular school places humans and animals on the same level when undertaking medical research should be no cause for surprise, since it flows from the very method of molecular biology. The primary objects of analysis are molecular factors, and these are approached as 'molecules' regardless of whether they are animal, plant or human – or indeed, of whether they are human or non-human matter.

What is central in their knowledge of the scientific object is not so much that they deny any validity to the concept of living – they do not deny that an 'animal is living'. Rather, it is a matter of blurring the boundaries between human/non-human or living/not-living. For the anthropologist this distinction involves choices which are culturally significant. This blurring of the boundaries has been made possible by the power, in experimental terms, of the molecular school. This simplification is the bedrock of molecular biology and has led to genetic determinism.[10]

Conclusion

As in any experimental research, the anthropologist knows that his field-work cannot be entirely mapped out in advance. It is built up as data is collected. Such has been the case in this study of the French scientific community. From an ethnographic survey of the breeders' know-how, I was naturally led in the direction of an animal genetics laboratory involved in selection, where I was able to consider researchers' practices and discourses. In order to assess the general validity of my conclusions, I carried out a comparative study (based on their statements and publications) in the field of surface sciences, fundamental researches at the interface of the three hard sciences (physics, chemistry and, increasingly, biology). A comparison between these three surveys (surface sciences, geneticists and breeders, quantitative and molecular geneticists) has revealed a number of common characteristics in the conceptions that actors have of their respective branches. Let us summarise the most striking points concerning each case study.

The new tendencies in 'surface sciences' indicate changing representations, involving an upheaval in hard sciences, in terms of hierarchies and the relative strengths of entrenched positions. Surveys of semiconductor surfaces have only recently opened up to new themes inspired by molecular technology. This delay may be accounted for by the fact that this sub-branch of surface sciences is, in the main, the privileged field of physicists whose representation of organic molecules has, for a long time, been 'negative' (considered as 'impure' on a 'pure' semiconductor surface). The 'organic' has finally been removed by the physicists from the 'impure' category (in which it was never included by the physical chemists). This shift may be accounted for by the fact that physicists are now moving in the direction of a common representation, where biology plays the dominant role, stimulating research on biomimetic devices. It will be interesting to see, in the future, whether physicists will gain access to 'biological' and 'chemical' fields of knowledge (where they seek to publish, which conferences to attend, etc.).

The second example analysed two professional categories – quantitative geneticists working on animals, and cattle breeders – working in selection. This study highlights the distance between the statements by the parties concerned as well as the anthropological analysis to which their practices and relationships may be subjected. Both categories described themselves as 'closed-in', but it is clear that both belong to the same network and information is exchanged between them. On the one hand, time-honoured selection practices are formulated mathematically and built into science by the quantitative geneticists; on the other, crucial statistical information provided by quantitative genetics are well assimilated by breeders. Both share a conception of the animal in terms of its productivity performance, and their aesthetic standards are similar. However, whether the animal is close to hand (as in the case of the breeder) or at a remove (from the geneticist at work in the lab), contrasting representations do arise, materialising in terms of the breeders' criticism of the quantitative geneticists' reductionism.

The third example demonstrates how scientists, belonging to the same animal biology department of a large institute of applied research, tend to erect barriers between two sub-branches, quantitative and molecular genetics. However, it also shows that cooperation between the two remains a possibility. The division appears as a purely local fact and stands in stark opposition to the development of 'integrated biology', characteristic of our post-genomic era. It points to a change in the local balance of forces – in favour of molecular geneticists – which is linked to developments in European scientific and economic policies. The shift has been in favour of medical research (consequently, of molecular biology) and away from intensive agriculture (in which quantitative genetics has been one of the prime movers). There again, molecular geneticists, when locally claiming their primacy (which would of course be contested in other places[11]) can always rely on the highly reductionist vision (one gene, one disease) as it is upheld by the media.

An analysis of these three case studies clearly shows that the notion of science as 'entrenched' is largely fictional, despite the statements and representations of the scientists themselves when referring to their position. In challenging the alleged boundaries between the laboratories and society at large, STS heavily underlines this notion as fictional. Researchers attempt to set up barriers not only with the rest of society (case study one) but also within the scientific area (case studies one and two). This form of entrenchment allows each community to present its activity as the most legitimate. The assumption of superiority is expressed in terms of negative comments about beliefs and methods of other (sub-)disciplines: 'organic impurity' casting chemistry and biology as of lesser status than physics; quantitative genetics being 'outmoded' in comparison with molecular genetics. These negative representations are likely to evolve under the pressure of altered balances of forces as between hard sciences (for example the thrust of nanosciences and biology, in the first case study). However, the third case study indicates that the negative representation of others may simply be an expression of changes within the local balance of forces fostered by a given political situation. In this sense, it does not necessarily decrease the importance of the scientific sub-area concerned, especially at a time when integrative biology is developing.

Notes

1 A question to which B. Latour's answer is negative. In his introduction to the French version of *Laboratory Life: The Construction of Scientific Facts*, Latour argues that there is no difference between a science researcher and the native of an exotic country speaking an unknown language. The investigator, however, does not have the option of calling upon an interpreter.

2 In this evolution, involving a degree of militancy, whereby science is forced into democracy, scientific extension work has been playing a leading role in France in recent years. The places where science debates with the public at large are institutions (such as the *Cité des Sciences et de l'Industrie* or the *Palais de la Découverte*) involved in scientific extension work. Increasingly, however, debates take place in non-academic surroundings such as bars or 'science cafés', as they grow in popularity over the years, which goes to show that science is indeed spreading outside the walls of its laboratories. Well known scientists are prepared to make contact with the general public, organise lectures, and produce documentary films on TV.

3 Consider such crises as the Chernobyl disaster, blood tainted by the AIDS virus, animal food contaminated by the BSE prion.

4 Such as local groups, neighbours, chicken breeders, wine growers and professional groups.

5 Michel Callon illustrates this process with a particularly enlightening case study relating to research on spinal amyotrophy. Had it not been for the careful census taken of families including one child with the disease; for the hosts of general practitioners who were unsure of their diagnosis; above all, for the volunteer parent members of the French association against myopathy; had it not been for the systematic collection of cells and their stocking in banks, the extraction and analysis of DNA, it would have been impossible to identify the responsible gene and to conclude that the disease was due to an alteration of the SMN gene.

6 Surface physicist's reluctance to study 'the organic' may also be accounted for by the fact that the transition from classic solid state physics (where most surface physicists were trained) to molecular electronics involves a 'cultural revolution'. The familiar theoretical basis of metal and semiconductor physics ceases to apply, or to apply only in part. Other approaches must be considered – as indeed they were by a pioneering group of physicists thirty years back – but so far they failed to be included in the textbooks of today, or only to a very limited extent.

7 A good example of this embeddedness is Telethon – a TV show to raise money for research on myopathy – which is the 'one-gene-one disease' catchword.

8 This has been done since at least the eighteenth century.

9 The molecular branch is involved within international consortiums, and pig sequencing involves frequent cooperation with the Danes and the Chinese.

10 In molecular genetics, prime importance is given to the genetic 'programme', as F. Jacob coined the phrase in the 1950s, embedded in the genes so that the organism is not too heavily dependent on its environment.

11 For many historians of sciences, molecular biology is today more a technique than a science (Morange 2003), or, in any case, it has not the capacity of explaining complexity in the post-genome era (Fox-Keller 2000).

References

Aviram, A. and Ratner M. A. (1974) 'Molecular rectifiers', *Chemical Physics Letters*, 29, 277–83.

Bent S. F. (2002) 'Organic functionalization of group IV semiconductor surfaces: principles, examples, applications, and prospects', *Surface Science*, 500, 879–903.

Callon, M. (1991) 'Techno-economic networks and irreversibility', in J. Law (ed.), *A Sociology of Monsters: essays on power, technology and domination*. London: Routledge.

Callon, M., Lascoumes, P. and Barthe, Y. (2001) *Agir dans un monde incertain*. Paris: Seuil.

Chateauraynaud, F. (1991) 'Forces et faiblesses de la nouvelle anthropologie des sciences', *Critique*, 529–30: 459–78.

Cranney, J. (1996) *INRA. 50 ans d'un organisme de recherche*. Paris: INRA.

Douglas, M. (1966) *Purity and Danger: an analysis of the concepts of pollution and taboo*. London: Routledge & Kegan Paul.

Fox Keller, E. (2002) *The Century of the Gene*. Cambridge, MA: Harvard University Press.

Gibbons, M., Limoges, C., Nowotny, H., Schwartzman, S., Scott, P. and Trow, M. (1994) *The New Production of Knowledge: the dynamics of science and research in contemporary society*. London: Sage.

Gieryn, T. F. (1995) 'Boundaries of science', in S. Jasanoff, G. E. Markle, J. C. Petersen and T. Pinch, *Handbook of Science and Technology Studies*. London: Sage.

Grasseni, C. (2003) *Lo sguardo della mano*. Bergamo: Bergamo University Press.

Johansson, M. (2003) 'Plenty of room at the bottom: towards an anthropology of nanoscience', *Anthropology Today*, 19(6): 3–6.

Kevles, D. J. (1987) *The Physicists: the history of a scientific community in modern America*. Cambridge, MA: Harvard University Press.

Knorr-Cetina, K. (1995) 'Laboratory studies: the cultural approach to the study of science', in S. Jasanoff, G. E. Markle, J. C. Petersen and T. Pinch, *Handbook of Science and Technology Studies*. London: Sage.

—— (1999) *Epistemic Cultures. How Sciences Make Knowledge*. Cambridge, MA: Harvard University Press.

Latour, B. (2004) *Politics of Nature: how to bring sciences into democracy*. Cambridge, MA: Harvard University Press.

Latour, B. and Woolgar, S. (1986) *Laboratory Life: the construction of scientific facts*. Princeton, NJ: Princeton University Press.

Martin, E. (1998) 'Anthropology and the cultural study of science', *Science, Technology & Human Values*, 23(1): 24–44.

Minvielle, F. (1998) *La sélection animale*, Paris: PUF (Que sais-je?).

Morange, M. (2000) *The History of Molecular Biology*. Cambridge, MA: Harvard University Press.

Pestre, D. (1997) 'La production des savoirs entre académies et marché', *Revue d'économie industrielle*, 79: 163–74.

Procoli, A. (2004) 'Le temps et la construction du regard sur l'animal de rente. Ethnographie des pratiques et récits des éleveurs bretons', *Cahiers d'économie et sociologie rurales*, 72: 91–112.

Rabinow, P. (1999) *French DNA: trouble in purgatory*. Chicago, IL: University of Chicago Press.

Traweek, S. (1988) *Beamtimes and Lifetimes*. Cambridge, MA: Harvard University Press.

Yoshinobu, J., Tsuda, H., Onchi, M. and Nishijima, M. (1987), 'The adsorption states of ethylene on Si(100) c(4×2), Si(100) (2×1), and vicinal Si(100)9: electron energy loss spectroscopy and low-energy electron diffraction studies', *Journal of Chemical Physics*, 87, 7332.

11 Genomics and the transformation of knowledge

The bioinformatics challenge

Henrik Bruun

Introduction

The significance of computer science and bioinformatics for biological and biomedical research (hereafter 'bioscience'[1]) is rapidly increasing. Functional genomics, proteomics, metabolomics and many of the other new research platforms are based on the use of bioinformatics tools for the storage, manipulation and analysis of data. This chapter focuses on the competence-related effect of the increasing role of bioinformatics: How should research be organised? What kind of education should be given to students? What knowledge needs to be integrated and how can such integration be achieved? These and related issues are discussed with reference to recent developments in bioscience. A particular research platform, the use of DNA microarrays as a tool for the study of gene expression, is presented and analysed as a paradigmatic example of the interdisciplinary challenges that 'the new biology' faces. The chapter ends with a proposal for new research topics within the framework of a broader research programme on genomics and transformation in the production of knowledge.

The rise of bioinformatics

Interdisciplinary collaboration is not a new phenomenon in the history of biology (Bechtel 1993; Fujimura 1996; Kay 2000; Morange 1998). Many of the most important discoveries of the twentieth century were results of interaction across the boundaries of disciplines. Jim Watson and Francis Crick's discovery of the double helix structure of the DNA molecule is probably the most spectacular example. Watson had an education in biology, while Crick's background was in physics. The two were not unique in their collaboration. Crick in particularly was typical of the period: a physicist who looked for challenges in the field of biology. Physicists and technicians from the world of physics played a crucial role in the emergence of molecular biology. Their influence was not immediate, however, in the sense of reducing biology to chemistry or physics. It seems that the knowledge transfer took place at a more general, methodological level.

Michel Morange (1998), a noted historian of biological science, has suggested that the most important contribution of the physicists was to transform biology into an operational science in which scientific concepts are expected to correspond to simple, experimental operationalisations. Another, perhaps even more important, influence was their conviction that 'the secret of life was not an eternal mystery, but was within reach' (Morange 1998).

Today, bioscience is subjected to a new trajectory of epistemic transformation. The background is that decades of research on DNA, genes and proteins have produced extensive knowledge about their characteristics. The international genome projects are constantly generating new knowledge on the DNA structure of different organisms. Less is known, however, about the function of genes and about the patterns of interaction between genes, proteins, genes and proteins, and the environment of the organism. Therefore, the next step, according to a growing consensus among researchers, is to focus on the dynamic aspects of biochemical processes: signalling pathways, regulatory pathways, molecular machines, organelles and, ultimately, the cell as a whole (some even speak about the organism as a whole). Methodologically, this implies a shift from a 'local' to a 'global' perspective: instead of studying single genes or proteins, attention is shifted to how they interact (Duyk 2002; Lander 1999). The use of bioinformatics tools to handle large amounts of data is a condition for such research. It is precisely these two aspects – the global perspective and the manipulation of large amounts of information – that are said to constitute the core of the 'new' biology (Kallioniemi 2002). The situation is comparable to the early days of interaction between biology and physics, but now with computer science as the disciplinary counterpart.[2]

The big genome projects, initiated during the 1990s, were instrumental for establishing bioinformatics as a key element in contemporary bioscience. On the one hand, electronic databases were necessary for storing the vast amounts of information that the sequencing projects produced. Traditional archives would have been highly impractical, even from the simple perspective of information storage. A book with the whole DNA sequence of, for instance, the fruit fly, *Drosophilia Melanogaster*, would, according to estimates, comprise 27,000 pages (letter size 8, no marginal). Such a book would not only be impractical to use: much of its information would be inaccessible to the scientific community. The importance of databases lies not only in their storage capacity, but also in the new forms of information retrieval that they make possible (MacMullen and Denn 2005). The multitude of databases – DNA databases, SNP databases, EST databases, protein databases, bibliographic databases and so on – have become a constitutive (Bruun and Langlais 2003) element of the 'new bioscience'. The latter would be unthinkable without them. An early sign of the scientific significance of databases was Bert Vogelstein's utilisation of an EST database to isolate a human mismatch repair gene called MLH1. This gene, which was discovered in 1994, turned out to be one of the keys to understanding the disease mechanism of colon cancer (Davies 2001).

On the other hand, bioinformatics is also an important tool for the processing and analysis of data and in this capacity had a key role in the efforts to sequence the human genome. Craig Venter and his company's (Celera Genomics) shotgun sequencing technique was based on slicing the human genome into small, randomly organised DNA fragments, which were then reassembled into a running sequence by computer algorithms. In 1999, Celera read up to 40 million bases per day. Assembling such an amount of information was not easy, and there is no doubt that this would have been impossible without advanced bioinformatics tools.

Challenges for education and research

The new bioscience contains several challenges for the scientific community. As a result of the introduction of high-throughput techniques into the heart of the explorative phases of research, basic research is becoming even more capital intensive. Increasingly, the competitiveness of laboratories will be dependent on mechanical automation, instrumentation and systems for processing laboratory data (Bassett *et al.* 1999). These changes raise a set of questions about financing, competence and organisation. In this context, I want to discuss the latter two. The emerging research platforms pose new demands on competence. The new approaches are broad and require a heterogenic scientific basis. Mastery of just one research platform will not be enough in the future. The global perspective implies that the results from one type of investigation, such as the study of gene expression, are integrated with research that is based on other platforms. Expression studies, to continue the example, will therefore, to an increasing extent, be combined with research in proteomics, metabolomics, cell biology and other relevant fields, the long-term goal being, for instance, the creation of models of complex phenomena such as cellular pathways and disease mechanisms (Kallioniemi 2002; Mäkelä and Porkka 2002). This is not a prediction of the future; the integration is already happening, and the various tools of bioinformatics – everything from databases to specialised software for data processing and modelling – have a key role in this development. As a result, research organisations need to develop strategies for responding to the situation.

My own on-going research suggests that the development of organisational competencies, along the lines discussed above, is not without problems – at least not in Finland, which has been described as a model case of the information society (Castells and Himanen 2001). The study I am referring to is based on interviews with Finnish scientists involved in research that uses cDNA – or oligonucleotide microarrays (from now on 'DNA microarrays'). The DNA microarray, which is used to study gene expression, among other things, was developed in the mid-1990s by researchers at Stanford University and has since become one of the core platforms of the new bioscience (Duggan *et al.* 1999; Keating and Cambrosio 2003). The great advantage of the DNA microarray is that it allows for an analysis of changes in the expression of thousands of genes in a single experiment, whereas earlier research had to advance one gene at a time (Brown and Botstein 1999). The DNA microarray experiments generate very large sets of information and researchers are therefore completely dependent on bioinformatics tools for storing, processing and analysing data (Bassett *et al.* 1999). At the same time, however, most bioscientists lack formal competence in bioinformatics, computer science, statistics or mathematics. As a result, many researchers experience the bioinformatics component as challenging and even downright problematic.

A common way to solve competence problems is to introduce a division of labour, which in this case would mean that data processing and analysis is externalised to professional computer scientists or bioinformaticians. In fact, a large part of biological research relies on a modular organisation of labour, designating restricted tasks to people with the appropriate expertise. However, my interviews with Finnish scientists, and the literature in the field, suggest that it is difficult to separate the biological component from the bioinformatics component in gene expression experiments. Knowledge about the opportunities and limitations of the data analysis techniques is needed already at the stage of experimental design (Churchill 2002). Correspondingly, biological knowledge is crucial at the analysis stage after the experiment: it is needed to separate the biologically relevant results from those that only have statistical relevance (Kohane *et al.* 2003).

Arguably, the dilemma of competence requirements and division of labour is acute, not only for gene expression research, but more generally for all bioscience that is based on heavy use of bioinformatics. How should laboratories and departments respond to this? Should they give extension training to the bioscientists and require formal bioinformatics competence when recruiting new researchers? Or should they start encouraging people with a computer science degree or a mathematics degree to apply for research student positions? Or despite what was said above, is it still possible to respond to the problem by a division of labour, that is, by externalising some of the tasks to either in-house or external bioinformaticians? A fourth strategy could be to do little or nothing along these lines, with the expectation that the automation of research instruments and procedures will proceed rapidly and that the need for specialised bioinformatics competence will be relatively restricted.

There are no self-evident answers to these questions, and practices vary from laboratory to laboratory. Perceptions of the future vary significantly. Some researchers predict that more than half of the staff of future laboratories will consist of bioinformaticians, mathematicians, statisticians, etc. Others, at the opposite end of the continuum, have argued that there will be little need for such people at the level of the research group or the laboratory, because much of the work for which they are needed today will be automated in the near future. Considering the key role that bioinformatics is expected to play in the emerging new biology, and the prevailing uncertainty about how to translate this knowledge to long-term competence strategies, it is surprising that there has been so little debate about the issue raised here. Most of my interviewees testified that the integration of biology and bioinformatics actually causes problems at the ground level of research. The frustration is felt not only by biologists, but also by bioinformaticians. In a recent interview, a researcher belonging to the latter group summarised his feelings about the collaboration with biologists with the following complaint: 'It is hard to explain the complexity of these things to those who don't know anything about databases and information technology.'

Similar questions can be posed at the level of education. In a recent viewpoint paper in *Science*, William Bialek and David Botstein (2004) complain that further development in the understanding of biological systems is hampered by the fragmented teaching of natural sciences in American universities. They argue that the natural science disciplines have bifurcated into two cultures; one that is mathematically and quantitatively oriented, and one that is not. Mathematics, the physical sciences, chemistry and engineering belong to the former, while biology (with the exception of a few specialised areas[3]) and medicine constitute the latter. Bialek and Botstein observe that American biologists and physicians generally receive just one year of training in mathematics and the physical sciences and that they are taught separately and differently from the students majoring in these subjects. As a result, many students in biology and medicine 'have too little education and experience in quantitative thinking and computation to prepare them to participate in the new world of quantitative biology' (Bialek and Botstein 2004: 788). According to the authors, the problem is not only that the students lack deep enough knowledge about mathematics and computing, but more fundamentally that they do not internalise the mathematical understanding of the world that will be necessary in future biological research. Bialek and Botstein (2004) conclude that 'there is an enormous challenge in raising a generation of scientists who are equally at home with this quantitative mode of thought and with the complexities of real organisms'.

In Finland, the cultural divide seems to be even deeper than in the US. There is hardly any compulsory mathematics, physics or computer science in the curricula of biological and medical education. Studies in biochemistry are required by some of the programmes. All in all, however, it seems that students have a great freedom of choice, which probably means – assuming that there is some truth to the claim about a cultural divide – that they do not tend to select the mathematical and quantitative subjects. This is illustrated by the University of Kuopio, one of the Finnish life science centres, where only one out of five students in biology or biomedicine attends the course on bioinformatics (and there is just one such course).[4] Note that the course in question is not about making algorithms and programming, which is what Bialek and Botstein have in mind, but in the use of standard software and tools. Another illustration of the cultural divide is provided by some of the computer science graduates I have interviewed, who are writing their doctoral dissertations as members of a bioscientific research group. From the perspective of bioscience, their work is valuable if it provides new tools, such as software, for biological or biomedical analysis. From the computer science perspective, however, writing software is not enough for a dissertation. As one student puts it: 'Even if my software led to the identification of an important cancer gene, it would be of little scientific value for the computer science community.' Balancing the different requirements, and finding synergetic strategies through which the requirements of both traditions can be fulfilled at the same time, will be a necessity for bridging the cultural divide. If Bialek and Botstein are correct, these students should be seen as pioneers whose situation mirrors the future of biological education and research.

The example of DNA microarrays

The DNA microarray platform is a good starting point for getting more detailed insight into the contemporary attempt to integrate computer science and bioscience (see Kohane *et al.* 2003 for a comprehensive and accessible introduction). As stated earlier, the DNA microarray technology was developed in the mid-1990s by researchers at Stanford University[5] and has since then become one of the key technologies for the new biology. The microarrays, or biochips as they are also called, are slides (for instance, microscope glasses) that have been spotted with thousands of PCR products (genes) or, alternatively, with chemically synthesised long oligonucleotides. In the cDNA microarray, spots generally correspond to a particular gene in some organism. In oligo microarrays – normally with 50 to 70 bases in each spot – each gene is represented by several spots. Microarrays are produced both commercially and by microarray core facilities in academic and government institutions. Some of the best-known commercial producers are Affymetrix and Agilent Technologies.

DNA microarrays can be used for a number of purposes: the comparison of gene expression in two or more samples; the detection of single nucleotide differences between genomic samples; and the characterisation of genes within an organism through comparative genomic analysis. Potentially important fields of application are drug development and the diagnosis of clinically relevant diseases. In the following review and discussion I will restrict myself to the use of DNA microarrays in gene expression studies.

The rationale for studying gene expression is that it tells us something about the function of the genes, 'their association with a particular biological or clinical feature' (Kohane *et al.* 2003). A more straightforward way to study the phenotypic effects of genes would be to focus on their end products, the proteins. After all, it is the proteins, and not the genes, that have a direct structural and functional significance for cellular and inter-cellular processes. What is more, there is no direct linkage between gene expression levels and the biological significance of proteins. A large change in gene expression does not necessarily mean that the concentration of the protein changes significantly, and even if such a change was to occur, there is no guarantee for it being biologically significant. Significance is affected by a number of factors that are independent of gene expression (Kohane *et al.* 2003). So gene expression analysis should be seen as an indirect way of studying the function of genes. Yet many consider it to be the best and most practical technology available at the moment (Brown and Botstein 1999). In the future, the focus may shift from gene expression to protein concentration and protein interaction. Proteomic assays are, in fact, being developed at the moment, and the first maps of protein interaction in an organism have been published (Giot *et al.* 2003).

The purpose of this section is to show that statistical analysis and the use of bioinformatics tools have become an internal part of microarray research and that biologists who avoid extending their competencies into these areas will have a hard time using the research platform. As already stated in the previous section, knowledge about the opportunities and limitations of the data analysis techniques is needed already at the stage of experimental design, at the same time as biological knowledge is crucial for analysis after the experiment. This is now demonstrated at a more detailed level. I begin with a description of the microarray experiment. Readers who are familiar with the experimental design issues in this field can proceed directly to the subsection on data processing.

The DNA microarray experiment

At the heart of DNA microarray-based research is the microarray experiment, in which cDNA from two biological samples is hybridised with the DNA on the microarray (the DNA is located in the spots of the microarray; remember that in cDNA microarrays each spot represents a gene).[6] The purpose is to detect differences in the gene expression levels of the two samples. The experimental design can be illustrated with an example (the description that follows is based on Brown and Botstein 1999; Cheung *et al.* 1999; Duggan *et al.* 1999; Kohane *et al.* 2003; Wong 2003). Let us assume that we are doing an experiment with two human samples. Then we need a DNA microarray that is spotted with human genes. Today, there are microarrays that cover all human genes and transcripts that are known. But why do we need two samples? The microarray technology is based on a *comparison* of gene expression in different samples, and not on measurement of absolute expression levels. Thus, the point is to see how the expression levels of genes differ between the samples. The selection of samples is a key decision in the design of the experiment. One of the samples should be a reference sample, while the other should represent the condition or feature that we are interested in (a target sample). So if we are interested in some particular human disease, we could compare the gene expression between a (reference) sample from a person who does not carry the disease, and a (target) sample from someone who does carry it. In advanced experiments, several microarrays with different target samples are used. Here, however, it is enough to consider the simplest case: one reference sample and one target sample.

The activity of the genes in each sample is shown by the amount of mRNA transcripts (the products of gene expression). In principle, one could use these transcripts in the microarray experiment. There are, however, several technical reasons for transforming the mRNA to cDNA before performing the experiment. The relevant sequence information is preserved through this transformation, which means that the cDNA from a particular sample gene is complementary ('c') to the DNA that represents the same gene in the microarray spot. When poured on the microarray, and under proper conditions, the sample cDNA will hybridise with ('attach to') the DNA on the spot in question.

The experiment is prepared by attaching two distinct fluorescent markers to the cDNA from the samples, one to the reference sample cDNA and the other one to the target sample cDNA. The two samples are then mixed with other chemicals into a hybridisation solution, which is pipetted on to the microarray. A hybridisation chamber is used for keeping the temperature constant and to avoid evaporation. The cDNAs compete with each other in the hybridisation process, and the success of a particular cDNA sequence is dependent on its abundance. Basically, a high level of hybridisation on some particular spot means that the corresponding gene (in either sample or in both) has produced a large number of transcripts: there is a high level of gene expression. A low level of hybridisation means the opposite.

The hybridisation levels of DNA from the two samples are distinguished from each other by the use of the fluorescent markers. After hybridisation, the microarray is beamed with a laser, which excites the markers and causes them to emit a signal that can be read by a scanner. The marker signals have different frequencies and can therefore be separated. However, as mentioned before, the analysis does not focus on absolute hybridisation levels, but on the difference in hybridisation. The question is not 'What is the expression level of the genes of the target sample?' but instead 'How much are genes up- or down-regulated when the target sample is compared with the reference sample?' The scanner produces an image of the microarray in which up-regulated spots are red and down-regulated spots green (intensity of colour reflects ratio of expression). Black spots signify that there was little or no difference in the level of expression.

The coloured microarray image can be used to check the general quality of the hybridisation, but apart from that it is the numerical values of the marker signals that matter. Each spot generates a series of data that represent various aspects of the intensity of the marker signal from the spot. This information is the basis for all computational operations that follow, and it is generally represented in a matrix with one row for each spot on the array. The matrix for a microarray with ten thousand spots consequently contains ten thousand rows. Studies that are based on several experiments produce several such matrices. The next step is to process the data. In principle it would be possible to focus on the expression level change of a single gene, and that could be done manually by analysing the numbers in the table. The point of doing a microarray experiment is, however, that the behaviour of many genes can be compared. Most phenotypic features involve expression changes (up- or down-regulation) in several genes. Expression analysis can also be used for the genetic classification of conditions (such as cancer subtypes) on the basis of the expression pattern of groups of genes. The purpose of the study is then not to identify the mechanisms of a disease, but only the expression profile that identifies a particular subtype of the disease. Common for both causal and profiling studies is that computer software and knowledge about statistics are needed for the analysis.

Processing of data: preprocessing, analysis and data mining

The data processing starts with 'preprocessing'; that is, 'various analytical or transformational procedures ... need to be applied to the data before it is suitable for a detailed analysis' (Tuimala 2003). The purpose, in other words, is to refine the raw data, because in its raw state it is not meaningful. There are several preprocessing steps (Knudsen 2002; Kohane *et al.* 2003; Tuimala 2003): treatment of missing values; assessment of the correlation between the intensities of backgrounds and spots; calculation of expression change; checking the quality of replicates and treatment of bad ones; treatment of outliers; filtering of bad data; filtering of uninteresting data; normalisation; checking for normality and skewness; checking for the linearity of the data; and checking for spatial effects. Each step involves decision-making from the analyst: in some cases there are several statistical tools available, and in other cases decisions have to be made about where to draw the line between good and bad, or interesting and uninteresting data. Even if there are some standardised procedures for making these decisions, microarray experts warn researchers away from the belief that software can do the data analysis for them. 'This kind of approach for statistical analysis is simply erroneous, because the results coming out from the program can be statistically erroneously derived. In such cases, also the biological conclusion can be wrong. Statistical tests [often have strict] assumptions, which need to be fulfilled. Violation of assumptions can lead to grossly wrong results' (Tuimala 2003).

The importance of decision-making in the preprocessing of data can be illustrated with one of the steps, normalisation. The purpose of normalisation is to remove systematic variation in the gene expression ratios that have been measured. Such variation occurs when some factor other than the samples themselves introduces a systematic bias into the raw data. There are numerous possible sources of systematic variation: differences in marker efficiencies; scanner malfunction; uneven hybridisation; varying quality of the microarray spots; plate and reporter effects; varying quality of microarrays from different print runs; and varying skills and styles of different experimenters. The subsequent data analysis will be flawed unless this bias is removed from the data. The notion of normalisation derives from the idea of making the data more normally distributed (assuming that deviations from normality are caused by disturbing factors). In the microarray context, however, the term normalisation also refers to methods of standardisation and centralisation.

Normalisation generally starts with a log-transformation of all intensity-ratios, which makes 'the variation of intensities and ratios of intensities more independent of absolute magnitudes' (Tuimala *et al.* 2003). Log-transformation is just another way of representing the data; it gives a better visualisation of the distribution, but does not change the data. Centralisation, on the other hand, introduces such change. It manipulates the data so as to make it more normally distributed by centring the distribution over an expected mean, thus removing the bias that caused the deviation from normality. For linear data, *median centring* can be applied. This is a global method in the sense that the median of the complete set of data is subtracted from the log ratio of every spot. For non-linear data, local techniques have to be used. Non-linearity is a sign of spatial bias in certain parts of the microarray, and as a result normalisation needs to be performed locally and not globally. In this case, the mean or median has to be calculated separately for different parts (subarrays) of the array. There are several normalisation techniques: mean centring, median centring, trimmed mean centring, standardisation, Lowess smoothing, ratio statistics, variance analysis (the ANOVA method), spiked controls and dye-swap experiments. Specialised software such as GeneSpring suggests a normalisation scheme for the experiment. GeneSpring also warns the user if the normalisations performed do not make sense. However, according to microarray experts, there is little consensus about what normalisation method to use, and this means that it is the researchers themselves who, in the end, are responsible for the choices (Tuimala 2003; Tuimala *et al.* 2003). Enlightened decisions cannot be made without a basic understanding of the statistics underlying the various methods and techniques.

Preprocessing should result in reliable data. When this has been achieved, the next step is the analysis proper. There are several methods for analysing changes in gene expression (Knudsen 2002; Kohane *et al.* 2003; Vihinen and Tuimala 2003). A common goal is to find groups of genes that behave in a similar way, the assumption being that similarity in expression pattern is a sign of similarity in biological function (as we have seen). Such groups can be identified by different clustering techniques. Clustering 'organizes the data into a small number of (relatively) homogeneous groups' (Vihinen and Tuimala 2003). If successful, the cluster analysis groups the genes together in a biologically meaningful way. The problem is that there are several clustering techniques, and all give different results. There is no mathematical way to distinguish biologically relevant clustering from clustering that has statistical relevance only. Why should this be so? The challenge can be illustrated with the example of so-called *hierarchical clustering*.

The hierarchical clustering algorithm starts with considering every gene (that is, every expression pattern) to be a cluster in itself. It then joins the two clusters that are closest to each other (the genes whose expression patterns are the most similar) and forms a new cluster out of these. A new value is calculated for the distance between the new cluster and all the other clusters. From now on, the two genes have been replaced by their common cluster. The algorithm then goes on to do the same thing again: to join the two clusters that are closest to each other in the new data set. This process is repeated until all genes belong to some larger cluster. Although this might sound rather straightforward, the researcher has to make important decisions that will affect the outcome of the clustering procedure. The reason is that there are several ways to measure the distance between clusters. First, there are different techniques for measuring the distance between the expression patterns of two genes. Second, there are several, distinct ways to measure the distance between two clusters that contain more than one member. Distance measures can be calculated on the basis of (i) the shortest distance, (ii) the longest distance, or (iii) the average distance between members of two clusters. All three methods are statistically legitimate, but will give different clustering outcomes. What is more, statistics alone cannot provide grounds for what distance measure will yield the most interesting results from the perspective of biological function: 'The decision about the applied similarity measure depends on the biological question you are interested in' (Vihinen and Tuimala 2003).

In addition to decisions to be made within the framework of particular clustering techniques, the researcher also has to decide what clustering technique to use (presuming that clustering is what is desired). Hierarchical clustering is just one of the techniques available. Other common techniques for grouping the data are self-organising maps (SOM), k-means clustering, and principal component analysis (PCA). They all give different results, and, just as in the case of hierarchical clustering, call for key decisions by the researcher. The k-means clustering technique, for instance, starts by distributing a specified number – k – of centroids into the data (randomly chosen genes become centroids). Each gene is then assigned to the centroid that is closest to it. All centroids now define their own cluster (all the genes that are assigned to it). Next, the k-means algorithm uses the positions of the cluster members to calculate a new position for each centroid (it moves the centroid to the real centre of the cluster). This operation changes the distance between genes and centroids, which means that a new check of the cluster membership of genes must be carried out. Genes continue to belong to the original cluster as long as the latter's centroid remains the closest one. If, however, one of the other centroids has overtaken the centroid of the original cluster and thus become the closest one to a particular gene, the latter's membership changes: genes are always assigned to the closest centroid. The process continues until new repetitions result in little or no change in the grouping of the data. The result should be a clustering in which the dispersion within clusters is minimised. The problem, however, is that the number of clusters – k – has to be specified *a priori* and that differences in the initial conditions (the random distribution of centroids) might affect the outcome (the clustering). There are, unfortunately, 'no objective rules for determining the correct number of gene expression clusters' (Vihinen and Tuimala 2003). Also, there are no known algorithms to determine when the absolute minimum of dispersion within clusters has been reached. The results of a particular clustering procedure can be validated by repeating the analysis (new initial conditions are automatically created in each repetition). It is, however, up to the researcher to decide when to stop this validation and how to respond to differences in the outcome of repetitions.

To sum up, both the data preprocessing phase and the data analysis phase of DNA microarray experiments require decision-making that affects the reliability of the data and the outcome of the analysis. Even though there are some recommendations for how to proceed in specific cases, it is, in the end, the researcher who makes the crucial decisions. It should be obvious that such decision-making requires familiarity with the different techniques and, at least to some extent, the statistical principles upon which they rely. The need for such knowledge might sometimes be concealed by the smoothness with which different analyses can be carried out with specialised software. Yet the decisions made in the preprocessing and processing of data are extremely important, because 'subsequent analyses and data mining rely on the partition obtained by clustering' (Vihinen and Tuimala 2003). To repeat, the point of clustering is to identify groups of genes that share some biological function. Clustering as such can never be used as a proof of shared function; it is, rather, a necessary step in a longer investigation. The clustering analysis is generally followed by a laborious process of data mining from various databanks; looking for sequence similarities in the promoter regions of the genes in some particular cluster; looking for information about the functional characteristics of the genes; seeking information about the diseases that the genes have been associated with, etc. Much of this information (such as sequence information) is directly available from the numerous databanks that exist. Equally important, however, are articles and abstracts with information about the genes. Articles can be mined on the web with the help of specialised mining tools (MacMullen and Denn 2005). All this work might be in vain if biologically insignificant clusters are taken as a starting point. On the other hand, the data mining process might also feed back into data processing, inspiring the researcher to make changes in the clustering procedure.

The interdependencies between biological questions and data processing, and the iterative nature of research – the movement back and forth between data analysis and data mining – have important implications for the organisation of work in microarray experiments. Biologists need to have a basic knowledge in statistics and bioinformatics in order to understand what they are doing, and any externalisation of the analysis is dubious to say the least. The analytical procedures as such are generally not difficult to perform, thanks to the user-friendly interfaces of specialised software. The difficult thing is to make the decisions that we have discussed. Many of those decisions depend on the biological question of interest and can therefore not be left to a computer scientist or even a bioinformatician. If interdisciplinary collaboration is desired, the division of labour cannot be modular – the bioscientist being responsible for the biology component, and the bioinformatician being responsible for the computer science component – but must instead be integral (Bruun *et al.* forthcoming; Bruun and Sierla forthcoming) in the sense of both collaborators going beyond their disciplinary frameworks in order to work in an integrative manner. The support from a bioinformatician or computer scientist is perhaps not that important in conventional processing of data – assuming that the bioscientist has learnt how to do this – but can be truly invaluable to identify limitations in available algorithms, methods and software, and to develop new tools that are better equipped to help the bioscientist finding answers or testing hypotheses. This again requires that the bioinformatician has a computer science competence that goes beyond the mere utilisation of existing bioinformatics tools. On the other hand, Bialek and Botstein's point (2004), discussed in the previous section, was precisely that the bioscientists themselves should have the ability to do this kind of work, too.

New research topics

My goal in this chapter has been to demonstrate, at different levels of specification, that genomics indeed does imply a transformation in knowledge production, and that this change needs to be analysed in further depth by social scientists. My particular perspective has been to focus on transformations with educational, cognitive, epistemological and practical research implications for institutions, organisations and the scientists themselves. I think that such an approach to the new bioscience is legitimate and promising, despite its lack of a direct focus on the ethical, social, political and economic issues that constitute the core of contemporary social science research on genomics and society. Inspired by the conference that led to this book,[7] I label my own line of research as a research programme on *genomics and transformation in the production of knowledge*. There is no conceivable end to the issues that could be studied within such a programme. Here is a list of a few of the questions that I find interesting, and that I am currently addressing, or hope to address in future projects:

- How have educational and science-producing organisations responded to the challenge of new research platforms, including bioinformatics? Are there national differences in these responses?
- How is interdisciplinary collaboration organised and carried out within and between the various research platforms of present-day genomics?
- Are integrative difficulties – such as difficulties in integrating knowledge or effecting collaboration across disciplinary boundaries – resulting in delays in the diffusion of new research platforms? Again: are there national differences?
- Is it true that bioscientists in general find mathematics, statistics and computer science to be particularly difficult subjects? Why is that?
- Does the introduction of computer science in bioscience imply that analytical procedures are automated; that is, that the role of the researcher is diminished? Or does it lead to the opposite, a pronouncement of the role of the researcher in analysis?

The set of research topics can also be extended, so as to integrate some of the social and economic perspectives of contemporary studies of science, technology and innovation:

- Does the increasing interdisciplinarity of bioscientific research affect the business models of science-based companies in the field?
- Does the increasing interdisciplinarity have implications for public perceptions of science, public participation in decision-making about science and, more generally, communication between scientists and the public?
- How do the recent developments affect power structures within the scientific community, or between the scientific community and other parts of society?

Acknowledgements

I thank all the interviewees for their participation and positive attitude. I particularly thank Katja Kimppa and Richard Langlais for their encouraging support. The research was financed by the Academy of Finland (project 49974) and Tekes National Technology Agency (decision 40022/04).

This text is based on a shorter viewpoint paper (language: Swedish) published in *Svenska Läkaresällskapets Handlingar* 1/ 200.

Notes

1 For the sake of simplicity, I will use the word 'bioscience' to mean biological and biomedical research that takes the biochemical processes of cells and organisms as its starting point.
2 This chapter emphasises the role of computer science. In reality, biology and medicine seem to be moving towards an increasingly complex pattern of interaction with *several* other disciplines, including, among others, chemistry, mathematics, computational science, physics and engineering.
3 The exceptions are population genetics, structural biology and some areas of neuroscience.
4 This number is an approximation given by a researcher at the university.
5 The history of the technology is, of course, much longer. DNA microarray technology is based on a separation of single-stranded DNA in a way that prevents the strands from reassociating with each other, but permits hybridisation to complementary RNA or cDNA. This technological concept was first applied successfully by Gillespie and Spiegelman in 1965 (Southern *et al.* 1999). Other important background developments were the advents of DNA cloning, Southern blotting, Dot-blotting and differential display PCR (Debouck and Goodfellow 1999; Phimister 1999; Southern *et al.* 1999). As one of the early developers of the DNA microarray puts it: 'There is a sense in which microarray hybridization might be thought of as a simple scaling up (in numbers), miniaturization (in size and sample requirements) and automation of hybridization measurements that have been standard for many years' (Brown and Botstein 1999). On the other hand, he continues to argue that the microarray demonstrates a qualitative difference from earlier technologies, because of the large amounts of data that microarray experiments produce: 'When the body of expression data is large enough, and only then, the patterns of systematic features become apparent and we begin to build an integrated picture of the whole system' (Brown and Botstein 1999). See also Keating and Cambrosio (2003) for an analysis of the biomedical context for the invention of the microarray technology.
6 In a DNA microarray experiment, the following equipment is needed: a spotted microarray; two biological samples to be investigated and kits for extraction of RNA from the samples; two fluorescent markers (generally Cy3 and Cy5); reverse transcriptase for producing cDNA from RNA; a reactive group for coupling the fluorescent marker to the cDNA, and purification chemicals (according to the protocol used) for removing nucleotides that did not couple with markers; a hybridisation solution and a hydrophobic cover slip or a hybridisation chamber, which keeps the temperature constant and the hybridisation solution from evaporating; a thermally stable, humidified environment in which the hybridisation can take place; salt buffers for washing the microarrays after hybridisation; a laser scanner for reading the fluorescently labelled microarrays; software for preprocessing data; and, finally, software for analysing and visualising the data.
7 'Genomics and Society', CESAGen's first International Conference, 2–3 March 2004, Royal Society, London.

References

Bassett, D. E., Eisen, M. B. and Boguski, M. S. (1999) 'Gene expression informatics: it's all in your mine', *Nature Genetics,* 21: 51–5.
Bechtel, W. (1993) 'Integrating sciences by creating new disciplines: the case of cell biology', *Biology and Philosophy,* 8(3): 277–99.

Bialek, W. and Botstein, D. (2004) 'Introductory science and mathematics education for 21st-century biologists', *Science,* 303 (6 February): 788–90.

Brown, P. O. and Botstein, D. (1999) 'Exploring the new world of the genome with DNA microarrays', *Nature Genetics,* 21: 33–7.

Bruun, H. and Langlais, R. (2003) 'On the embodied nature of action', *Acta Sociologica,* 46(1): 31–49.

Bruun, H. and Sierla, S. (forthcoming) 'Distributed problem solving in software development: the case of an automattion project', *Social Studies of Science.*

Bruun, H., Langlais, R. and Janasik, N. (forthcoming) 'Knowledge networking: a conceptual framework and taxonomy', *VEST: Journal for Science and Technology Studies.*

Castells, M. and Himanen, P. (2001) *Suomen tietoyhteiskuntamalli* (The Finnish model of information society). Helsinki: WSOY.

Cheung, V. G., Morley, M., Aguilar, F., Massimi, A., Kucherlapati, R. and Childs, G. (1999) 'Making and reading microarrays', *Nature Genetics Supplement,* 21(1): 15–19.

Churchill, G. A. (2002) 'Fundamentals of experimental design for cDNA microarrays', *Nature Genetics,* 32: 490–5.

Davies, K. (2001) [2002] *Cracking the Genome: inside the race to unlock human DNA.* Baltimore, MD and London: The Johns Hopkins University Press.

Debouck, C. and Goodfellow, P. N. (1999) 'DNA microarrays in drug discovery and development', *Nature Genetics Supplement,* 21(January): 48–50.

Duggan, D. J., Bittner, M., Chen, Y., Meltzer, P. and Trent, J. (1999) 'Expression profiling using cDNA microarrays', *Nature Genetics Supplement,* 21(January): 10–14.

Duyk, G. M. (2002) 'Sharper tools and simpler methods', *Nature Genetics,* 32: 465–9.

Fujimura, J. (1996) *Crafting Science: a sociohistory of the quest for the genetics of cancer.* Cambridge, MA, and London: Harvard University Press.

Giot, L., Bader, J. S., Brouwer, C., Chaudhuri, A., Kuang, B., *et al.* (2003) 'A protein interaction map of *Drosophila melanogaster*', *Science,* 302(5 December): 1727–36.

Kallioniemi, O. (2002) 'Geenisiruista biosiruihin. Uuden biotekniikan haasteet ja mahdollisuudet' (From gene arrays to bioarrays. The challenges and opportunities of the new biotechnology), *Duodecim,* 118: 1149–56.

Kay, L. E. (2000) *Who Wrote the Book of Life? A History of the Genetic Code.* Stanford, CA: Stanford University Press.

Keating, P. and Cambrosio, A. (2003) 'Real compared to what? Diagnosing leukemias and lymphomas', in M. Lock, A. Young and A. Cambrosio (eds), *Living and Working with the New Medical Technologies.* Cambridge: Cambridge University Press.

Knudsen, S. (2002) *The Biologist's Guide to Analysis of DNA Microarray Data.* New York: John Wiley & Sons.

Kohane, I. S., Kho, A. T. and Butte, A. J. (2003) *Microarrays for an Integrative Genomics.* Cambridge, MA, and London: The MIT Press.

Lander, E. S. (1999) 'Array of hope', *Nature Genetics Supplement,* 21(1): 3–4.

MacMullen, W. John and Denn, S. O. (2005) 'Information problems in molecular biology and bioinformatics', *Journal of the American Society for Information Science and Technology,* 56(5): 447–56.

Morange, M. (1998) [2000] *A History of Molecular Biology*. Cambridge, MA, and London: Harvard University Press.

Mäkelä, T. and Porkka, K. (2002) '"Omiikat" tulevat – yksi geeni ei enää riitä' (The 'omics' are coming – one gene is not enough any more), *Duodecim*, 118: 1146–8.

Phimister, B. (1999) 'Going global', *Nature Genetics Supplement* 21(1): 1.

Southern, E., Mir, K. and Schchepinov, M. (1999) 'Molecular interactions on microarrays', *Nature Genetics Supplement* 21(1): 5–9.

Tuimala, J. (2003) 'Preprocessing of data', in J. Tuimala and M. M. Lainne (eds), *DNA Microarray Data Analysis*. Espoo: CSC – Scientific Computing Ltd.

Tuimala, J., Saarikko, I. and Laine, M. M. (2003) 'Normalization', in J. Tuimala and M. M. Lainne (eds), *DNA Microarray Data Analysis*. Espoo: CSC – Scientific Computing Ltd.

Vihinen, M. and Tuimala, J. (2003) 'Cluster analysis of microarray information', in J. Tuimala and M. M. Lainne (eds), *DNA Microarray Data Analysis*. Espoo: CSC – Scientific Computing Ltd.

Wong, G. (2003) 'Introduction', in J. Tuimala and M. M. Lainne (eds), *DNA Microarray Data Analysis*. Espoo: CSC – Scientific Computing Ltd.

12 Science, media and society

The framing of bioethical debates around embryonic stem cell research between 2000 and 2005

Jenny Kitzinger, Clare Williams and Lesley Henderson

Introduction

The media are key arena through which ideas about bioethics are played out. What rhetorical work is being performed in this context? How do media practices impact on how controversies are framed? And how does coverage change over time? This chapter examines such questions in relation to one of the most controversial aspects of new biotechnologies – the use of embryos in stem cell research. We examine how the embryo is defined, imagined, visualised and represented in such controversies, and examine the strategies adopted by leading protagonists in the debate across a five-year period.

We start by analysing the heated debate about shifting regulation that played out in the UK during the year 2000. We then follow this up by revisiting the analysis in relation to the scientific 'breakthroughs' that followed in 2004 and 2005. Our analysis demonstrates how both sides in the dispute mobilise metaphors and use personification to recruit support; and how they promote different ideas about the embryo's significance, size and social embeddedness and present competing narratives about its origins, destiny and 'death'. The role of visual representation is key here. It does not follow the usual pattern whereby, in the abortion debate, those 'on the side' of the foetus display its image while those who are 'pro-choice' shy away from this. In the stem cell debate the pattern is inverted, highlighting the role of technologies of visualisation in defining what counts as human.

Our analysis also demonstrates how, in spite, or even *because*, of the apparently 'balanced' nature of media coverage, it systematically disregards more fundamental challenges to science and curtails discussion of broader social and political issues. We go on to show how, in spite of many continuities in rhetorical strategies between 2000 and 2004–5, the changing scientific context and the shifting form of the news event (for example the 'science breakthrough' story) can impact on which discursive repertoires are mobilised and which gain most prominence. In conclusion we reflect on the methodological implications of our research, and how the findings might inform efforts to support more diverse debates around science and society.

A brief review of the theoretical and policy background

Innovative health technologies such as stem cell research and the development of embryos for 'therapeutic cloning' may have the potential to diagnose, treat and possibly even prevent illness and disease. However, they also raise new risks and give rise to ethical, legal and social concerns. Many commentators have highlighted how such technologies are redefining the boundaries of medicine, and the relation between health technologies and 'the social', as well as the relations people have with their own bodies and with each other (Franklin 1997; Edwards 1999; Webster 2002). Such analyses complement an extensive literature which explores how new reproductive technologies might impact upon the status of the embryo or foetus and hence on women (e.g. Spallone 1989; Stacey 1992; Pfeffer 1993; Rose 1994; Casper 1998; Williams *et al*. 2001; Kent 2000; Hartouni 1997).

These debates are not confined to abstract sociological/philosophical academic theorising; they are also subject to extensive policy discussion and legislative decision-making. There is, of course, a long history of such engagement around abortion, but the debate has recently accelerated in relation to new technologies and research. In Britain during the 1980s, scientific research on human IVF embryos generated intense controversy (IVF refers to in vitro fertilisation). This was exemplified by the setting-up of a committee in 1982, chaired by the philosopher Mary Warnock, to examine the ethical, legal and social implications of developments in the field (Mulkay 1997). This was followed, in 1990, by the Human Fertilisation and Embryology Act which regulated the practice of IVF and the creation, use, storage and disposal of any resulting embryos. Under the Act, research on embryos older than fourteen days, or whenever the 'primitive streak' appears (when the embryo first develops cells that go to make up the spinal nerve), was prohibited. Before this time research was allowed but required a licence from the Human Fertilisation and Embryology Authority (HFEA). Licences were only granted for research in limited areas mainly to do with fertility, reproduction and congenital disease.

In 1998, following further developments, in particular the cloning of Dolly the sheep, the HFEA and the Human Genetics Advisory Commission undertook a public consultation on human cloning. The report recommended that consideration should be given to two further areas for which the HFEA might issue research licences: the development of therapy/treatments for mitochondrial disease and for diseased or damaged organs or tissues. A group chaired by the Chief Medical Officer, Professor Liam Donaldson, was then set up to review the area. The Donaldson report was published in August 2000. One of its principal recommendations was to expand the ways in which embryos could be used to include research aimed at increasing understanding about human disease and disorders and their treatment. The report recommended that such research should be permitted on embryos created either by IVF or by cell nuclear replacement (CNR), subject to the controls in the 1990 Act (CNR relates to the process of inserting the nucleus of an adult cell into a donated egg from which the original nucleus has been removed, a process often referred to as cloning). The government subsequently drafted regulations to turn these recommendations into law but allowed MPs a free vote (i.e. one not determined by party membership). The Human Fertilisation and Embryology (Research Purposes) Regulations 2001 were passed by the House of Commons on 19 December 2000 (Select Committee on Stem Cell Research Report, 13 February 2002.) This set the scene for the UK to become a 'world leader' in this area of research, closely trailing South Korea in expertise and 'breakthroughs'. There were four key events which re-ignited media interest following on from the legislation of 2000. In February 2004 Korean scientists announced that they had created the first cloned human embryo via CNR. The following August a British team was granted the first licence for similar work in the UK. In May 2005 this was followed by announcements that the British team had created their first cloned embryo. This announcement was timed to coincide with the declaration that the South Korean team had successfully used cloning to derive stem cell lines genetically designed to match a group of patients. Although, as this chapter goes to press the scientific papers published in *Science* by the South Korean team have been officially retracted and Professor Hwang has been dismissed from his post. He is currently on trial on three charges: embezzlement, fraud and violating a bioethics law.

Our research

This chapter starts from the recognition that embryos are socially, culturally and politically constructed, and that the construction of the embryo varies depending on who is attributing the meanings, and what the work goals are (Casper 1998). We start by examining how this operated within the national UK press and TV reporting during the two key events during 2000 outlined above: the publication of the Donaldson Report and the subsequent vote in the House of Commons. We then reflect on how these constructions played out in subsequent coverage of 'breakthroughs' in 2004 and 2005.

We focus on the debate as it played out in the mass media because the mass media are a crucial source of public information about health and about medial research (Miller *et al.* 1998). They are also an arena through which policy battles are fought, and are often the focus of intense lobbying by competing sources (Henderson and Kitzinger 1999; Philo 1999). There is no necessary correlation between public opinion, public discourse and public policy (Condit 1999; Kitzinger 2002). However, the media do have a strong influence on what, and how, things come to be defined as public issues. The 'framing' of stories, the selective presentation of one set of themes, oppositions, associations, 'templates', 'facts' and claims, rather than others, is a critical part of this (Conrad 1997; Hansen 2000; Kitzinger 2000; Petersen 2001).

Our initial analysis was based on a sub-sample from a comprehensive archive of reporting about all aspects of human genetic research for the year 2000 in all national UK newspapers and main TV news bulletins (Kitzinger *et al.* 2003). Each item within the original archive was indexed on to computer by date, headline, name of journalist, etc. and coded for the main story focus, the type of visual images used, who was quoted and what potential medical or ethical, social and legal risks were raised in the reporting. For the purposes of the present chapter, this database was scanned for all articles about stem cell research. This identified two periods of intense coverage: 13–30 August 2000, around the release of the Donaldson Report, and 19–21 December 2000, around the parliamentary vote. These two time frames were thus selected for analysis in the first instance.

During these two periods there were a total of 55 newspaper items (including news reports, editorials and feature articles) and eight TV news bulletins about stem cell research. In addition to the main indexing already established, these items were then subjected to more detailed scrutiny. We analysed how the embryo was described, including the use of different terminology and visual images, and all references to its significance, status and size. We also examined how the embryo was positioned in the narrative: its origin and potential and relationship to others. Close attention was given to the ways in which journalists or their sources defined 'life', and their use of metaphors. We also considered the overall framing of the debate and the presentation of the potential beneficiaries of stem cell research. Finally the data were systematically re-examined to clarify apparent gaps – including, for example, to identify the ways in which women appeared (or did not) in the debate. All exceptions to the dominant patterns in coverage were also re-examined.

For the final section of this chapter, similar sampling strategies and analytical methods were adopted in order to examine how the media covered the new events around stem cell research that emerged in 2004 and 2005. On this occasion we were also able to include local/regional press coverage, creating an archive of over 100 media reports for further analysis. This analysis is contextualised by interviews with journalists and their sources, such as the key stakeholders who seek to inform the public debate around this issue.

Findings

An overview of the media coverage in 2000: the focus on the embryo and the binary structuring of the debate

The stem cell debate in the year 2000 was framed as, above all, a controversy about the status and potential of the embryo (rather than other potential controversies, such as the validity of the science *per se* or the context in which it was being realised). *The Daily Telegraph's* main science report, for example, displayed a large image of embryonic cells with the simple headline '*This* is what the human cloning row is about' (*The Telegraph*, 16 August 2000, our emphasis). *The Guardian* carried a similar image (a 22×14 cm lurid green photograph of a clump of cells floating against a black background). The headline reads 'MPs agonise over matter of life and death. Embryo debate. Pleas from the disabled and ill, charges of Nazism' (*The Guardian*, 16 December 2000). TV news bulletins used background logos of a cell (magnified many times, sometimes filling the entire studio wall) and their reports were illustrated with images of cells dividing or shots of vats containing frozen embryos. After the Parliamentary vote, one bulletin showed the vats while the voiceover declared: '*This* is what tonight's vote was about'. As tongs were used to remove tubes from the vat, the voiceover went on to explain: 'Each of these tubes contains frozen human embryos ... To some they are the first stage of human life which we interfere with at our peril. To others they are small clusters of cells which could offer hope to thousands living with devastating disease' (*BBC News*, 19 December 2000).

The media presented the debate as a dispute between two contrasting perspectives. On one side were those who felt embryonic stem cell research was an abuse of embryos which set dangerous precedents (e.g. for reproductive cloning). On the other side were those who argued that the benefits outweighed any such ethical dilemmas or risks (if indeed such risks were seen to exist at all). This binary opposition was a central organising pillar structuring media coverage. Channel 4 News, for example, opened with the question: 'Is it a miracle cure or Frankenstein science?' (C4, 16 August 2000); *Newsnight* asked: 'Is this the stuff of dreams or nightmares?' (BBC2, 16 August 2000). News footage showed confrontations in the House of Commons between the Labour Health Minister and the Conservative Shadow Health Minister, or included organised studio debates between those on different sides of this divide. The press organised similar confrontations on their pages – offering guest writers 'head-to-head' columns to dispute the case 'for' and 'against'.

The battle lines were drawn between, on one side, scientists, Labour politicians and 'patients', and, on the other side, religious spokespeople, Conservative politicians and anti-abortionists. Roughly equal time or space was allocated to each set of protagonists. Table 12.1 shows the different people given space on the TV news bulletins and the position they took on embryonic stem cell research. Table 12.2 presents parallel information for the press coverage.

Having briefly outlined the main form of the reporting, the following section presents detailed analysis of the concepts, images, terminology, metaphors and narratives used by the proponents and opponents given a platform in this debate.

Table 12.1 TV news sources: showing number of bulletins featuring people in these categories *supportive of* or *opposed to* embryonic stem cell research (13–30 August 2000 and 19–21 December 2000)

Category of person (as identified on screen)[1]	Number of bulletins featuring this category of person speaking in support of embryonic stem cell research[2]	Number of bulletins featuring this category of person speaking against embryonic stem cell research
Labour MPs	5	0
Professor Donaldson	4	0
Other scientists/Drs (Professors Winston/Higgins)	2	0
Patients/Patients' support groups	5	0
Ruth Deech HFEA	1	3
Conservative MPs	0	2
Pro-'Life' spokesperson	0	2
Genetic watchdog groups	0	2
Helen Watt ('clinical researcher and Roman Catholic')	0	1
Jacqueline Laing (Professor of Law, London Guildhall University)	0	1
Dr Nicholsor (Editor of the *Bulletin Medical Ethics*)	0	1
Dr Bruce (Director of Society, Religion and Technology project for Church of Scotland)	0	1

Notes:
1 These categories are based on the introductions provided on screen/in the article in any particular case. Thus, for example, Peter Garrett was sometimes introduced as the Director of Life and sometimes as a representative of 'The Movement Against Human Cloning'. People often operate through different organisations, and self-presentation is an important lobbying strategy and a discursive act in itself. The categories we have used in coding any individual always echo the media framing in that case rather than superimposing other categories.

2 This is the number of bulletins including footage of this type of speaker, not the number of speakers.

Table 12.2 Newspaper sources: showing number of articles containing quotes from people in these categories *supportive of or opposed to* embryonic stem cell research[1] research (13–30 August 2000 and 19–21 December 2000)

Categories	Number of articles containing quotes supportive of stem-cell research from this category[2]	Number of articles containing quotes critical of stem-cell research from this category
Scientists/Drs	22	0
Labour MPs	10	0
Professor Donaldson	6	0
Government spokesperson	4	0
Patients/Patients' support groups	4	0
Lib-Dem MPs	2	0
Lord Alton (cross-bench peer, Pro- Life campaigner)	0	3
Conservative MPs	1	14
Religious figures	0	13
Pro-Life groups	0	13
Genetic watchdog groups	0	3
Other	3	2

Notes:
[1] There were only five quotes which could *not* be easily assigned to the 'pro' or 'anti' embryonic stem cell research category,- further demonstrating the polarised way in which the debate was presented. These five quotes have been excluded from the table.
[2] This is the number of articles containing this type of quote, not the number of 'quotes'.

Contrasting discourse around the embryo: fluorescent frogspawn or a young human being?

Opponents and proponents of embryonic stem cell research promoted very different understandings of the embryo. Differences revolved around key dimensions, such as the interpretation of the 14-day cut-off point, references to the embryo's size, use of physical descriptions or images, and the way in which the embryo was characterised in relation to human qualities and connections. Diverse rhetorical strategies were also employed to locate the embryo's biography (from womb to birth/from Petri dish to stem cell line) and to define the very meaning of life and death. Each of these aspects is outlined below.

The 14-day time limit

Pro-Donaldson speakers repeatedly referenced the time limit (that is, fourteen days after fertilisation), after which embryos could not be used for research. This was used as a mantra to guarantee that moral boundaries would be maintained. The cut-off point was presented as a technical, scientific 'truth' defining the moment beyond which embryos deserved greater protection. By contrast, those *opposed* to using embryos (however young) presented the embryo in very different terms. They either ignored or explicitly refused the significance of the 14-day rule: 'Human life is inviolably sacred, both before and after any arbitrary 14-day deadline' (Cardinal Winning, *Express*, 17 August 2000). Whereas opponents defined the embryo at any stage in terms of its human potential, some proponents implied that, before 14 days had elapsed, the cells did not constitute an embryo at all. They made efforts to distinguish between 'proper' embryos and things they called the 'early-stage embryo' or 'blastocyst' (*Guardian*, 17 August 2000). In doing so they were building on pre-existing initiatives in the run-up to the government's Human Fertilisation and Embryology Bill when efforts were made to distinguish the 'pre-embryo' (Mulkay 1997). To present a moment in time as a 'technical' or 'scientific' fact is itself a rhetorical strategy. As Spallone points out:

> [T]he idea of the beginning of an embryo at the 15th day after fertilisation acquired power, not on the basis of natural facts, but because it was empowered by a complex of social, cultural, technical and political factors.
>
> (Spallone 1999: 3)

The significance of size

A second, related strategy used by proponents of embryo stem cell research was to emphasise the microscopic size of the pre-14-day-old embryo in order to underline its lack of human status or qualities. At the original press launch Professor Donaldson talked of 'a tiny ball' (BBC 2 *Newsnight*, 16 August 2000), and the issue of size was reiterated across subsequent coverage. The embryo to be used in stem cell research was 'still the size of a pinhead' (*The Guardian*, 17 August 2000) and 'smaller than this full stop' (*The Telegraph*, 16 August 2000). By contrast, opponents of such research spoke and wrote as if the size of the embryo was irrelevant. In fact, there was only one example of any opponent referring to size. This was when Cardinal Winning underlined the embryo's vulnerability as 'a very small human being' (Cardinal Winning, *Daily Mail*, 18 August 2000).

The battle of images and the role of magnification

A third intertwined strategy adopted in support of stem cell research was to detail the actual appearance of the embryo at this stage. Ironically this included pen-portraits based on *magnifying* technologies and the use of photographic *enlargements* of the 'tiny ball of cells'. The pre-14-day embryo looks nothing like its 12-week-old counterpart used by anti-abortionists to display perfectly formed fingers and toes. In the war of images it is *proponents* of stem cell research who invite us to visualise the embryo. Vastly magnified images of balls of cells that look rather like extra-terrestrial alien blobs were used to underline the fact that these cells should not, indeed *could* not, be recognised as human. A large picture in the *The Telegraph*, for example, was captioned: 'The human blastocyst consists of about 100 cells, [and] lacks a brain, heart or any recognisable feature' (16 August 2000).

All these strategies combined to emphasise the non-personhood of the embryo and its lack of consciousness or feeling. Embryos are 'tiny, unfeeling breeze blocks' (*Sunday Times*, 20 August 2000), 'no more a "person" than a scrape of skin from a grazed knuckle' (*Express*, 16 Aug 2000). A memorable image was conjured up by one *Sunday Times* columnist. He explained that: 'Only after 14 days do those 100 or so indeterminate cells begin to shape into the embryo of human life'. Before this point, the cells are merely '"stem cells" ... stuck together like fluorescent frogspawn' (*Sunday Times*, 20 August 2000). By contrast, opponents of embryonic stem cell research never attempted to describe the actual appearance of the embryo. Instead they invoked symbolic imaginings, repeatedly emphasising the embryo's potential and individuality.

Embryos as human persons

Opponents of embryonic stem cell research often used words such as 'person' in relation to embryos and portrayed them as 'young human life' with a right to 'dignity' (Jacqueline Laing, BBC2 *Newsnight*, 16 August 2000). They were also much more likely to prefix the term 'embryo' with the word 'human' (three times as often as proponents). Indeed, the word 'human' echoed through their comments in many forms – with multiple references to 'human life', 'humanity', 'dehumanising', 'human beings' and 'human rights' (e.g. Helen Watt, *The Guardian*, 17 August 2000).

The language used by opponents of embryonic stem cell research also implied notions of social obligation and connection. Where proponents of stem cell research used clinical terminology about embryonic cells, their opponents evoked parent–child relations with words such as 'cherish' and 'nurture'. Embryos were 'human beings at the most vulnerable stage of their life' (Life spokesperson, *Financial Times*, 17 August 2000); 'something to be cherished and protected' (Helen Watt, *Guardian*, 17 August 2000). The embryos were attributed with childlike qualities, thus demanding *more*, rather than less, adult consideration. They were, for example, located as the ultimate representatives of childhood innocence. A Vatican spokesman described embryonic stem cell research as a gross violation which would 'stain the blood of innocents' (Vatican spokesman, *Daily Mail*, 18 August 2000). (For a discussion of 'innocence', see Kitzinger 1988.)

Such disputes were not only played out through descriptions of the embryo, but also through the implicit and explicit trajectories or *biographical narratives* within which the embryo was embedded. The next section examines this more closely.

Embryonic narratives: source, destiny and social context

The embryonic stem cell controversy evoked competing stories about the origin, destiny and social context of the embryo to be used in such research. The Donaldson report, and subsequent legislation, concerned embryos created *either* by IVF (originally created for infertility treatment) *or* by cell nuclear replacement (created for research). Close attention to the accounts framed by different protagonists shows that, during the year 2000, proponents of stem cell research emphasised the former source of embryos while opponents implicitly emphasised the latter. *Proponents* declared that the cluster of cells would be 'left over' or 'surplus', 'the unwanted by-products of infertility' (*The Guardian*, 16 August 2000). *Opponents* spoke of embryos being deliberately 'created and nurtured ... for experiments in the course of which they will be killed' (Cardinal Winning, *The Sunday Telegraph*, 20 August 2000).

Each side also conjured up different images of the embryo's destiny. For proponents of stem cell research we are talking about cells which would otherwise be 'destined to be discarded' (*Daily Mail*, 24 August 2000). For their opponents, however, the embryo is an entity which would, in the 'normal' course of events, become a person. Whereas one side positioned the embryo in Petri dishes, frozen vats or clinical waste disposal units, the other implicitly presented the embryo nestled in a womb (although this womb, and the woman it belonged to, was never explicitly acknowledged).

According to Lord Alton, for example, each embryo is 'a new entity which, *if left alone,* will flower into a human being' (*Sunday Express*, 13 August 2000, our emphasis).

Defining 'Life' and 'death'

Running through the above debate there is, of course, a struggle over the definition of human life itself – a struggle which has a long history in abortion debates and is condensed in the very nomenclature of anti-abortion groups such as 'Life'. (This claiming of the word 'life' is not, however, undisputed. The Newcastle stem cell research team, by contrast, are located in 'The Centre for Life' and often pose for news camera in front of their logo, 'Life'.) Stem cell research also adds a new twist to this long-standing controversy. To the question: 'when does life begin?' it adds the question: 'when does life *end* and how do we define death?'

Opponents of stem cell research spoke of embryos being 'killed', but such terminology was rigorously avoided by those on the other side of the debate. Instead they spoke, for example, of 'dismantling' the embryo as if it were a mechanical rather than a living object (*Telegraph*, August 16 2000). When challenged they explicitly refused to accept their opponents' terminology. Such a dispute was explicitly played out in the following exchange between the Channel 4 news anchor and the MP, Ian Gibson, chair of the parliamentary office of science and technology and active proponent of stem cell research.

> Anchor: But, none the less, it is about beginnings, the creating of human life.
> Gibson: Well, you say human life. I see life as a continuum, sperm are alive, eggs are alive, so life and death are very difficult to define.
> C4 *News*, 16 August 2000)

Interestingly, the 'demise' of an embryo was, in any case, usually referenced by proponents of stem cell research in another context entirely: IVF research. It was in this context that proponents located embryos as being 'discarded' or 'left to perish' (*Sunday Express*, 13 August 2000). (The last word is often used to describe food that has been left to 'go to waste'.) This meant not only that stem cell research could be presented as less 'wasteful' of spare IVF embryos, but that stem cell research could even be subtly presented as a form of rescue. Indeed, through this lens, the research could actually ultimately confer a form of 'immortality' (Lord Winston, *BBC2 Newsnight*, 16 August 2000). As Waldby notes, opponents of such stem cell research perceive the life of the embryo as biographical, in contrast to advocates, who view the life of the embryo as 'a form of biological vitality. From this point of view the embryo is not killed. Rather its vitality is technically diverted and reorganized' (Waldby, 2002: 313).

However, it should be noted that, although unwilling to accept that embryonic stem cell research involves 'killing' anything, proponents of such research did acknowledge the strong 'feelings' or 'opinions' held by opponents. They often agreed that embryos deserved 'some respect', but they placed the embryo's well-being in the balance against the well-being of suffering individuals. It is here that the mobilisation of patients to represent potential beneficiaries of stem cell research became crucial.

Comparing the embryo to the patient: 'It's either me or an egg'.

At the press launch, Professor Donaldson declared: 'Stem cell research opens up a new medical frontier with enormous potential' (*C4 Evening News*, 16 August 2000). This was picked up by reporters who informed readers that stem cell research could, for example, 'relieve the suffering of millions' (*Daily Mail*, 17 August 2000). This was not just argued in abstract. Viewers and readers were invited to identify with specific individuals. Where embryos were presented by supporters of stem cell research as free-floating, disembodied, non-sentient and anonymous cells – abstracted from all social context – patients who might benefit from such research were profiled very differently. Potential beneficiaries were introduced as firmly embodied individuals, experiencing intense physical and emotional sensations, and deeply embedded in family relations, especially relations with children. Yvette Cooper, the Public Health Minister, for example, was quoted inviting us to consider 'the woman with Parkinson's who struggle with speech so she cannot sing nursery rhymes to her children. The grandfather who cannot enjoy his grandchildren growing up because of the devastation of stroke' (*Daily Mail*, 20 December 2000). In addition to such vignettes, actual patients were profiled in newspaper articles and often appeared in person on TV news. One news bulletin, for example, opened with an interview with Graham Kaye, a man with Huntingdon's disease. He was filmed at home surrounded by his family: 'I'm dying of this disease', he stated, offering viewers a stark choice: 'It's either me or an egg' (*BBC1 9 O'clock News*, 16 August 2000).

It is this 'me or an egg' choice which proponents of stem cell research insisted upon. Viewers and readers were invited to weigh the suffering of real people against any emotional reactions or ethical concerns they might have about the embryo (or what Graham Kaye, above, defined as an 'egg'). The conclusion was meant to be self-evident. As one broadsheet editorial commented: '[F]or most of us, the person who counts higher is the sentient human being before us, not the potential one in the Petri dish' (*Independent on Sunday*, 20 August 2000).

Those objecting to embryonic stem cell research could gain little oppositional purchase against such accounts, and did not usually attempt to question the benefits of stem cell research. Instead they emphasised that it was the *source* of the stem cells they wished to criticise: 'A good end doesn't make good an action that in itself is bad', declared the Vatican (*Telegraph*, 25 August 2000). They also tried to argue that adult stem cells would be a much more appropriate source, and that: 'We are being duped into believing that we can conquer disease like Alzheimer's. ... only by cloning, cannibalising and killing human embryos' (Cardinal Winning, *Sunday Telegraph*, 20 August 2000). This brings us to the final area we wish to examine – the way in which stem cell research was framed by competing metaphors.

Characterising the researchers: frontier pioneers or savage cannibals?

Metaphors are a powerful way in which ideas are framed and concepts communicated (Lakoff and Johnson 1980). Metaphors were woven throughout this debate – most strikingly in relation to how the science, and scientists, were represented. Stem cell research was repeatedly characterised by supporters as a new 'frontier'. The research, they asserted was 'civilised' (Robert Kay MP, *The Times*, 20 December 2000) and the research scientists were 'pioneers' (e.g. *C4 Evening News*, 16 August 2000). In contrast, opponents of therapeutic cloning evoked a different kind of boundary crossing – not the bold adventures of frontier pioneers but an illegitimate invasion. They talked of a sort of piracy. Lord Alton, for example, asked, rhetorically, 'By what right do we *plunder* a unique, tender new life – and *raid* it for spare parts' (*Sunday Express*, 13 August 2000, our emphasis). Some went further and used metaphors that involved the breaching of the most fundamental taboos and moral boundaries of all. They labelled stem cell research as a form of 'cannibalism' (e.g. *Sunday Express*, 13 August 2000; *Sunday People*, 13 August 2000; *Daily Mail*, 21 December 2000). Although these metaphors appear to be in fundamental opposition, they share an underlying logic. Both, in part, draw their power from a deeply racialised notion of civilisation versus primitive barbarity, both leave the concept of 'progress' unquestioned.

Marginalised discourses

The above analysis highlights the competing (and sometimes strangely complementary) rhetoric used by each side in the embryonic stem cell debate. In the following analysis we briefly examine some perspectives which were marginalised or excluded all together. We demonstrate how the focus on the embryo, the stock casting of proponents/opponents and the setting up of a simple, oppositional, framework side-lined three crucial issues.

Lack of attention to the therapeutic gap or medical risks

Among all the claims for the bright new future offered by embryonic stem cell research, opponents were seldom quoted as drawing attention to the potential gap between hope and reality. The fact that the potential benefits of stem cell research lay largely in the future was usually used, by proponents of such research, to assert the imperative of hope and the need for further work; it was rarely used to question whether such research would actually succeed. Only two out of the 55 newspaper articles in our sample for the year 2000 included quotes raising the 'therapeutic gap' as a challenge to the value of the research in the first place. The possibility that there might be any risks in stem cell-based therapies was similarly marginalised. Only four newspaper articles included quotes mentioning potential dangers (for full discussion of the rhetorical construction of the 'imperative of hope', see Kitzinger and Williams 2005).

Ignoring wider social and political contexts

The second issue which was largely sidelined was how 'choice' would operate for patients in the future and how medical innovations would actually be delivered within the global health economy. Although patients were represented as parents, children and grandparents, their location within any broader social context was vague. As Kerr points out:

> The focus on individual choice or nuclear family units takes attention away from the wider communities and societies in which people are rooted, assuming an independence and insulation from wider commercial and discriminatory pressures which is not borne out in people's everyday experiences of health and disability.
>
> (Kerr 2002: 10)

The global context of such research and treatment was even more obscured in the coverage. There were some positive statements which implied that stem cell research would have global relevance: it might, for example, 'end an organ donor crisis which means that millions of adult and child patients *worldwide* die or are forced to rely on clumsy mechanical replacement' (*Daily Express*, 16 August 2000, our emphasis). Quite how this would play out given existing relations between North and South was not discussed. We located just four references, in passing, to 'commercial pressures' in the press brought up by *opponents* of the research and one whole sentence by a proponent of stem cell research, Ian Gibson. In the conclusion of an article he wrote for the *Guardian*, he declared: 'This is a global debate, which is necessary to ensure benefits extend across the planet on an equal basis, and that multinational corporations do not engineer the debate to reap financial profits' (*Guardian*, 15 August 2000). A further brief discussion emerged for a few seconds on *Newsnight*. Here Richard Nicholson (editor of the *Bulletin of Medical Ethics*) stated that 'this research will only benefit the wealthy, it's very expensive treatment for very few people, it will do nothing for mankind as a whole' (BBC2 *Newsnight*, 16 August

2000). His 'opponent', Lord Robert Winston, agreed that they both felt this was an important issue but asserted that the research would lead to cheaper treatments. However, after a brief exchange, the anchor brought them back 'on track' with the question: 'That ball of cells ... does it deserve no respect?' Any further discussion of global health provision was abruptly curtailed.

It seems that concerns about the wider political/economic context of biomedical research and interventions might be a shared concern for protagonists on both sides of the embryonic stem cell debate. However, the focus on the embryo and the oppositional structure of the debate in the media did not allow this issue to be explored.

The absence of feminist perspectives

The third area of debate excluded from the media coverage is the issue of women's status and treatment within this scientific endeavour. Stem cell research raises concerns from a feminist/women's health perspective because of women's role as a source of eggs or embryos. Concerns include questions about informed consent, health risks associated with the artificial stimulation of egg production, and the whole issue of women's relationships with their bodies and with the medical or scientific establishment in a context of gendered and global inequalities. However, during the debate as it played out in the year 2000, such issues were barely mentioned.

As subjects, women were often represented in a passive or fragmented way, with few explicit references to women as social actors (see also Mulkay 1997). There was little sense of any participation in procedures which might be 'done to them'. They were simply wombs into which clones might be implanted. Women's bodies were also rendered literally redundant and invisible. We have pointed out how those opposing stem cell research could ignore the role of women in gestation – talking, for example, of embryos which, if 'left alone', would somehow develop into babies. (An embryo literally alone will, of course, never survive.) This theme carried through into visual representations of eggs and embryos as independent entities. Computer-generated graphics on the TV news showed eggs flying in from one side of the screen like an asteroid from outer space, or embryos being extracted from frozen vats with no indication of their origins (e.g. *BBC2 Newsnight*, 16 August 2000; *C4 News*, 19 December 2000). Where women were shown in these diagrams on television they were often drawn as faceless, naked and transparent (e.g. *BBC2 Newsnight*, 16 August 2000).

There was just one clear example of women's involvement and active participation being addressed in any of the media coverage in the year 2000. This was Leah Wild's *Guardian* column documenting the progress of her IVF treatment. Commenting on the stem cell debate, she remarked: 'In all this furore, there is one voice missing: mine. The majority of embryos for research come from people like me, as the unwanted by-products of infertility treatment' (*Guardian*, 16 August 2000). She went on to write about the deep sense of connection with the embryos and their potential that donors might experience even while supporting such research. This exceptional example of such a point of view highlights its exclusion from the rest of the debate and hints at some of the complexities that were thereby ignored.

A reflection on context and implications: how journalistic practices impact on media coverage

Our analysis so far has highlighted how opposing discourses about the embryo were projected to assert competing ethical positions and policy recommendations. Although proponents of stem cell research attempt to present their perspective as 'scientific' and 'neutral', and their opponents as 'emotive', this is a false dichotomy. In fact, both sides use language, images and narrative structures to invite us to imagine, identify with, and have emotional responses to the embryo (or patients) in competing ways. Neither side presents simply technical, unambiguous 'facts', although both try to claim some higher scientific, objective or moral truth. We hope that our analysis has helped to map out and deconstruct some of their strategies for persuasion.

Simply analysing the discourses circulating within (or absent from) the mass media does not, however, give the whole picture; nor does it provide a fully adequate basis for intervention and innovation. The discourses presented on television and in newspapers depend, in part, on what is actually said and done by those trying to promote their perspective (media 'sources'). Understanding their strategies and deconstructing their rhetoric is an important part of the analysis. At the same time it is important to recognise that these points of view are never presented 'raw'; they are mediated through news institutions, values and conventions involving editorial/journalists' choices, such as which story to cover, whom to interview and how to edit and frame the debate (see Henderson and Kitzinger, in press; Petersen 2002; Schlesinger 1987, 1990).

There are several clear media factors which impacted on the stem cell debate as it operated in the year 2000. The first key issue to note is that news reporting is 'event oriented' and responds to high-status official sources. Both the Donaldson report and the subsequent parliamentary vote fitted into standard news values and were thus almost guaranteed coverage. It follows that, as both events focused on the status of the embryo, this would help to define the terms of the subsequent debate as it played out in the media.

The second point to note is that standard 'hard' news values, such as those outlined above, are now increasingly complemented by 'soft' human interest stories with greater entertainment value and high appeal to audience identification (Henderson and Kitzinger 1999). In the stem cell debate this was very much in evidence in the attention and prominence given to personal accounts from patients. This gave the proponents of stem-cell research extra leverage with the media. It is hard to give a clump of embryonic cells the same human-interest value.

The third point we wish to highlight is the way in which news production and reporting is gendered. Male journalists in our sample outnumbered female journalists ten to one. In addition, men were four times more likely to be quoted as a source in the press, and given twice as much airtime on the TV news coverage. This sort of gender imbalance is so routine as to be almost unremarkable, and in itself, of course, need not *determine* the nature of the 'voices' or arguments given space. However, such imbalance may contribute to the marginalisation of concern about women's health issues and the placing of feminist criticisms outside the frame of reference. The opportunities offered by columns such as Leah Wild's personal account, or in editorial spaces such as the 'Women's Page', shows how different types of discussion may be allowed 'at the edges', raising issues which are excluded from routine 'hard news'.

The fourth, and most fundamental, point we wish to highlight here is the implications of the way in which the news media frame issues as two-sided controversies, with stock characters 'for' and 'against'. This default framing technique raises fundamental issues about the ability of the news media to tackle complex issues in fresh ways. The oppositional approach may be embedded in source organisations – for example the structure of British politics – but it is also explicitly sought by newspaper journalists and TV reporters in pursuit of drama and ease of understanding. Time and space limitations encourage sound bite exchanges; controversy is assumed to make better copy than consensus. Indeed, sources we spoke to complained that any attempt to introduce more nuanced debate was often written off by journalists. As one source told us: 'They wanted tension, they wanted argy bargy. You would see journalists' disappointment if you didn't argue enough. ... Often the media press me to say something extreme and, when I won't, they go on to pro-life groups' (interview with JK). In his experience, journalists sought out general polarised arguments. Although this presentational form may appear to represent 'balance', in fact such didactic and dyadic framing contributes to the exclusion of more nuanced debate (see also Smart 2003).

Some of the lengthier TV studio footage hints at how discussion might have developed in different ways. It allowed some proponents and opponents of stem cell research very briefly to develop dialogue around areas of agreement and to raise broader issues for discussion. Analysis of such examples shows, however, that protagonists themselves considered there was 'not enough time' to pursue the issues in this way and that such conversations were quickly redirected by the TV interviewers. (See, for example, the brief discussion of broader political context between Nicholson and Winston cited on pages 217–8). Clues to the restraining influence of standard media values and practices are also evident in the fact that, when we re-scrutinised the data to locate any exceptions to the general pattern of coverage, we found that these appeared in relatively unusual formats with less journalistic mediation. So, for example, Ian Gibson's statement about global politics (cited on page 217) appeared in a guest piece he wrote himself. Similarly, the unusual account of ambivalence and a highly 'connective' notion of the embryo and sense of their human potential, combined with a pro-stem cell position, appeared in a personal-story column by a rare woman contributor, Leah Wild (see pages 218–9).

Revisiting subsequent coverage: an analysis of the 2004–5 stem cell 'breakthroughs'

Any media studies project will benefit from a longitudinal perspective and revisiting earlier analysis to see how dynamics play out over time. The above analysis was first produced in 2002. Since then, four major news events have propelled the stem cell story back into the public arena. In 2004 Korean scientists announced that they had apparently created the first cloned human embryo and a British team was granted the first licence for similar work in the UK. In 2005 this was followed by announcements that the British team had created its first cloned embryo and the (now challenged) claim from the South Korean team that it had successfully derived patient-compatible stem cell lines. Examining the continuities and discontinuities with reporting from 2000 can help to refine our analysis.

Similarities in reporting between 2000 and 2004–5

In many ways, reporting of the events of 2004–5 displays many of the features evident in the earlier debate. Again, the primary moral debate focuses on the status of the embryo, and the argument is presented largely as a simple binary opposition between scientific progress and religious or anti-abortion sentiment (for example, the *Telegraph* subheading 'Scientists celebrate milestone for medicine. Pro-Life groups fear misuses of new technique', 13 February 2004). Alternative perspectives, e.g. those of feminist or humanist critics, are marginalised. Concerns about wider social and political issues are also largely silenced – at least as far as the mainstream news formats are concerned, although, again, we found some room for diversity in formats such as guest-authored pieces.

The discourses around the embryo, employed by each side in 2004–5, also closely mirror those deployed in 2000. Supporters of stem cell research repeatedly focus on the microscopic nature of the embryo – no larger than a 'speck of dust' (*Independent*, 13 February 2004) or 'a grain of sand' (*Express*, 13 February 2004). Those in favour of embryo stem cell research also repeat their insistence that the object of research lacks any recognisable human features; indeed, they seem to have refined their strategy here and are clear about the misunderstandings they are targeting. Ian Wilmut, for example, creator of Dolly the sheep, emphasises the need to 'correct' people's perceptions:

> Many people think of an embryo as a little foetus, a tucked-up person, knees up to the chest. But this is such a small ball of cells that you can't see it without a microscope.
>
> (*Sunday Times*, 15 August 2004)

As in the 2000 debate, we can also see how proponents of embryonic stem cell research envisage 'life' in rather different ways than the anti-abortionists. Far from killing anything, stems cell research, they argue, avoids 'wasting' the embryo (Christopher Shaw, C4 *7 O'clock News*, 19 May 2005). As we witnessed in 2000, different terminology is also sometimes introduced to name the dividing cells as something other than the emotive 'embryo' with its implicit link to foetuses and babies. There is the occasional use of the term 'blastocyst', and one journalist even describes the cells simply as a 'growth', using language associated with tumours ('Why cloning can benefit us all', *The Journal*, 17 August 2004).

The arguments used by those *opposed* to the developments in stem cell research in 2004–5 reiterate many of their concerns from earlier years. They repeatedly assert the embryo's humanity, describing it not as 'microscopic' but as 'tiny' in a way which emphasises the imperative to protect this 'tiniest member of the human race' (Josephine Quintavalle, Core, Pro-Life group, *Channel 5 News*, 11 August 2004). Opponents not only attribute to the embryo 'personhood', they also sometimes ascribe it gender. For example a *Daily Mail* columnist refers to 'a five-day-old female embryo' in her piece entitled 'We've bargained away our humanity for the illusion of a world without pain' (*Daily Mail*, 23 May 2005). Stem cell research is deemed to make the embryo a victim to be 'dismembered' (Helen Watt, *Sun*, 13 February 2004), 'mutilated' (Jack Scarisbrick, *Express*, 13 February 2004) or 'pulled apart and then flushed away' (Melanie Phillips, *Daily Mail*, 23 May 2005). The embryo is embedded in traditional family relations (discarded embryos are represented as 'siblings') and its normative destiny to become a child is routinely evoked. The *Daily Mail*, for example, illustrates its front page coverage of the first Korean breakthrough with a picture of the cloned cells beginning to divide. This shows 'the very beginning of a process which, nine months later, *should* lead to the birth of a child' ('Playing God?', *Daily Mail*, 13 February 2004, our emphasis).

So far, so familiar. However, analysis of the coverage of 'breakthroughs' in 2004–5 also reveals some interesting differences, both in rhetorical emphasis and in the range, and framing, of certain arguments.

Differences in reporting between 2000 and 2004–5

Four core areas of change are worth highlighting in comparing how the debate played out in 2000 and in 2004–5. The key differences involve the emphasis on discourse of nation, the structuring of reports around 'breakthroughs', new rhetorical twists on the question of size, adaptations to the ever-moving goal of 'cures', and the shifts in discursive strategy to take into account the new emphasis on CNR rather than spare IVF embryos.

Foregrounding discourses of nation and place

The first point we wish to highlight concerns ideas about nation. The debate about UK legislation during 2000 did involve ideas about national identity and the place of the UK in a global economy. However, this became even more explicit in the years that followed. In 2004–5 the location of the main research events, in South Korea and in north-east Britain (an area of economic deprivation), intensified the way in which the debate was inflected through discourses about nationality, national legislation and local/national economic interests. (See CheKar and Kitzinger, in press, for detailed discussion of this point.)

Structuring reports around 'scientific breakthrough' events

A second point worth highlighting is the changing nature of the news events themselves. The shift from debating legislation in 2000 to the actual initiation of experiments in 2004–5 had two effects. On the one hand, it added an urgency and a sense of horror to some of those opposed to this work. However, it also shifted the way in which journalists were writing up their stories. Although the moral debate presented in 2004–5 took the same *form* as it had earlier, it was often given *less prominence*. In 2004–5, journalists were drawing on press releases from laboratories and research funding bodies which emphasised the details and the 'promise' of the science. Three out of the four news events in this field in 2004–5 were 'breakthroughs' (the fourth case involved the application for a licence). Moral questioning was not, therefore, the primary *raison d'être* for most of the stories. Such questioning therefore was sometimes relegated to a paragraph at the end, or a news anchor's aside, demoted from the earlier 'balanced' discussion of the debate around legislation. BBC1 News, for example, reporting on the South Korean and UK breakthroughs in May 2005, simply mentioned that the breakthroughs made some people feel 'queasy' and included a brief interview with a concerned health specialist (BBC1, *1 O'clock News*, 19 May 2005). Coverage of some events by Channel 4 News only included one source concerned about the implications of the research, showing a few seconds of US President Bush, 'a powerful enemy' of the research, declaring 'Life is a creation, not a commodity' (Channel 4, *7 O'clock News*, 19 May 2005).

A new rhetorical twist on size

A third point we wish to highlight here is that in 2004–5 the issue of size is given extra rhetorical effect by contrasting the tiny size of the object of research with the 'giant leap' of the research achievement. BBC1 News, for example, showed images of cloned cells while the voiceover intoned: 'smaller than a pinhead, but a gigantic achievement' (BBC1, *1 O'clock News*, 19 May 2005). The dramatic comparison between the *minute* object of research, and the great size of the scientific achievement enacted upon it, underlines the 'breakthrough'. It also, of course, implicitly evokes another earth-shattering event – the small step for man, giant leap for mankind of the first Moon walk.

Words of caution

A fourth point to highlight is that, with one notable exception (see *C4 News*, 19 May 2005), the predictions of cures in 'five or ten years' that proliferated in 2000 were often *not* updated in the 2004–5 reporting. The breakthroughs were identified as 'milestones' and 'landmarks', but talk about breakthroughs was also sometimes now accompanied by calls for caution about predicting imminent medical benefits. BBC1 News, for example, informed viewers that these breakthroughs might 'revolutionise medicine' and eventually 'cure the incurable', but also offered words of caution, stressing that it was important not to encourage 'false hopes', warning that the Newcastle team had already been contacted by patients asking for help, and that this was much too early in the process (BBC1, *1 O'clock News*, 19 May 2005). The 'economy of hope', and how it operates over time in the context of the sociology of expectations, is a fascinating aspect of this debate which we do not have time to explore further here (but see Kitzinger and Williams 2005 for further discussion).

Perhaps the most important point, in the context of this chapter, however, is to revisit the rhetorical strategies used about the embryo. So far we have drawn attention to the continuities, but it is also vital to reflect on differences of emphasis in the two time periods we have analysed. In 2000 the creation of CNR cloned embryos (rather than the use of 'spare' IVF embryos) was only a theoretical possibility. In 2004–5 it became a reality. This led to a shift in emphasis by both proponents and opponents.

Shifting debates around the CNR embryo

In 2004–5, the proponents of the science could no longer simply emphasise the 'spare IVF' source of research material. Instead they had to confront the debate about the deliberate creation of embryos for stem cell research, and to find way of arguing in defence of this technique. There was also a need to develop new public messages in order to facilitate the access to raw materials needed for this work. In particular, scientists now realised they needed access to fresh eggs. In 2004–5, therefore, it is interesting to find proponents of embryo stem cell research beginning to shift their public statements. For example, it was only in 2004–5 that we found examples of proponents of embryonic stem cell research highlighting the potential medical dangers of using IVF embryos ('What slippery slope?', Christopher Reeve, *Guardian*, 13 February 2004). Some scientists also started to argue that CNR might raise less, rather than more, ethical dilemmas (because of the potential to use the egg and cell from the same woman without involving a third party). And, in some cases, the very term 'embryo' was disputed as a relevant term for a CNR creation because 'it is important to point out that these are not strictly embryos, so it is debatable whether the "right to life" argument applies. They are stimulated to behave like embryos' ('Cloning decision is right', *Glasgow Herald*, 12 August 2004).

Opponents, for their part, were able to develop more in-depth critiques of this form of research endeavour. First, it is interesting to note how they increasingly mobilised manufacturing imagery to evoke the commoditisation of the embryo in the CNR process. While those who were pro stem cell research tended, as noted earlier, to use mechanistic language to describe the *destruction* of the embryo – talking, for example, about it being dismantled (e.g. *The Times*, 22 May 2005), it was those who were against the research who now emphasised mechanistic metaphors. But they used such metaphors to talk about the *creation* of the embryo – referring, for example, to them as being 'manufactured' (Julia Millington, ProLife Alliance, quoted in the *Herald*, 20 May 2005). The same metaphor, applied to different stages of a research process, can clearly have different implications.

In the 'breakthroughs' of 2004 and 2005 CNR was also sometimes implicitly, and occasionally explicitly, presented as particularly abhorrent because it disrupted 'normal' family relations. Jack Scarisbrick, National Chairman of LIFE (the anti-abortion body), opened his attack on therapeutic cloning with this declaration:

> All laboratory cloning of human beings is wrong because it involves manufacturing a new kind of human being: one produced asexually and without parents in the traditional sense ... [It is] a violation of the natural order.
>
> (*Express*, 13 February 2004)

The Daily Mail columnist, Melanie Phillips, protested that 'The links between sex, procreation and parenthood have progressively been snapped while we have stripped unborn life of meaning and respect', adding that the clone is 'the ultimate single-parent child' (23 May 2005).

In fighting against the developments in 2004–5, some of the rhetoric went even further and presented CNR therapeutic cloning as worse than reproductive cloning and even as worse than abortion. Indeed Melanie

Phillips declared that abortion was morally more acceptable than cloning because cloning 'means deliberately creating a life solely in order to destroy it' (Melanie Phillips, *Daily Mail*, 23 May 2005), while another commentator declared: 'Reproductive cloning is bad but it does not have a 100 per cent mortality rate, whereas in therapeutic cloning all the embryos die' (Dr Helen Watt, Director of the Catholic Church's Linacre Centre for Health Care Ethics, quoted in 'Scientists celebrate a milestone for medicine; Pro-life groups fear misuse of new technique', *Telegraph*, 13 February 2004).

The shift of emphasis in the science to the CNR embryo that was evident during 2004–5 is also having another effect. The issue of CNR, combined with the possible need for 'fresh' eggs, is making it more difficult to avoid questions about the source of experimental material. Although proponents sometimes try to emphasise that they would use 'unwanted' eggs, it rapidly became clear that some would also be seeking eggs specially donated for the purpose of research. This is necessary both because of the high failure rate in CNR and because the experiments both in South Korea and in the UK seemed to suggest that 'fresh' eggs, no more that two hours old, were necessary to maximise success.

At the time of writing, the status of women as the source of eggs is beginning to emerge in a few reports. Rumours are circulating about whether or not South Korean scientists followed ethical protocols in obtaining eggs. In addition, in an apparent strategy of 'rhetorical diversification', anti-abortion groups are increasingly appealing to notions of women's rights to press home their case, expressing concern that 'women may one day be used to "farm" donor eggs' (*Guardian*, 13 February 2004) or that such work will 'make the exploitation of some women more likely. Cloning involves exposing women to dangerous fertility drugs in order to collect sufficient eggs to use in the cloning process' (LIFE, quoted in the *Telegraph*, 20 May 2005). There is also growing activism among women's health activists and those concerned with distributive justice. Leading campaigner Judy Norsigian, from 'Our Bodies Ourselves' (also known as the Boston Women's Health Collective), for example, describes herself as 'a supporter of most embryonic stem cell research', but declares substantial concerns about the use of CNR embryos. She has been a prominent figure in the call for a moratorium on such research in the US. Although she is often cast by the media as a 'strange bedfellow' with the Catholic Church, this is a positioning she vehemently rejects. 'This is a colossal myth that it's a matter of the pro-choice, pro-science votes on the one side and the religious social conservatives on the other side', she told us (Interview with JK). There are, she points out, legitimate health and safety reasons for objecting to some forms of stem cell research, and these reasons are generally not the primary concerns of the anti-abortion lobby which focuses on quite different distinctions. 'Although there are those who have deliberately confused this issue, sometimes conflating embryo cloning research with ALL embryo stem cell research, it is important to keep the two sepa-rate and to insist that health concerns for women don't take a back seat' (Norsigian 2005: 702). Clearly, CNR can raise different issues for cam-paigners than the use of 'spare' IVF embryos. This suggests that new sta-keholders may come to the fore as the science develops. Watch this space.

Since this chapter was drafted the South Korean team has admitted it paid for eggs, and that some eggs came from junior colleagues – an event that, momentarily, highlighted the issue of women's position in such research. In turn, this ethics scandal has been overtaken by allegations of scientific fraud.

Conclusion

This chapter has explored the interplay between scientific developments, news events, stakeholders' strategies and journalistic practices in the evo-lution of public debate around a specific social issue. Through a combina-tion of quantitative and qualitative analysis, placed in the context of source–journalist relations and an attempt to track coverage longitudinally, we have highlighted some of the complex dynamics that can shape media coverage and the public profile of a controversy.

Many policymakers are committed to encouraging wide-ranging public debate about medical and scientific innovations (see, for example, 'Human Genetics', The House of Commons Science and Technology Select Com-mittee 2005). In some ways the release of the Donaldson report back in 2000 was engineered precisely to facilitate such debate prior to the parlia-mentary vote. However, our analysis shows that, although a major con-troversy was aired, many fundamental questions remained unaddressed. The coverage left the existing system of medicine and the scientific enter-prise itself largely unchallenged. Although appearing to represent the range of conflict, certain positions, including the position of ambivalence, were largely silenced. Although including quotes from an apparently wide range of 'balanced' sources, some voices were systematically marginalised. The situation five years on shows many similar limitations, although some issue are beginning to gain new forms of prominence and rhetorical strategies are not static.

Here we agree with many other critics, that there is a need to create different types and fora for discussion if we are really to develop inclusive public democratic debate, and that there is an urgent need to include more diverse views. (See arguments put forward by, among others, Kerr *et al.* 1998; Barnes 1999; Shakespeare 1999; Rose 2000; Williams *et al.* 2002.) This is essential in order to achieve more accountable and sophisticated policymaking, and to move beyond the notion of simply 'educating' the public about science. Rather, people's 'lay expertise' should be drawn upon to inform debate about what directions science should – or should not – be taking. Siding with either the pro-embryo research lobby, with its generally uncritical acceptance of biomedical research agendas, or with the pro-life, anti-embryo research lobby, is an inadequate set of possibilities (Casper 1998; Franklin 1999; Morgan and Michaels 1999). There is a need for more discussion which addresses feminist perspectives, examines the mutual construction of the socio-technical (Webster 2002) and seeks to develop sociologically informed bioethics (DeVries and Conrad 1998). There are possibilities to pursue such aims partly (but not exclusively) via the mass media. However, if this is to happen, then there is a need both to be critical of existing discourses and to engage constructively with the type of obstacles, challenges and opportunities within media practices and processes that have been outlined in this chapter.

Acknowledgements

The initial work on the case studies in 2000 was funded by the Wellcome Trust (Project Number: 058105). Lesley Henderson compiled the media archive and Clare Williams and Jenny Kitzinger wrote the analysis. The subsequent analysis of 2004–5 was conducted and written up by Kitzinger and made possible by the support of the ESRC Centre for Economic and Social Aspects of Genomics (CESAGen). A version of the first part of this chapter was originally published in *Sociology of Health and Illness*, 25 (7) 2003.

References

Barnes, B. (1999) 'Acceptance: science studies and the empirical understanding of science', *Science, Technology and Human Values*, 24: 376–83.

Bosk, C. (2000) 'The sociological imagination and bioethics', in C. Bird, P. Conrad and A. Fremont (eds), *The Handbook of Medical Sociology*. Englewood Cliffs, NJ: Prentice Hall.

Casper, M. (1998) *The Making of the Unborn Patient*. New Brunswick, NJ: Rutgers University Press.

CheKar, C. and Kitzinger, J. (in press) 'Science, patriotism and constructions of nation and culture', *New Genetics & Society*.

Condit, C. (1999) *The Meanings of the Gene*. Madison, WI: University of Wisconsin Press.

Conrad, P. (1997) 'Public eyes and private genes: historical frames, news constructions, and social problems', *Social Problems*, 44: 139–54.

DeVries, R. and Conrad, P. (1998) 'Why bioethics needs sociology', in R. DeVries and J. Subedi (eds), *Bioethics and Society*. Englewood Cliffs, NJ: Prentice Hall.

Edwards, J. (1999) 'Explicit connections: ethnographic enquiry in north-west England', in J. Edwards, S. Franklin, E. Hirsch, F. Price and M. Strathern (eds), *Technologies of Procreation*. London: Routledge.

Franklin, S. (1997) *Embodied Progress: a cultural account of assisted conception*. London: Routledge.

—— (1999) 'Making representations: the parliamentary debate on the Human Fertilisation and Embryology Act', in J. Edwards, S. Franklin, E. Hirsch, F. Price and M. Strathern (eds), *Technologies of Procreation*. London: Routledge.

Hansen, A. (2000) 'Claims-making and framing in British newspaper coverage of the "Brent Spar" controversy', in S. Allan, B. Adam and C. Carter (eds), *Environmental Risks and the Media*. London: Routledge.

Hartouni, V. (1997) *Cultural Conceptions: on reproductive technologies and the remaking of life*. Minneapolis, MN: University of Minneapolis Press.

Henderson, L. and Kitzinger, J. (1999) 'The human drama of genetics: "hard" and "soft" media representations of inherited breast cancer', *Sociology of Health and Illness*, 21: 560–78.

—— (in press) 'Orchestrating a science event', *New Genetics & Society*.

Holland, S., Lebacqz, K. and Zoloth, L. (eds) (2001) *The Human Embryonic Stem Cell Debate*. Cambridge, MA: MIT Press.

House of Commons Science and Technology Select Committee (1995) *Human Genetics*. London: House of Commons.

House of Lords Select Committee on Stem Cell Research (2002) *Report*. London: The Stationery Office.

Kent, J. (2000) *Social Perspectives on Pregnancy and Childbirth for Midwives, Nurses and the Caring Professions*. Buckingham: Open University Press.

Kerr, A. (2002) 'Sociologies and bioethics of genetics: tensions and potentials'. Paper presented at 'Society for Human Biology' symposium, London, 20–23 September.

Kerr, A., Cunningham-Burley, S. and Amos, A. (1998) 'Drawing the line: an analysis of lay people's discussions about the new genetics', *Public Understanding of Science*, 7: 113–33.

Kitzinger, J. (1988) 'Defending innocence: ideologies of childhood', *Feminist Review*, 28: 77–87.

—— (1998) 'The gender-politics of news production: silenced voices and false memories', in C. Carter, G. Branston and S. Allan (eds), *News, Gender and Power*. London: Routledge.

—— (2000) 'Media templates: patterns of association and the (re)construction of meaning over time', *Media, Culture and Society*, 22: 64–84.

—— (2002) 'Media influence revisited: an introduction to the "new effects research"', in P. Cobley and A. Briggs (eds), *The Media: an Introduction*, 2nd edition. London: Longman.

Kitzinger, J., Henderson, L., Smart, A. and Eldridge, J. (2003) *Media Coverage of the Social and Ethical Implications of Human Genetics*. Final report for The Wellcome Trust.

Kitzinger, J. and Williams, C. (2005) 'Forecasting science futures: legitimising hope and calming fears in the stem cell debate', *Social Science and Medicine*, 61(3): 731–40.

Lakoff, G. and Johnson, M. (1980) *Metaphors We Live By*. Chicago, IL: Chicago University Press.

Miller, D., Kitzinger, J., Williams, K. and Beharrell, P. (1998) *The Circuit of Mass Communication: media strategies, representation and audience reception in the AIDS crisis*. London: Sage.

Morgan, L. and Michaels, M. (eds) (1999) *Fetal Subjects, Feminist Positions*. Philadelphia: University of Pennsylvania Press.

Mulkay, M. (1997) *The Embryo Research Debate*. Cambridge: Cambridge University Press.

Norsigian, J. (2005) 'Stem cell research and embryo cloning: involving laypersons in the public debate', *New England Law Review*, 39(3): 701–8.

Petersen, A. (2001) 'Biofantasies: genetics and medicine in the print news media', *Social Science and Medicine*, 52: 1255–68.

—— (2002) 'Replicating our bodies, losing our selves: news media portrayals of human cloning in the wake of Dolly', *Body and Society*, 8: 71–90.

Pfeffer, N. (1993) *The Stork and the Syringe: a political history of reproductive medicine*. Cambridge: Polity Press.

Philo, G. (ed.) (1999) *Message Received*. Harlow: Addison, Wesley and Longman.

Rakow, L. and Kranich, K. (1991) 'Women as sign in television news', *Journal of Communication*, 41: 8–23.

Rose, H. (1994) *Love, Power and Knowledge: towards a feminist transformation of the sciences*. Cambridge: Polity Press.

—— (2000) 'Risk, trust and scepticism in the age of the new genetics', in B. Adam, U. Beck and J. Van Loon (eds), *The Risk Society and Beyond*, London: Sage.

Santos, B. and De Sousa, C. (1995) *Towards a New Common Sense: law, science and politics in the paradigmatic transition*. London: Routledge.

Schlesinger, P. (1987) *Putting Reality Together*. London: Methuen

—— (1990) 'Rethinking the sociology of journalism', in M. Ferguson (ed.), *Public Communication*. London: Sage.

Shakespeare, T. (1999) 'Losing the plot? Medical and activist discourses of contemporary genetics and disability', *Sociology of Health and Illness*, 21: 669–88.

Smart, A. (2003) 'Reporting the dawn of the post-genomic era: who wants to live forever?', *Sociology of Health and Illness*, 25(1): 24–49.

Spallone, P. (1989) *Beyond Conception: the new politics of reproduction*. London: Macmillan.

—— (1999) 'How the pre-embryo got its spots'. Paper presented at 'Genetics and Genealogy' Conference, Potsdam, 4–6 July.

Stacey, M. (ed.) (1992) *Changing Human Reproduction: social science perspectives*. London: Sage.

Stormer, N. (2000) 'Prenatal space', *Signs*, 26: 109–44.

Theriot, N. (1993) 'Women's voices in 19th century medical discourse: a step toward deconstructing science', *Signs*, 19: 1–31.

Waldby, C. (2002) 'Stem cell, tissue cultures and the production of biovalue', *Health*, 6: 305–23.

Webster, A. (2002) 'Innovative health technologies and the social: redefining health, medicine and the body', *Current Sociology*, 50: 443–57.

Williams, C., Alderson, P. and Farsides, B. (2001) 'Conflicting perceptions of the fetus: person, patient, "nobody", commodity?', *New Genetics & Society*, 20: 225–38.

—— (2002) 'Too many choices? Hospital and community staff reflect on the future of prenatal screening', *Social Science and Medicine*, 55: 743–53.

13 'Natural forces'

The regulation and discourse of genomics and advanced medical technologies in Israel[1]

Barbara Prainsack

Societal turbulences: deliberating and regulating genomics and advanced medical technologies

In the last decades, debates on genomics[2] and advanced medical technologies have led to heated controversies throughout the Western world. Technologies such as embryonic stem cell (ESC) research and human cloning have instigated prolonged public debates in which sharp societal antagonisms and the fragility of the societal consensus on 'core values' have become visible. In most Western countries, an increasing number of experts from both the life sciences and the social sciences have started to discuss the ethical permissibility, the limits, and the potential benefits of these new research strategies. Many experts diagnose a certain kind of 'gut feeling' prevalent in many Western countries that there is something 'wrong' with these medical technologies, where they are perceived as being likely to change the meaning and the boundaries of our understanding of life. For example, Leon Kass, the former chairman of the US President's Council on Bioethics, regards 'revulsion' as 'the emotional expression of deep wisdom, beyond reason's power fully to articulate it' (Kass 1997). In many places it is perceived to be the task of governments to 'protect' individuals and individual rights (such as the 'rights' of the embryo, the rights of the mother-to-be, etc.) or societies from the dangers that are deemed to be inherent in these research strategies and new technologies.

The Israeli bioethics discourse stands in sharp contrast to bioethical discourses in other Western countries in two ways. First, by operating with the same terms (such as 'human dignity', 'sanctity of life', etc.) but often arriving at different conclusions, it threatens to unveil the particular Christian notion of the way supposedly 'secular' terms are used in bioethical debates in large parts of the Western world. Second, Israel's permissive approach towards genomics and advanced medical technologies in general seems to contradict the existence of a general 'gut feeling' that there is something wrong with research in fields such as ESC and human cloning. Some of the most emotionalised issues in other Western countries are simply not regarded as controversial in Israel, where the regulation of new medical technologies has never caused any large public debates. What often follows is the assumption of the absence of a moral debate on medical research and technologies in Israel, or of the existence of an 'immoral' bioethics discourse. My argument will be that this is not the case, but that the Israeli discourse is based on the discursive creation of a particular risk setting, as well as on a particular understanding of risk with regard to the regulation of genomics. This context of risk situates the unhindered advancement of medical research and technologies, as well as their use, in the context of a 'natural' process to grant the continuity of the collective. In order to understand the microphysics of power inherent in those risk settings, we need to turn our attention towards a field which is sometimes neglected by social science work on genomics: we need to look at politics.

Embracing life: genomics in Israel

In May 2001, the Prime Minister of the German *Land* of North Rhine-Westphalia, Wolfgang Clement, travelled to Israel to explore the possibilities of a potential future collaboration between the University of Bonn and the University of Haifa in embryonic stem cell research. What he had in mind was the importation of human ESC lines from Israel to Germany, because whereas Germany's *Embryonenschutzgesetz* (Embryo Protection Law) prohibits harvesting ESC in Germany, it does not entail any regulations with regard to the importation of ESC lines. Clement's endeavour did not only result in harsh criticism from the Christian churches but also from within his own Social Democratic party, as well as from large parts of the German public. The German weekly *Die Zeit* called the negotiations between Clement and the researchers in Haifa an 'explosive deal'[3] (Bahnsen 2001), and the Christian Democratic Party found Clement's conduct 'egregious' (*Spiegel online* 2001). How could Israel, the Germans wondered, a nation which has experienced the traumatising effects of inhuman conduct, of atrocious persecution and mass murder, be capable of 'disregarding human dignity' by allowing almost unrestricted research on human embryos (see Schnabel 2001)? After all, as (the then) German President Johannes Rau phrased it, '[i]f we deem something to be unethical, it is because it is unethical and immoral always and everywhere. With regard to fundamental ethical questions, there is no geography of the permitted and the prohibited' (Rau 2001). Israeli journalist Tamara Traubman joined the debate about the morality or immorality of Israeli bioethics with the question: 'So today the Germans are the moral ones and we're the Nazis?' (Traubman 2004).

Were the German media portrayals correctly conveying the message that there are many 'immoral' practices going on in Israel? Not only with regard to ESC research, but also concerning genetic testing and human cloning, Israel has one of the most permissive legal frameworks of all countries regulating its biotechnology. The main difference, though, lies at the level of discourse. It is precisely the public and semi-public discourses preceding and surrounding the regulation of genomics which are marked by a very positive attitude towards research and technologies that are extremely controversial in other parts of the world. Why do Israelis seem to embrace these newly emerging medical technologies, whereas they give people the shivers in other parts of the world? Why does the Israeli bioethics discourse differ so significantly from bioethics discourses elsewhere?

Israeli regulations: Human cloning and germ line therapy

While there are currently no laws explicitly dealing with so-called research cloning (or 'therapeutic cloning') in Israel, the 1999 Prohibition of Genetic Intervention Law (State of Israel 1999; revised in 2004) prescribed a five-year moratorium on:

1 Human cloning: the creation of a complete human being, chromosomally and genetically absolutely identical to another person or fetus, living or dead;
2 Germ line therapy: causing the creation of a person by use of reproductive cells that have undergone a permanent intentional genetic modification.

As far as the creation of a person 'by use of reproductive cells that have undergone a permanent intentional genetic modification' is concerned, the Minister of Health can, upon the recommendation of the advisory committee, permit certain types of genetic intervention, given that he or she is of the opinion that human dignity would not be prejudiced. Any person breaching the provisions of the law is liable to two years imprisonment.[4] As we have seen, the Prohibition of Genetic Intervention Law does not formulate provisions on research cloning.[5] In addition to this, the law should be seen in the context of a discussion which does not object to reproductive cloning *in principle*, but rather regards it as a matter of the right timing: '[T]here is no upfront condemnation of an entire area of scientific inquiry' (Revel 2005: 118). If the procedure is unsafe, and if society is not fully aware of its potential implications, the technology should be banned temporarily.

The Prohibition of Genetic Intervention Law was extended (with minor changes) for another five-year period in March 2004. This decision was preceded by a conflict between a group which coalesced around the Knesset committee's chairwoman, Melli Polishuk-Bloch of *Shinui*[6] on the one hand, and some academic experts involved in the Knesset committee deliberations on the other. While the group around Polishuk-Bloch was in favour of turning the temporary ban on reproductive cloning into a permanent one,[7] most of the scientists and philosophers who participated in the policymaking process preferred a more permissive regulation. One bioethicist explained that, whereas the preamble of the 1999 Prohibition of Genetic Intervention Law had stipulated that the five-year temporary ban of reproductive cloning should be used to learn more about the potential social and ethical implications of reproductive cloning, it had now turned out that there were no valid ethical or moral arguments against human reproductive cloning *per se*; consequently, there was no reason to come up with a permanent ban (interview with bioethicist).

Carmel Shalev, a legal scholar, feminist and former director of the Gertner Institute for Epidemiology and Health Policy Research in Tel Hashomer near Tel Aviv, in her background paper for the United Nations Ad Hoc Committee on an International Convention against Reproductive Cloning, asks:

> Why should particularly cloning not be included as a method of choice for infertility treatment? Even if undergone for health reasons – for example, in the case of a male carrying a dominant gene that affects 100 per cent of his offspring – why should we object?[8]
>
> (Shalev 2002: 5)

The Bioethics Advisory Committee of the Israeli Academy of Sciences and Humanities supports this view:

> Even reproductive cloning may one day be a safe technology for which there might be important individual medical implications ... Therefore, some may see the cloning of the embryo as part of the necessary research for such distant goals in the future.
>
> (IASH 2001)

The Israeli Academy of Sciences and Humanities, stem cells, and again: cloning

Currently, there is no explicit legislation on stem cell research in Israel.[9] In 2001, the Bioethics Advisory Committee of the Israeli Academy of Sciences and Humanities, a non-governmental body comprised of established Israeli scientists and scholars, issued a comprehensive report on the ethical implications of stem cell research. The report also included recommendations for legislation (IASH 2001). It aimed at balancing religious rationales, ethical and humanistic considerations and scientific objectives. Furthermore, it elaborated on different religious (Jewish, Christian, Muslim) views regarding the status of the embryo, and on provisions in other countries as well as on the international level.

Drawing upon Jewish Law, which does not ascribe human dignity to an embryo outside of the uterus (this will be discussed in more detail later), the use of surplus embryos from *in vitro* fertilisation (IVF) procedures, as well as the use of embryos created through somatic cell nuclear transfer for research purposes, is declared ethically permissible. IVF embryos suitable for implantation should not be used for research, unless it is clear that they will not be needed for IVF. Ova should not be fertilised solely for research purposes.[10] The Academy report is of no immediate legal relevance, but it attracted considerable attention in Israel and abroad.

At the level of practice, according to one of Israel's celebrities in stem cell research, Professor Josef Itskovitz-Eldor,[11] the sources of embryos used for ESC research in Israel are threefold: first, surplus IVF embryos are used with the informed consent of the donors. When surplus embryos are stored after an IVF cycle, the couple is routinely asked at certain intervals whether they want these surplus embryos stored, or whether the medical institution should discard them. If the couple chooses the latter option, explains Itskovitz-Eldor, only then will they be informed that there is also another possibility: donating the embryos to research. This procedure currently represents the main source of embryos for ESC research in Israel.

Second, Itskovitz-Eldor's hospital also uses 'abnormally fertilised oocytes', namely, ova that have been fertilised in a way that renders the resulting embryos unsuitable for implantation in an IVF procedure. One of these 'abnormal fertilisations' would be so-called polysperming, in which more than one sperm enters and fertilises the ovum and creates, instead of a regular embryo, an embryo with a double or triple set of DNA. According to Itskovitz-Eldor, stem cells obtained from such embryos are especially useful for medical research, since the resulting stem cell lines contain the same genetic defect as the embryo and thereby provide opportunities to explore diseases and test pharmaceuticals on the tissues grown from them.

The third source of embryos for ESC research are so called 'non-retrievable embryos' from pre-implantation diagnosis procedures (PGD). In this procedure, which always takes place within the framework of IVF treatment, fertilised embryos are examined prior to implantation and screened for particular disease carrier genes which run a high risk of being passed on to the couple's offspring. Only the embryos free from the disease carrier gene will be implanted in the woman's womb, while all others will be discarded (or used for research).

Research cloning is currently not performed in Israel, because the local Helsinki Committee has not yet approved any requests for performing this procedure. An additional obstacle represents the shortage of donor ova, which at this point can only be legally obtained from women undergoing IVF. Most of these women, explains Itskovitz-Eldor, need their ova for themselves. Scientists have been pushing for a legalisation of ova donation independent of IVF treatments (interview with Itskovitz-Eldor; see also Judy Siegel 2001).

Tissue banking and genetic testing

In general, tissue banks and biobanks (especially in the form of combining biological materials with personal and demographic data) are regarded very positively in Israel. In December 2002, the Bioethics Advisory Committee of the Israeli Academy of Sciences and Humanities issued a report on 'Population-Based Large-Scale Collections of DNA Samples and Databases of Genetic Information'. It suggests the establishment of 'publicly funded DNA collections and genetic databases … as a common resource for medical research in Israel. The public funding could include financial support from government and research granting funds in Israel, as well as philanthropic or charity funding' (IASH 2002: 7). However, the concept of privately funded and privately owned biobanks does not generally encounter hostility in Israel either. 'Money is a legitimate incentive', says Avinoam Reches, chairman of the Israeli Neurological Association, and Head of the Medical Ethics Section in the Israeli Medical Association (interview with Reches). A member of Israel's newly established National Bioethics Council explains the Israeli position as follows: '[Our] principle behind it [is]: Unlike Iceland or Estonia, we don't want the government to consider the DNA as a national resource. We want private companies to do this [to run large-scale biobanks, BP]' (interview with bioethicist). The Bioethics Advisory Committee of the Israeli Academy of Sciences and Humanities therefore suggests the creation of independent monitoring agencies (IASH 2002) to safeguard the compliance of biobanks with legal and ethical norms.

One publicly funded biobank, the so-called National Laboratory for the Genetics of Israeli Populations,[12] was established upon the initiative and with financial support of the Israeli Academy for Sciences and Humanities in 1994. The collection, which consists of more than 2,000 immortalised human cell lines from individuals in Israel, is located at the Sackler School of Medicine at Tel Aviv University. The objective of the biobank is to facilitate research on 'complex diseases … and studies comparing disease-associated factors in different ethnic backgrounds', as they are, according to the director of the laboratory, 'more likely to yield meaningful results' (Gurwitz *et al.* 2003: 96). Currently, the maintenance costs of the Laboratory are paid for by Tel Aviv University, and a small segment of the cost is covered by income through sample sales.

The only large-scale, privately funded and owned biobank which gained a lot of media attention was IDgene Pharmaceuticals Ltd, which aimed to discover the genetic basis of common diseases (diabetes mellitus types I and II, schizophrenia, Parkinson's, Alzheimer's, asthma, breast and colon cancer) in Ashkenazi Jews (about 2.5 million residents in Israel), and to develop diagnostic markers and pharmaceuticals. Although IDgene had to overcome a number of obstacles in the process of its establishment, these were procedural problems rather than obstacles related to a negative attitude of either governmental bodies or the bioethics community to DNA banking as such.

Furthermore, genetic testing is also widely embraced in Israel (see Hashiloni-Dolev 2004). The traditional reason for this phenomenon is that, due to the demographic composition of Israel's population, in particular the high rate of endogamy within Ashkenazi-Jewish communities, the incidence of inherited genetic diseases (such as Tay Sachs, cystic fibrosis, Fragile-X syndrome, Gaucher's disease, or breast cancer) is higher in Israel than in other parts of the Western world (Zhang *et al.* 2004). For example, the BRCA 1 mutation known as 185delAG is found in approximately 2–2.5 per cent of the Ashkenazi-Jewish population (Warner *et al.* 1999) and in 20 per cent of all Ashkenazi women who are afflicted by breast cancer before the age of forty (Fitzgerald *et al.* 1996) or forty-two (Offit *et al.* 1996). The alleged genetic 'homogeneity' of Ashkenazi Jews is commonly attributed to two phenomena: first, the 'founder effect', understood as the loss of genetic variation due to endogamy; and second, the so-called 'genetic drift', the inter-generational change of gene frequencies due to chance, instead of natural selection (not due to external stimuli). Ashkenazi Jews, who make up more than 80 per cent of world Jewry, are believed to descend from about 1,500 Jewish families dating back to the fourteenth century.

However, the higher prevalence of inheritable genetic diseases does not fully explain the large number of people who undergo genetic testing in Israel,[13] especially in and around the field of reproduction (Hashiloni-Dolev 2004). Israel has implemented screening policies[14] for various population sectors which are often conceptualised in terms of ethnic origin (Zlotogora and Leventhal 2000; Gross 2002; Zlotogora and Chemke 1995; Shahrabani-Gargir *et al.* 1998; Brownstein *et al.* 1991; Prainsack and Siegal 2006). These policies aim at both preventing the marriage between two carriers of the disease, and at diagnosing and aborting affected embryos and foetuses. Medical genetics is a recognised medical speciality in Israel, and eleven clinical genetic centres serve a population of only six million.

Leading Jewish ethicists are supportive of a wide use of screening programmes. For example, 'Rabbi J. David Bleich indicates that the elimination of Tay Sachs disease is, *of course*, a goal to which all concerned individuals subscribe' (Rosner nd, emphasis added). Indeed, (prenatal) screening programmes for Tay Sachs have been in place in Israel since 1986 (Broide *et al.* 1993). 'The discarding of the affected zygotes is not considered as abortion since the status of a fetus or a potential life in Judaism applies only to a fetus implanted and growing in the mother's womb' (Rosner nd). IDgene's founder, Ariel Darvasi, gives a similar picture when he refers to the initial phase of building up his population DNA collection: 'We have met with a few leading rabbis, and they were all very much in favour of what we are doing' (quoted from Joshua Siegel 2001).

'Natural forces': the necessity of keeping the collective body in check

How can our initial puzzle be solved? What makes the Israeli bioethics discourse so special? In the remainder of this chapter I will argue that the permissive Israeli approach towards regulating genomics is not rooted in the marginalisation of 'morality' in the bioethical discourse, but in a particular perception of risk, which transcends the field of genomics and bioethics and connects to a situation of violent conflict and general political crisis.

'Risk', a central concept in bioethics discourses, represents an important means of both governing and self-governing individuals. It is 'a strategy of making events and situations governable and introduces a calculative rationality for governing the conduct of individuals, populations, and collectives' (Gottweis 2003: 5; see also Gottweis 1998). It confronts 'patients' with a repertoire of 'factual problems' and tailored ways of solving them, thereby predetermining ways for individuals to constitute their own selves. The political aspects of risks include 'creating new dangers and "empowering" individuals to confront them', as Vaz and Bruno (2003: 282) put it. Medically suggested ways to avoid the materialisation of risk scenarios prompt people to adhere to particular regimes of self-surveillance and to stick to certain 'healthy' ways of life. Risk scenarios include both individuals and the total population. By 'learning' about our individual risks, we are led to establish our identities in accordance with what is necessary to minimise them. We adopt particular lifestyles and adapt the order of our value preferences accordingly. The perception of individual risks prompts us to translate the aim of risk reduction into our own actions and commitments, manifested in the ways we constantly invent and re-invent our bodies. Collective risks have a similar effect on us: we participate in the common endeavour of reducing the risks by incorporating this goal into our personal lives.

'Risk' with regard to genomics is perceived and understood in certain ways throughout the Western world: as a risk of being afflicted with a disease which is already 'discernible' in one's genes; as a risk of inheriting 'bad genes' from a parent; or as a risk of others infringing one's 'genetic privacy' by leaking information to third parties.[15] Within the framework of genomics, risk is the concept that connects the material body, through the concept of genes, enzymes and blood cells, to the immaterial concept of disease. A broader understanding of risk with regard to genomics typically concerns the risk that genomics poses to society.

In Israel, the understanding of risk in the context of genomics is very different. Before discussing this in more detail, I will first outline the traditional explanation for the permissive Israeli approach towards genomics and advanced medical technologies: religion.

Jewish religion explains a lot in regard to the particularities of the Israeli bioethics discourse ...

One reason for the permissive regulatory framework of biotechnology in Israel is that Jewish religion does not regard as morally challenging many technologies which are perceived as highly problematic in societies with a Christian cultural background. For example, in Jewish Law *ex utero* embryos are not regarded as comparable in any way to an implanted embryo, not to mention a fully fledged human being.[16] Genetic materials derive their legal status from belonging to a human body. If they do not belong to one, as it is the case with embryos in a Petri dish (which do not count as their own 'human body' either), their *halachic* status is comparable to that of gametes (sperm and oocytes), which need not be 'protected' from medical research.

Another feature of Jewish religion is that 'playing God' is not something which is regarded as reprehensible in general terms. Since human beings are created in God's image, they are not only entitled, but mandated to create. The capacity to participate in God's creation includes the command to improve it. In this context, applying changes to God's creation is regarded as legitimate, if it is done in a considerate and responsible manner.[17]

A central reason for the way in which Israelis have embraced medical technologies and research that could broaden fertility options is the immense importance of procreation in Jewish religion. The divine command to 'be fruitful and multiply' is regarded as binding for every male Jew, and it is often named the 'first *mitzvah*'[18] of the Torah. In cases where couples are carriers of disease genes, where Jewish ethics is concerned about the burden that such procedures can impose on individuals, genetic screening is the favoured option compared with not procreating at all (Rosner nd).[19]

The strong emphasis on procreation is paralleled by the principle of absolute sanctity of life in Judaism, which manifests in the imperative to protect and sustain (born) human life at almost any cost. The basis of the sanctity of life argument lies in the conviction that every individual is created in God's image. Life must be preserved, regardless of its quality (see also Cohen-Almagor 2001).

Acknowledgement of the importance of procreation and of the sanctity-of-life principle in Judaism contributes to our understanding of why genomics are so widely embraced in Israel. As mentioned above, Israel has the largest number of IVF clinics per capita in the world.[20] Judaism generally 'supports medical care and advancement if it can lead to improved health care' (which is also true, as a general principle, for orthodox communities; see Cox 2002: 15), as it traditionally puts strong emphasis on healing. If a non-living entity (such as an *ex utero* embryo) is capable of serving research which aims to find new cures for fatal diseases, then it is rather uncontroversial that this research should be conducted. In this respect, religion is rather an inducement for research than a barrier (interview with Shapira).

Indeed, the religious segment of Israeli society seems to be less insistent on a dense regulation of research than are secular people. As Tel Aviv University philosopher Asa Kasher notes, the scientific community has never had any problems getting along with orthodox Health Ministers (the current Israeli Health Minister, Yacor ben Yizri, belongs to the Pensioners' Party. Both his predecessors belonged to the ultra-religious *Shas* Party. Interview with Kasher).[21]

A lot ... but not everything

Although religion, as we have seen above, has considerable explanatory power for understanding the differences between Israeli bioethics discourses and bioethics discourses in other parts of the Western world, it cannot explain everything. Significantly, it does not account for the secular, non-religious segment of the Israeli population. Why are they not driven by the same concerns and objections against ESC research and human cloning as secular people in the United States and in continental Europe, where there are significant movements to halt that kind of research? And why is it the case that governmental regulations are in accordance with religious laws in fields such as ESC research and human cloning, whereas religious impera-tives are compromised in other bioethical fields?[22]

Contested Israeli bodies

In order to understand the particular context of risk in which the Israeli bioethics discourse is embedded, we also need to look at politics. Israel is in the midst of a political crisis, suffering from everyday bloodshed, from internal power conflicts, and from an increasing pressure from the inter-national community to compromise on its claims to land and sovereignty over all territories. The longer the conflict takes, the more difficult it becomes to justify the current status quo: Israel regards itself as a demo-cratic state and as a Jewish state at the same time. As long as Israel's population maintained a Jewish majority as an unchallenged fact, the insistence on maintaining the Jewish character of the state seemed legit-imate. However, population estimates predict that Jews will be out-numbered by non-Jews within another two decades (*Israel Today* online 2002).[23] Consequently there is a growing need to safeguard the Jewish nature of the state by maintaining that majority. The 'demographic crisis' has become a prominent topic in newspaper headlines, magazines, and prime-time television shows. According to a survey carried out by the Tami Steinmetz Center for Peace Research at Tel Aviv University, 67 per cent of the Israeli public 'strongly or moderately fear ... the scenario of a de facto binational state west of the Jordan' (Yaar and Hermann 2003: 2–3).[24] The image of a collective body in danger of 'annihilation' (*Israel Today* online 2002) was created. This Israeli collective body – the discursively created image of the collective entity (see Weiss 2002) – appears as a Jewish body, despite the fact that about 20 per cent of all Israeli citizens are non-Jews. Collective identities are created along the lines of 'us' – the Jews, and 'them' – the non-Jews. The latter are not part of the collective body. Interfaith marriages, as well as inter-religious adoptions, are legally vir-tually impossible (in the case of adoption, an exception is made for chil-dren from abroad who undergo conversion to Judaism when they are adopted by Israeli parents) (Kanaaneh 2002: 44; Israeli Adoption Law 1971). Even the Israeli law on surrogate motherhood upholds a strict line of separation between Jews and Arabs (State of Israel 1996).[25]

The effective policing of the boundaries of the Jewish collectivity (Kahn 2000: 72) is discursively framed as 'staying alive' as a nation and/or as a people, rather than as an endeavour in which 'racist' rationales are the ultimate end. Remaining Jewish, for the collective body, is a necessary means to its survival. In December 2002, Prime Minister Ariel Sharon declared *aliya*, Jewish immigration from the *diaspora*, to be 'Israel's most important and central goal'. Furthermore, Sharon declared that it was 'necessary to bring an additional 1 million Jews to Israel within 10–13 years', and that it was '*everyone's duty* to further the absorption of immigrants in the country' (State of Israel Cabinet Communiqué, 29 December 2002, emphasis added).

Achieving a high birth ratio in the Jewish population sector is another way of sustaining the collective body. In autumn 2002, orthodox Welfare Minister Shlomo Benizri (*Shas*) decided to reinstate the so-called Public Council for Demography and entrusted it with creating 'conditions that will make it easier for people to want to have children and ensure that Israel retains its Jewish character' (Arutz-7 2002).[26] As Susan Martha Kahn states in her book on *Reproducing Jews* in Israel, 'the overwhelming desire to create Jewish babies deeply informs the Israeli embrace of reproductive technology' (Kahn 2000: 3). Daniel Elazar also frames the demographic problem in terms of reproduction, summarising the words of one of Israel's leading demographers, Roberto Bacchi: '[T]he Jewish people as a whole [is] still suffering as a result of the loss of the reproductive capacity of a major segment of its population' (Elazar 1987).

The great emphasis which Jewish religion and culture places upon procreation has always fostered the goal of increasing the Jewish birth rate. Furthermore, the Israeli healthcare system facilitates access to assisted reproduction techniques for all citizens and subsidises the techniques generously. A particular system of child allowances, which had been in effect until the budget cuts of summer 2003, had raised the monthly income sharply for families with six or more children. As Meira Weiss diagnoses, 'Israeli society is still obsessed with fertility' (Weiss 2002: 2).

Genomics as a 'natural' necessity to sustain the collective body

In light of imminent danger to the existence of the collective body, one could assume that drastic means are being employed by the government in order to ensure its continuity. Interestingly, this is often not the case. Whilst, as indicated above, fostering immigration has been declared as Israel's prime goal,[27] 'adding' to the strength of the collective body by raising the birthrate (with the help of artificial reproduction techniques) and by securing its health (through modes such as genetic testing), are not governmentally imposed policies. What we find here is an intriguing combination of technologies of the self and technologies provided (but not prescribed) by the government. On the one hand, people create themselves and their identities in accordance with discursive truths ('we need more babies', 'we need to be strong') by internalising these truths and translating them into their own preferences and commitments (Rose 1996). In this sense, all actors involved in deliberating the legal framework of genomics and advanced medical technologies in Israel, as well as patients and clients, are 'governing themselves' more than they are governed by external actors.

On the other hand, government technologies complement the technologies of the self. Individual bodies, through the discursively created truth of the endangered health of the collective body, adopt patient behaviour and 'overuse' medical services and medical technologies (interviews with Shalev and Itskovitz-Eldor). Not only does Israel have the highest ratio of IVF clinics per capita in the world, *in vitro* fertilisation treatments are also largely subsidised by public health insurance (the cost of treatment is practically fully covered until the second live born child of the couple – or single woman – receiving treatment; see Rabinerson *et al.* 2002). A large variety of prenatal genetic testing technologies is performed (despite the fact that religious authorities prohibit the performance of abortions unless the health of the mother is in danger; see Raz 2004), and ESC research and human cloning for research are regarded as promising research fields which should not be inhibited. The presence of governmental techniques complementing the individual's quest for survival and procreation (on both an individual and a collective level) is well illustrated by the words of a senior worker in a women's health clinic in Tel Aviv: 'This is a government service. *The government is supplying the sperm* and we are given the mandate of deciding who is fit and who is not [through social and psychological evaluations of candidates for artificial insemination performed by a social worker and a psychiatrist or psychologist; BP]' (Kahn 2000: 28–9).

Members of the staff of fertility clinics are simultaneously the extended 'arms of the state' and the 'fathers', thereby providing the reproductive activities of single mothers for the collective with the attributes of the 'natural' sexual reproduction of a couple. As it should be 'natural' for people to protect the lives of their fellow human beings in case of sickness and disease, keeping the collective body alive is also discursively created as a 'natural' endeavour, almost a 'natural force'. In contrast to many other Western countries, where many perceive invasive and/or advanced medical technologies as an illegitimate interference with nature, in Israel it is the other way round. The advance of medical science and its use is seen as the 'natural course of events'. To hinder medical research would be perceived as inhumane and 'against nature'. This notion is very well conveyed by an Israeli legal expert – a woman – who was involved in the debates preceding the law on surrogate motherhood (for information on the law, see Benshushan and Schenker 1997). Whereas for some (with different cultural backgrounds), infertility might be seen as something which is given by God and should be accepted, or something that a couple should 'bypass' by adopting a child, to her the need to find a way for infertile couples to produce genetically related offspring seemed most obvious. The experience of the plight of infertile couples provoked the desire to help them by working out a way in which they could at least have some genetic relation to their child-to-be.[28] 'I saw their suffering, and I felt I had to do something', she told me. 'I thought it was unfair. I have kids of my own; why shouldn't they be able to have a child of their own, too?' (interview with legal expert). Other bioethicists and policymakers shared her perception and expressed it in similar terms: 'Isn't the ability to bear children the core content of humaneness?' (interview with bioethicist). 'Banning human cloning would be against human dignity' (interview with lawyer and bioethicist).

These examples direct us to yet another characteristic of the 'natural' duty of everybody to sustain and keep the collective body alive. My interviewee saw it as her 'natural duty' to help. Reproduction, as Weiss and other authors suggest, is not just a personal affair in Israel, it is also a collective matter. 'Collectivism became the "civil religion" of Israel, the larger frame of reference through which other issues and problems ... are all defined and accounted for' (Weiss 2002: 6). It is 'natural' to join the quest for the survival of the collective body by contributing one's own share, by submitting oneself to genetic testing and artificial reproduction procedures, and not do anything to hinder the advancement of medical research. Giving life to new individual bodies and finding cures for the sick assists the survival of the collective body. This is portrayed as a 'natural' activity, in terms of allowing nature to run its course, while *not* securing a high Jewish birth rate would be 'unnatural'. The term 'natural growth', which is used to refer to population growth due to births within particular settlements in the West Bank (and formerly in Gaza), is equally applicable to the condition of the general population: 'I don't call them settlements, but rather communities', declared Prime Minister Sharon, referring to settlements in the territories, and he went on: 'We also won't force young mothers to have abortions' (Arutz-7 2003).

In this light, advancing medical research and broadening options for reproduction are portrayed as 'natural' needs of the collective. As the protocol of a Knesset committee prior to issuing the Prohibition of Genetic Intervention Law summarises the concerns of Israeli scientists: 'Politicians must not, and cannot forbid science from advancing. ... Science will keep moving forward, *that is its nature*' (quoted from Ben-Or 2000: 764, emphasis added). The 'natural growth' of Israel's population is paralleled by the 'natural growth' of medical research.[29]

'Human dignity' and 'rationality' are the discursive instruments through which, not only religious imperatives, but also the need to respond to the demographic threat translate into the imperative 'to help infertile couples' and single women to create genetically related offspring, and to engage in medical research that uses human embryos to find cures for diseases. Furthermore, the 'natural need' for individuals to procreate and for societies to advance medical research cannot be legitimately banned. Sabotaging this 'natural need' is wrong, 'irrational' and immoral. Therefore Haggai Meirom, who initiated a law proposal to the Knesset which aimed at banning all forms of human cloning in Israel, was told that 'he was prohibited to interfere with science' (interview with policymaker). Interpreted in this context, Meirom's 'problem' was his failure to acknowledge these natural needs of individuals and society and to acknowledge that nobody was entitled to interfere with it. In his initial law proposal, Meirom had failed to make 'healthy decisions' for both individuals and the collective. When Meirom was 'made aware' that he was 'entering the field of science' (interview with bioethicist), he finally agreed to 'correct' his proposal (interview with policymaker; interview with Meirom) and enable a temporary ban of the narrowly defined technique of reproductive cloning through somatic cell nuclear transfer.[30]

Conclusion

The Israeli discourse on genomics and advanced medical technologies goes beyond formulating the claim that Jewish religion simply does not pose as many obstacles to medical research as other religions do. The centre of the Israeli discourse is not the claim that religion does not contradict science. What we observe here is the existence of a *non-controversy*: all narratives, religious and political ones, serve the same goal, maintaining the continuity of the collective body which is under threat. The permissive attitude towards genomics and advanced medical technologies in general in Israel cannot be explained by the absence of a moral discourse in this field, but rather by a different discursive framing of risk. Risk is more related to a notion of a Jewish collective body at risk of 'annihilation' (*Israel Today* online 2002) – a risk that affects the collective entity of Israeli Jews – than, for example, the individual risk of having one's 'genetic privacy' infringed by society. The notion of individual risk of becoming afflicted with disease is embedded in the notion of 'sickness' of, and danger to, the collective body. Individual activities taken to reduce risks to their individual bodies is supported by the imperative to reduce risks to the collective. This leads to a situation in which scepticism towards technologies such as genetic testing, stem cell research, and even reproductive cloning is simply 'inconceivable'. The idea of the existence of a substantial conflict between religious or ethical considerations on the one hand, and medical objectives on the other, is not traceable in the discourse; it can be heard, but it cannot be understood. What *can* be observed is the discursive creation of the application and use of genomics and advanced medical technologies as a 'natural force': it is 'natural' to do what is necessary to 'sustain life'. The particular understanding of what kind of life it is, to which collective entity it is attached to, and which materiality or non-materiality it consists of, is where politics comes into play. Life is attached to the collective Jewish body which must be kept alive in the midst of political pressure and violent conflict. In this sense, the Israeli bioethics discourse is no less 'moral' than bioethics discourses elsewhere: and it is no less political.

Notes

1 This chapter is the result of a discourse-analysis approach to data obtained in several rounds of interviews with about twenty-five bioethicists and policy-makers in Israel in the years 2001–5, and to governmental materials and semi-official documents. I am grateful to my interviewees and colleagues in Israel, without whom my research would have been a 'mission impossible'. Whereas I often give the full names of my interviewees, I 'anonymised' data wherever I felt that exposing the sources of my information would put them in a vulnerable position. Israel's bioethics community is very small, and people can be identified very easily if attributes such as location, academic discipline, etc. are provided. I am also grateful to Hendrik Wagenaar for his insightful and helpful comments.

2 'Genomics' is understood here not only as the study and manipulation of the sequence and function of genes, and of their interaction with each other and with other factors, but in a broader sense. It signifies genetic and genomic research and applications in the genomic era (which began with the Human Genome Project in the early 1990s). The scope of this chapter is limited to human genomics. For a discussion of the regulation of both human and agricultural genomics and biotechnology in Israel, see Prainsack and Firestine (2006).

3 All translations were made by the author, if not indicated otherwise.

4 The maximum penalty has gone up from two to four years of imprisonment in the 2004 version of the law.

5 A close reading of the 1999 version of the Prohibition of Genetic Intervention Law reveals that even reproductive cloning was only prohibited if the donor of the ovum was identical with the donor of the genetic material that was to be cloned. This resulted from the term 'human cloning' being defined as 'the creation of a complete human being, chromosomally and *genetically* absolutely identical to another person or fetus, living or dead' (emphasis added). Because a donor ovum would contain mitochondrial DNA which would add to the genetic information of the other donor's somatic cell nucleus if the donors were not identical, the cloned entity would not be 100 per cent *genetically* identical to any already existing person or fetus, and could therefore be created lawfully. In the 2004 version of the law, the problem was solved by deleting the word 'absolutely' when referring to genetical identity. Furthermore, the phrase 'creation of a complete human being' was replaced by a description of cloning process. For an English translation of the 2004 version of the law, see http://www.academy.ac.il/bioethics/english/documents/bioethics_law.htm (accessed 14 September 2005).

6 *Shinui* (literally: 'change') describes itself as a 'liberal, Zionist, Democratic Party, dedicated to creating a modern and pluralistic Israel' (http://www.shinui.org.il/elections/eng/). The use of the word 'Zionist' here is no indication of a 'hawkish' position in the Israeli–Arab conflict.

7 Most activists in favour of a permanent ban on cloning were driven by feminist and social justice considerations (for example, the fear of exploitation of women due to the need for ova; the fear of funding money going into cloning research which should be spent on curing diseases). A bioethicist 'from the other side', favouring only a temporary ban on reproductive cloning, commented on the arguments of the group around Polishuk-Bloch as follows: 'Things like the image of women as having to bear children, etc., [are] issues of interest but ones that are far from having the power of justifying a ban on research or treatment' (personal e-mail communication with bioethicist 2004). Polishuk-Bloch herself, as anecdotal evidence suggests, was primarily afraid of the international scientific community pointing their fingers at Israel if it maintained its permissive regulation. Furthermore, Polishuk-Bloch's party,

Shinui, is known for its anti-religious agenda, which sometimes manifests itself as an anti-religious 'reflex': everything propagated by religious parties and groups, such as maintaining the temporary instead of a permanent ban on cloning, is intuitively rejected by *Shinui*.

8 Meanwhile, Shalev has become more sceptical towards human reproductive cloning, largely drawing upon Habermas's reasoning (Habermas 2003). Half-jokingly, she admits that several longer periods of working in the US might have influenced her approach (personal communication with Shalev 2004).

9 Therefore, ESC research can be, and currently is, legally performed (for details, see Revel 2005). The only requirement is the existence of an approval of research designs by the Supreme Helsinki Committee in the Ministry of Health.

10 Although the Academy report does not mention it explicitly, the main reason for this is the *halachic* (*halachah*: Jewish Law) prohibition of using sperm for any purpose other than procreation (interview with Halperin; see also Feldman 1968, chapter 6: 'Improper Emission of Seed').

11 Itskovits-Eldor is the director of the Department of Obstetrics and Gynecology at the Rambam Medical Center in Haifa, and a prominent figure in stem cell research. He was involved in the isolation of the first stem cell line in 1998 (Thomson *et al.* 1998).

12 The Laboratory's website can be found at: http://www.tau.ac.il/medicine/NLGIP/nlgip.htm (accessed 22 September 2005).

13 For a discussion of the claim of a 'pathologisation' of Ashkenazi Jews due to an overrepresentation in genetic research, see Birenbaum Carmeli (2004) and Lehrman (1997).

14 The National Health Law of 1994 stipulates that all Israeli citizens are eligible for a particular range of free genetic testing and counselling services. Three kinds of genetic tests and counselling services are provided: first, amniocentesis for women who are at least thirty-five years old at the beginning of the pregnancy; second, carrier detection screening for Tay Sachs; and third, thalassaemia screening for high-risk groups (for more details, see Raz 2004). Furthermore, the number of 'private screening programmes' (paid for by patients) and the number of their users keeps increasing (Zlotogora and Leventhal 2000). It is expected that genetic testing for more than twenty-five common diseases will be widely available in Israel in the not too distant future (Shohat 2000).

15 For a discussion of 'risk commonly associated with genetic research', see Chadwick and Berg 2001.

16 As we have seen in Note 10, the only *halachic* problem is with fertilising ova with sperm explicitly for research purposes.

17 For a discussion of the role of 'playing God' in Jewish religion, see Heyd 1992 and Wahrman 2002.

18 A *mitzvah* (plural: *mitzvot*) is a divine commandment. Six hundred and thirteen *mitzvot* are derived from the Torah.

19 In the case of genetic testing, there is also another reason for the permissive attitude of Jewish religion: as Rabbi Azriel Rosenfeld argues, no process invisible to the naked eye is forbidden by Jewish Law. Since the manipulation of tissue on the genetic and molecular level entails such procedures, it can be concluded that they are not prohibited (Rosenfeld 1979).

20 Namely, one IVF clinic to every 250,000 inhabitants, which is four times as many as in the United States (Heyd 1993; Schenker 2003).

21 Nevertheless, '*the* Jewish approach' towards genomics and advanced medical technologies should be seen as an assembly of various voices rather than as a monolithic entity. Jewish ethics is not a matter of black and white, of either/or, but it always entails a weighing of interests which remains open to change whenever new necessities arise and/or circumstances change. Whereas this short

overview of religious teachings relevant to ESC research, cloning, and genetic testing represents the positions held by the (Jewish) Israeli majority to the best of my knowledge, there are also Rabbis who voice scepticism towards cloning techniques (for a discussion, see Golinkin 2001; Broyde 2001). There is also a segment within the 'ultra'-orthodox Jewish population which rejects assisted medical reproduction techniques in general, although they are a tiny minority.

22 One example of a divergence between *halachic* laws on the one hand and policies on the other is the regulation of end-of-life care. In 2002, the Steinberg Committee on the Dying Patient presented a report to the Minister of Health, in which it suggested prohibiting active euthanasia but allowing physicians to end 'needless' suffering of dying patients in certain cases through the use of articulated advance directives and ethics committees consultations (Siegel 2002; Ravitsky 2005; Amidror and Leavitt 2005). It thereby compromised the religious imperative to protect and sustain life regardless of its 'quality'.

23 Drawing from a study conducted by University of Haifa professor Arnon Sofer, the non-Jewish population in Israel plus Gaza and the West Bank is expected to have outnumbered the Jewish population in the same area by 2020 (8 million non-Jewish Palestinians in contrast to 6.6 million Jews) (*Israel Today* online 2002). Also within the pre-1967 borders (within the so-called 'Green Line'), the development is pessimistic for the Jewish majority: Arnon Sofer released demographic data claiming that the current population of more than 5 million Jews and 1.2 million Arabs will change to a ratio of 6.6 million Jews to 2.1 million Arabs (Muslims, Christians, Druze) in Israel 'proper'. Currently, up to 28 per cent of Israeli citizens are not Jewish. By 2020, a third of the total population could consist of non-Jews (Beaumont 2002). The Arab birth rate in Israel is approximately 2.5 times that of the Jewish birth rate, and the difference is steadily increasing (Harel 2002).

24 The survey (carried out in October 2003) contained a total of 574 interviewees (representative for Israel's population); sample error was identified by the researchers as 4.5 per cent in each direction.

25 Chapter B, 2.5 of the law states: 'The carrying mother is of the same religion as the intended mother; however, if all the parties to the agreement are not Jewish, the committee [which decides on surrogacy application on a case by case basis; BP] is entitled to deviate from the provisions of this clause in accordance with the opinion of the religious representative on the committee (of that religion)' (State of Israel 1996, quoted from Kahn 2000: 142–3).

26 *Arutz-7* is a national-religious radio 'pirate' station in Israel that has been declared illegal by the Israeli High Court of Justice. Nevertheless, *Arutz-7* still legally runs a daily electronic newsletter, from which the information referred to in the article has been taken.

27 The arrival of new Jewish immigrants from abroad is situated in a context of 'natural growth'. For example, a radio commentator (of a national-religious radio station) articulates his joy over the arrival of 250 new citizens at Ben Gurion Airport in Tel Aviv as follows: 'I have to say, I've been at these flights a few times, and it gets more exciting each time ... – with regular people, a little haggard after their flight, making it happen right here and now. It's like they are little *seeds* who have been kept safe for 2,000 years, only now to be planted in the soil of the Holy Land' (*Arutz-7* 2004. Emphasis added). The same argument of 'naturalness' can be found in numerous debates on demographic and political issues. An editorial in the *Jerusalem Post* opposes the redivision of Jerusalem because '[c]ities, unlike other political entities, *grow organically*, and to sunder one part from another is like sawing a branch off a tree, or a leg off a man. It's not something that, *in the natural course of things*, ought to happen' (*Jerusalem Post*, 10 June 2004. Emphasis added).

28 According to the Israeli surrogate motherhood law, the sperm must come from the husband, whereas ovum donation can be authorised in special cases. The egg, however, must not stem from the surrogate mother. For more information, see Weisberg (2005).
29 The emphasis on the 'naturalness' of the advancement of medical and bio-technological research should also be interpreted as a response to accusations from abroad that Israeli policies supported fields of research and technology which were allegedly 'unnatural' or even 'counter-natural', such as embryonic stem cell research, human cloning, or the intense use of medical technologies for reproduction.
30 For a more detailed discussion of the deliberation of the 'Prohibition of Genetic Intervention' Law in Israel, see Prainsack (2006).

References

Amidror, T. and Leavitt, F. J. (2005) 'End-of-life decision making in Israel', in R. H. Blank and J. C. Merrick (eds), *End-of-Life Decision Making. A Cross-National Study*. Cambridge, MA: MIT Press.
Arutz-7 (2002) 'Keeping Israel Jewish' (16 October). Available online at http://www.israelnationalnews.com (accessed 14 September 2005).
Arutz-7 (2003) 'PM Sharon: "I don't call them settlements"' (12 May). Available online at http://www.israelnationalnews.com (accessed 14 September 2005).
Arutz-7 (2004) '2,000-year-old seeds planted in the soil of the Holy Land' (4 August). Available online at http://www.israelnationalnews.com (accessed 14 September 2005).
Bahnsen, U. (2001) 'Heiße Ware aus Haifa' (Hot goods from Haifa), *Die Zeit*, 24: 31.
Beaumont, P. (2002) 'Israel fears invasion of immigrants', *The Observer* (Sunday, 16 June). Available online at http://observer.guardian.co.uk/worldview/story/0,11581,738548,00.html (accessed 14 September 2005).
Ben-Or, G. (2000) 'The Israeli approach to cloning and embryonic research', *Heidelberg Journal of International Law*, 60(3–4): 763–70.
Benshushan, A. and Schenker, J. G. (1997) 'Legitimizing surrogacy in Israel', *Human Reproduction*, 12: 1832–4.
Birenbaum Carmeli, D. (2004) 'Prevalence of Jews as subjects in genetic research: figures, explanation, and potential implications', *American Journal of Medical Genetics*, 130A: 76–83.
Broide, E., Zeigler, M., Eckstein, J. and Bach, G. (1993) 'Screening for carriers of Tay-Sachs disease in the ultraorthodox Ashkenazi Jewish community in Israel', *American Journal of Medical Genetics*, 47(2): 213–15.
Brownstein, Z., Friedlander, Y., Peritz, E. and Cohen, T. (1991) 'Estimated number of loci for autosomal recessive severe nerve deafness within the Israeli Jewish population, with implications for genetic counselling', *American Journal of Medical Genetics*, 41(3): 306–12.
Broyde, M. J. (2001) 'Cloning people and Jewish law: a preliminary analysis'. Available online at http://www.jlaw.com/Articles/cloning.html (accessed 14 September 2005).
Chadwick, R. and Berg, K. (2001) 'Solidarity and equity: new ethical frameworks for genetic databases', *Nature Reviews Genetics*, 2(4): 318–21.

Cohen-Almagor, R. (2001) *The Right to Die with Dignity. An Argument in Ethics, Medicine, and Law*, New Brunswick, NJ: Rutgers University Press.

Cox, E. J. (2002) 'Religious ethics of BRCA testing in the Orthodox Jewish community'. Master's thesis. Sarah Lawrence College, Bronxville, NY.

Elazar, D. J. (1987) 'Backing into a Jewish majority in Israel'. Paper at the Jerusalem Center for Public Affairs website. Available online at http://www.jcpa.org/dje/articles2/majority.htm (accessed 14 September 2005).

Feldman, D. M. (1968) *Marital Relations, Birth Control, and Abortion in Jewish Law*, New York: New York University Press.

Fitzgerald, M. G., MacDonald, D. J., Krainer, M., Hoover, I., O'Neil, R., *et al.* (1996) 'Germ-line BRCA 1 mutations in Jewish and non-Jewish women with early onset breast cancer', *New England Journal of Medicine*, 334: 143–9.

Golinkin, D. (2001) 'The case against cloning humans', *The Jerusalem Post Online Edition* (25 March). Available online at www.jpost.com (accessed 14 September 2005).

Gottweis, H. (1998) *Governing Molecules: the discursive politics of genetic engineering in Europe and in the United States*. Cambridge, MA: MIT Press.

—— (2003) 'Embryos for Europe: emerging strategies and institutional solutions'. Paper presented at the European Consortium for Political Research (ECPR) Joint Session of Workshops, Edinburgh, 28 March to 2 April.

Gross, M. L. (2002) 'Ethics, policy, and rare genetic disorders: the case of Gaucher disease in Israel', *Theoretical Medicine and Bioethics*, 23(2): 151–70.

Gurwitz, D., Kimchi, O. and Bonné-Tamir, B. (2003) 'The Israeli DNA and cell line collection: a human diversity repository', in B. M. Knoppers (ed.), *Populations and Genetics: legal and socio-ethical perspectives*. Leiden: Koninklijke Brill.

Habermas, J. (2003) *The Future of Human Nature*. Cambridge: Polity Press.

Harel, Y. (2002) 'The demographic apocalypse' (11 March). Available online at http://www.kahane.org/oped/mar/13.htm (accessed 14 September 2005).

Hashiloni-Dolev, Y. (2004) 'Looking for the perfect child: fertility control and genetic management in Germany and Israel'. PhD thesis. Tel Aviv University, Israel.

Heyd, D. (1992) *Genethics. Moral Issues in the Creation of People*. Berkeley, CA: University of California Press.

—— (1993) 'Artificial reproductive technologies: the Israeli scene', *Bioethics*, 7(2–3): 263–70.

IASH (The Bioethics Advisory Committee of the Israeli Academy of Sciences and Humanities) (2001), *Report of the Bioethics Advisory Committee of the Israeli Academy of Sciences and Humanities on The Use of Embryonic Stem Cells for Therapeutic Research*, (August). Available online at http://www.academy.ac.il/bioethics/english/PDF/Embryonic_Stem_Cells.pdf (accessed 14 September 2005).

—— (2002) *Population-Based Large-Scale Collections of DNA Samples and Databases of Genetic Information*, (December). Available online at http://www.academy.ac.il/bioethics/english/report2/content-e.html (accessed 14 September 2005).

Israel Today online (2002) 'Israel losing the demographic race', (24 October). Available online at http://www.israeltoday.co.il (accessed 14 September 2005).

Jerusalem Post (2004) 'Divide Jerusalem?', (10 June). Available online at http://www.jpost.com (accessed 22 September 2005).

Kahn, S. M. (2000) *Reproducing Jews. A Cultural Account of Assisted Conception in Israel*. Durham, NC: Duke University Press.

Kanaaneh, R. A. (2002) *Birthing the Nation. Strategies of Palestinian Women in Israel*. Berkeley, CA: University of California Press.

Kass, L. (1997) 'The wisdom of repugnance: why we should ban the cloning of humans', *The New Republic*, 216(22): 17–26.

Lehrman, S. (1997) 'Jewish leaders seek genetic guidelines', *Nature*, 389: 322.

Offit, K., Gilewski, T., McGuire, P., Schluger, A., Hampel, H., Brown, K., Swensen, J., Neuhausen, S., Skolnick, M., Norton, L. and Goldgar, D. (1996) 'Germline BRCA 1 185delAG mutations in Jewish women with breast cancer', *Lancet*, 347(9016): 1643–5.

Prainsack, B. (2006), '"Negotiating life": the regulation of human cloning and embryonic stem cell research in Israel', *Social Studies of Science*, 36 (2): 173–205.

Prainsack, B. and Firestine, O. (2006) '"Science for survival": biotechnology regulation in Israel', *Science and Public Policy*, 33(1) 33–46.

Prainsack, B. and Siegal, G. (2006) 'The rise of genetic couplehood? A comparative view of premarital genetic testing', *BioSocieties*, 1: 17–36.

Rabinerson, D., Dekel, A., Orvieto, R., Feldberg, D., Simon, D. and Kaplan, B. (2002) 'Subsidised oocyte donation in Israel (1998–2000): results, costs and lessons', *Human Reproduction*, 17(5): 1404–6.

Rau, J. (2001) '"Wird alles gut?" Für einen Fortschritt nach menschlichem Maß' ('Is everything going to be ok?' In favour of progress according to human standards). Speech given in Berlin, Otto-Braun-Saal der Staatsbibliothek, 18 May (in German). Transcipt available online at http://www.kna.de/doku_aktuell/rau_berliner_rede_2001.pdf (accessed 14 September 2005).

Ravitsky, V. (2005) 'Timers on ventilators', *British Medical Journal*, 330: 415–17.

Raz, A. (2004) '"Important to test, important to support": attitudes towards disability rights and prenatal diagnosis among leaders of support groups for genetic disorders in Israel', *Social Science & Medicine*, 59(9): 1857–66.

Revel, M. (2005) 'Ethical issues of human embryo cloning technologies for stem cell research', in S. Blazer and E. Z. Zimmer (eds), *The Embryo: scientific discovery and medical ethics*, Basel: Karger.

Rose, N. (1996) *Inventing Our Selves. Psychology, Power, and Personhood.* Cambridge: Cambridge University Press.

Rosenfeld, A. (1979) 'Judaism and gene design', in F. Rosner and D. Bleich (eds), *Jewish Bioethics*. New York: Sanhedrin Press.

Rosner, F. (nd) 'Judaism, genetic screening, and genetic therapy'. Available online at http://www.us-israel.org/jsource/Judaism/genetic.html (accessed 14 September 2005).

Schenker, J. G. (2003) 'Legal aspects of art practice in Israel', *Journal of Assisted Reproduction and Genetics*, 20(7): 250–9.

Schnabel, U. (2001) 'Ohne Mutter keine Menschenwürde. Die Reproduktionsmedizin steht in Israel hoch im Kurs, die Zusammenarbeit mit Deutschland auch' (No human dignity in absence of a mother. Reproductive medicine is highly valued in Israel, as is collaborating with Germany), *Die Zeit*, 24 (in German).

Shahrabani-Gargir, L., Shomrat, R., Yaron, Y., Orr-Urtreger, A., Groden, J. and Legum, C. (1998) 'High frequency of a common Bloom syndrome Ashkenazi mutation among Jews of Polish origin', *Genetic Testing*, 2(4): 293–6.

Shalev, C. (2002) 'Reproductive cloning: a human rights framework'. Background paper to the United Nations Ad Hoc Committee on an International Convention against the Reproductive Cloning of Human Beings (25 February). New York.

Available online at http://www.biopolitics-berlin2003.org/docs.asp?id = 77 (accessed 14 September 2005).

Shohat, M. (2000) 'The future of genetics: where are we going in the next forty years?', *Israel Medical Association Journal*, 2: 690–1.

Siegel, Joshua (2001) 'IDgene plumbs secrets of Ashkenazi gene pool – a population more useful for study than Icelanders?', *The Forward*. Available online at http://www.israelseed.com/press/press_news_detail.asp?id = 13 (accessed 14 September 2005).

Siegel, Judy (2001) 'Health minister approves bill aimed at reducing shortage of donated ova', *The Jerusalem Post online edition*, (20 March). Available online at http://www.jpost.com (accessed 14 September 2005).

—— (2002) 'Panel draws up guidelines for ending suffering of terminally ill', *The Jerusalem Post online edition*, (20 January). Available online at http://www.jpost.com (accessed 14 September 2005).

Spiegel online (2001), 'Import von Embryo-Zellen: Union und Kirchen machen Front gegen Clement' (Import of embryonic cells: [Christian Democratic] Union and churches oppose to Clement) (in German). Available online at http://www.spiegel.de/politik/deutschland/0,1518,137641,00.html (accessed 14 September 2005).

State of Israel (1996) *Embryo Carrying Agreements Law*. Excerpts in English are available online at the WHO International Digest of Health Legislation website: http://www3.who.int/idhl-rils/frame.cfm?language=english (accessed 14 September 2005).

—— (1999) *Prohibition of Genetic Intervention (Human Cloning and Genetic Manipulation of Reproductive Cells) Law, 5759–1999*. An English translation of the revised (2004) version of the law is available online at http://www.academy.ac.il/bioethics/english/documents/bioethics_law.htm (accessed 14 September 2005).

State of Israel (2002) State of Israel Cabinet Communiqué, (29 December). Available online at www.mfa.gov.il/mfa/go.asp?MFAH0mw70 (accessed 14 September 2005).

Thomson, J. A., Itskovits-Eldor, J., Shapiro, S. S., Waknitz, M. A., Swiergiel, J. J., Marshall, V. S. and Jones, J. M. (1998) 'Embryonic stem cell lines derived from human blastocysts', *Science*, 282: 1145–7.

Traubman, T. (2004) 'A perfect baby', *Ha'aretz*, (7 July). Available online at http://www.haaretz.com (accessed 14 September 2005).

Vaz, P. and Bruno, F. (2003) 'Types of self-surveillance: from abnormality to individuals "at risk"', *Surveillance & Society*, 1(3): 272–91.

Wahrman, M. Z. (2002) *Brave New Judaism: when science and scripture collide*. Hanover, MA: Brandeis University Press.

Warner, E., Foulkes, W., Goodwin, P., Meschino, W., Blondal, J., *et al.* (1999) 'Prevalence and penetrance of BRCA1 and BRCA2 gene mutations in unselected Ashkenazi Jewish women with breast cancer', *Journal of the National Cancer Institute*, 91(14): 1241–7.

Weisberg, D. K. (2005) *The Birth of Surrogacy in Israel*. Gainesville, FL: University Press of Florida.

Weiss, M. (2002) *The Chosen Body: the politics of the body in Israeli society*. Stanford, CA: Stanford University Press.

Yaar, E. and Hermann, T. (2003) Peace Index, October 2003, The Tami Steinmetz Center for Peace Research. Available online at http://spirit.tau.ac.il/socant/peace/peaceindex/2003/files/oct2003e.doc (accessed 13 June 2006).

Zhang, B., Dearing, L. and Amos, J. (2004) 'DNA-based carrier screening in the Ashkenazi Jewish population', *Expert Review of Molecular Diagnostics*, 4(3): 337–92.

Zlotogora, J. and Chemke, J. (1995) 'Medical genetics in Israel', *European Journal of Human Genetics*, 3(3): 147–54.
Zlotogora, J. and Leventhal, A. (2000) 'Screening for genetic disorders among Jews: how should the Tay-Sachs screening be continued?', *Israel Medical Association Journal*, 2(9): 665–7.

Interviews

Dr Ariel Darvasi, founder of IDgene, conducted on 4 April 2004 in Jerusalem (followed up on several occasions during 2004–5).

Rabbi Dr Mordechai Halperin, director of the Schlesinger Institute for Medical Halachic Research, Shaare Zedek Medical Center, Jerusalem, and of the Unit for Medical Ethics in the Israeli Ministry of Health, conducted on 2 September 2002 in Jerusalem.

Professor Josef Itskovits-Eldor, director of the Department of Obstetrics and Gynecology at the Rambam Medical Center in Haifa, conducted on 4 August 2003 in Haifa.

Professor Dr Asa Kasher, The Laura Schwarz-Kipp Chair in Professional Ethics and Philosophy of Practice, Tel Aviv University, conducted on 5 September 2002 in Kiryat Krinizi (followed up on several occasions during 2003–5).

Attorney Haggai Meirom, former Member of Knesset (12th, 13th and 14th Knesset, 1988–99), conducted on 11 August 2003 in Tel Aviv.

Professor Dr Avionam Reches, Neurologist at the Hadassah Hebrew University Medical Center in Jerusalem, Chairman of the Israeli Neurological Association and Head of the Medical Ethics Section in the Israel Medical Association, conducted on 31 March 2004 in Jerusalem.

Dr Carmel Shalev, former director of the Unit for Health Rights and Ethics, Gertner Institute for Epidemiology and Health Policy Research, conducted on 1 September 2002 in Tel Hashomer (followed up on several occasions during 2003–5).

Professor Dr Amos Shapira, The Buchmann Faculty of Law, Tel Aviv University, conducted on 1 September 2002 in Tel Aviv.

Anonymous interviews: all conducted in Israel between 2002 and 2005.

14 Survival of the gene?

21st-century visions from genomics, proteomics and the new biology

Ruth McNally and Peter Glasner

In *The Century of the Gene*, Evelyn Fox Keller argues that the Human Genome Project (HGP) has undermined the very conceptual foundations on which it was predicated, making it impossible to ignore the gap between the reductionism of the gene-centred paradigm and the complexity of living organisms (Keller 2000). As a consequence she suggests that the gene may be a concept past its time. Looking to the future, she predicts that the primacy of the gene as a core explanatory concept of biological structure and function was more a feature of the twentieth century than it will be of the twenty-first, and she anticipates the emergence of a new lexicon for the life sciences.

In the years since the publication of Keller's book, the HGP has been formally completed, several new global research initiatives, including the International HapMap Project and the Human Proteome Project, have been started, and a number of events and publications have been organised where leading scientists have communicated their visions of the future of the biosciences. A recurrent feature of these communications is their use of revolutions, exemplars, new eras and other features of 'paradigm talk'. This chapter examines whether, at this admittedly early stage of the twenty-first century, such paradigm talk is evidence of Keller's predicted demise of the gene. Taking its cue from Keller's focus on language, the method of analysis is based on close attention to words used (and not used) and the meanings attached to them.

The structure of the chapter is as follows. The first section is a brief etymological overview of the twentieth century from the perspective of genes and DNA, while the following section supplements this history with information about the interplay of DNA and proteins. The purpose of these two sections is to provide background information for the next three sections, which analyse three contrasting visions of the future: the genomic era; from genomics to proteomics; and the new biology. The final section compares and contrasts these visions and reflects upon Keller's thesis of the fate of the gene in the twenty-first century.

From the century of the gene to the omic era

> With 35,000 genes and hundreds of thousands of protein states to identify, correlate, and understand, it is no longer sufficient to rely on studies of one gene, gene product, or process at a time. We have entered the 'omic' era in biology.
>
> (Weinstein 2001)

According to Evelyn Fox Keller, the twentieth century was the century of the gene, a period neatly flanked by the 'rediscovery' of Mendel's principles of heredity at the beginning, and the completion of the first rough draft of the human genome at the end (Keller 2000; see also Moss 2004). The term 'genetics' was coined in 1906 by William Bateson, the 'apostle of Mendelism in England', to name the science whose goal is to resolve two apparently contradictory observations – that organisms not only resemble their parents, they also differ from them. 'Gene' was coined by the Danish biologist Wilhelm Johannsen in 1909 as a 'little word' to refer to the unit of heredity (Keller 2000: 2). Using the gene as the unit of analysis, genetics attempts to explain both the constancy of inheritance and its variation (Magner 1979). By adding 'ome' to 'gene' in 1920, Hans Winkler created the word 'genome' (or *genom*), defined as 'the haploid chromosome set, which, together with the pertinent protoplasm, was said to specify the material foundation of the species' (Lederberg and McCray 2001).

There are various interpretations of the etymological source of the suffix 'ome' in genome. In one version, it is attributed to the 'ome' from chromosome (coloured body). However, the oldest 'ome' in the dictionary is the 'biome', an ecological community of organisms and environments, which was coined in 1916 (Mennella 2003). In this, and in other 'ome' words which predated 'chromosome', such as rhizome, phyllome, thallome and tracheome, 'ome' signifies the collectivity of units of a system, where the etymological origin is the Greek '*oma*', signifying condition, or having the nature of. A third interpretation is from the Sanskrit concept of *OM* which 'signifies fullness, completeness ... it encompasses the entire universe' in its limitlessness (Lederberg and McCray 2001).

In 1953, James Watson and Francis Crick published their famous paper proposing a double-helical structure for the biomolecule DNA (deoxyribonucleic acid), based on Rosalind Franklin's X-ray crystallography images from the laboratory of Maurice Wilkins. Through the influence of Watson and Crick's model, the concept and materity of the gene became identified with the biomolecule DNA. The subsequent development of methods for the direct analysis and manipulation of DNA created the technological conditions of possibility for the focus of study to shift from individual genes to whole genomes (see Wheale and McNally 1988). The word 'genomics', meaning the study of linear gene mapping and DNA sequencing, first came into widespread use in 1987 when adopted as the title for a new journal founded by Victor McKusick and Frank Ruddle (Venter *et al.* 2003). Known as the 'genomics grandfather', McKusick is also credited with the first proposal, in 1969, to map the human genome, and in 1988 he became the founder president of the Human Genome Organisation (HUGO), the 'UN for the human genome' (Burke 2003; McKusick 1989). That same year, James Watson accepted the directorship of the US National Institute of Health's (NIH) Office of Genome Research (and later the National Center for Human Genome Research).

The century of the gene culminated in the formal launch in 1990 of the Human Genome Project (HGP), a global enterprise to map and decipher the DNA sequence of every gene of the human (and of several other) species. Planned as a fifteen-year project, the HGP was initially a public initiative. However, in 1998 it was announced that a new private company, Celera Genomics Corporation, would sequence the entire genome in three years, resulting in what came to be viewed as a race between the public and private endeavours (Zweiger 2000). The completion of the first draft of the human genome was announced ahead of schedule in February 2001, and declared a joint accomplishment of the public HGP (Lander *et al.* 2001) and private initiatives, most notably that of Celera (Venter *et al.* 2001) (see Glasner and Rothman 2004).

In the wake of the HGP and the other genome initiatives, the frontiers of biological and biomedical research are being renamed and reshaped as bioscientific research has been 'omicised'. Just about every entity that has ever been the focus of biological or biomedical research has been re-conceptualised as a collective ome with its corresponding 'omic'. Examples from Cambridge Healthtech Institute's '-omes and -omic glossary' include cellulomics, chromonomics, degradomics, ligandomics, metabolomics, microbiomics, peptidomics, phenomics, physiomics, promoteromics, transcriptomics and vaccinomics, and the omic which has arguably most self-consciously modelled itself on genomics, proteomics (Chitty 2003).[1]

Proteomics as the 'omic' progenitor

> Until a few years ago, the sheer number and variety of proteins sustaining each cell seemed so impenetrable that no one even thought to bestow a collective name of them. Well now they have: the proteome.
>
> (Cohen 2000: 38)

The word 'proteome' was coined by Mark Wilkins, an Australian postdoctoral student, as a shorthand for saying 'all the proteins encoded by a genome' (Wilkins *et al.* 1995). It first entered into public use in 1994, although some commentators trace the history of the study of proteomes back to 1975 when two scientists independently described a method (two-dimensional gel electrophoresis – 2DGE) for separating and characterising all the proteins in a sample (Righetti 2004).[2]

Derived from the Greek *proteos*, meaning primary, the word 'protein' was first suggested by G. J. Mulder in 1838 (Stent 1971: 32). Proteins are a diverse set of molecules that includes skin, muscle, hair and nails, blood proteins, and enzymes, hormones, antibodies and antigens. Collectively, proteins comprise the structures and perform the functions that constitute living organisms in sickness and in health. Proteins are made of approximately twenty different units called amino acids. A short chain of amino acids is called a peptide; a long chain is called a polypeptide. Proteins are commonly composed of long, folded polypeptides, each of which comprises a precise sequence of between 50 and more than 2,000 amino acids.

Proteins occupy a prominent position in the history of the gene because for the first half of the twentieth century most scientists believed that they, rather than DNA, constituted its chemical basis. In favour of DNA was evidence that it was a major component of chromosomes and that it was able to convey and alter bacterial heredity. However, against it was the view that the molecular structure of DNA was too simple to encode the requisite vast amount of hereditary information. DNA was known to contain four different nucleotide bases called adenine (A), thymine (T), cytosine (C) and guanine (G). The inaccurate 'tetranucleotide theory' from the 1930s proposed that DNA was comprised of repeated units (tetranucleotides) comprising one of each of the four nucleotide bases. As such, like the biopolymer starch, DNA would be the same no matter which organism it came from. Watson and Crick's model for the structure of DNA rejected this theory. According to the double helix model, DNA is composed of two chains of nucleotide bases, and it is in the precise sequence of nucleotide bases that hereditary information is inscribed.

The principles of the relationships between DNA and proteins are expressed in the 'central dogma' of molecular genetics and the 'genetic code'. According to the central dogma, there is a uni-directional flow of hereditary information from DNA, where it is encrypted and faithfully transmitted from cell to cell and from generation to generation, to proteins which are its material expression in the structures and functioning of living organisms. Encoded within the base sequence of DNA is the information that directs the assembly of amino acids into proteins. The genetic code, which is virtually universal throughout living organisms, was fully deciphered in 1966.[3]

The central dogma expresses the primacy of DNA over proteins. It elevates DNA to the status of 'master molecule', and underpins one of the most arresting rationales made for the HGP, namely that in acquiring knowledge of the entire DNA sequence of living organisms we would have access to the 'Book of Life'. However, although DNA was the first biomolecule to become the focus of 'Big Science' enterprise,[4] proteins were arguably the first to be the focus of an 'omic' vision. In the late 1970s, long before the HGP had even been proposed, N. Leigh Anderson and his father Norman at the Argonne National Laboratory in Illinois attempted to enrol

US government and scientific support for an initiative they called the Human Protein Index, which would catalogue and compare all human proteins using 2DGE. In fact, in a leap of the imagination that prefigured the twenty-first century turn to the 'omics', they considered the Human Proteome Index to be just part a larger project which would characterise all the molecular constituents of living cells (Zweiger 2000).[5] However, as Anderson later acknowledged, the proposal was too advanced for its time, and they failed to attract the US government funding they sought. It was not until the 1980s that the 'kairos'[6] was right for a large-scale informatics approach to biological research, at which time the focus was on DNA rather than proteins in the form of the HGP.

The HGP was funded on the basis of promises that knowledge of the DNA sequence would reveal our biology and our destiny. Instead, however, as Keller argues, the outcome of the HGP and other genome initiatives has been to highlight and make impossible to ignore the complexity of biological systems. In comparison with what was anticipated at the start of the project, at a mere 35,000, the number of genes[7] in the human genome has turned out to be surprisingly small (and going down; the official count in August 2005 was a mere (and amazingly precise) 22,118). Not only is the number of genes in the human genome embarrassingly small from an anthropocentric perspective (slightly lower than the number of genes in the puffer fish – *Tetraodon nigroviridis* – genome), it is also humbling from a genocentric perspective, being outnumbered by several orders of magnitude by the estimated number of protein variants in the human proteome.

One explanation for the magnitude of the proteome relative to the genome is what is called 'gene splicing' which amplifies the number of different proteins that can be translated from a given DNA sequence. In gene splicing, 'gene transcripts' (the intermediaries between DNA and protein) are cut and then recombined so that a coding DNA sequence gives rise to a number of different protein 'isoforms'. So common is this process it seems that most proteins are the product of splicing between gene transcripts. It has been estimated that, through gene splicing, the 22,500 'genes' give rise to 100,000 protein variants (Krulwich 2001). Another factor which adds to the diversity of the proteome is that a functioning protein is more than a sequence of amino acids. Most proteins are modified by the addition of other chemical groups, such as phosphates and sugars, modifications which are critical to a protein's structure and function. In excess of 1,000 protein variants have been described for some individual DNA coding sequences. Moreover, over a lifetime and even at any point in time, an organism contains not one but many proteomes because the complement of proteins present and their relative abundance varies between youth and old age, from cell type to cell type, and between healthy and diseased tissue (Twyman 2004). On the face of it, it would appear that the diversity of the human proteome exceeds the explanatory power of the genetic code. Just two months after the publication of the first drafts of the human genome, the idea of a human proteome project resurfaced at a conference in Virginia, USA, entitled 'Human Proteome Project: Genes Were Easy' (Steinberg 2001). The proposal evoked a variety of opinions over the strategy that should be adopted for such a project, what its goals should be, which standards, technologies, instruments and protocols to use, and whether large proteomics projects should be undertaken by companies, or in public laboratories, or both. Eric Lander, a leading scientist in the HGP, even questioned whether such a project made sense, given the enormity of the proteome (Steinberg 2001). None the less, the 'Human Proteome Project' was taken forward by the newly-formed Human Proteome Organisation (HUPO).[8]

The thesis in Keller's book is that, in exposing 'clear and demonstrable gaps between the many different attributes that had historically been assumed to inhere in' the gene (Keller 2000: 70), the HGP has radically undermined its conceptual usefulness. She argues that 'to an increasingly large number of workers at the forefront of contemporary research, it seems evident that the primacy of the gene as the core explanatory concept of biological structure and function is more a feature of the twentieth century than it will be of the twenty-first' (Keller 2000: 9). Of the future she makes three predictions. The first is that biologists will have an expanded array of conceptual tools; the second is that the new post-genomic era will include numerous elements that defy categorisation as animate or inanimate; and the third is that, although the term 'gene' may have become a hindrance to biological understanding, biologists are not likely to stop talking about genes (Keller 2000: 9–10). The following sections look for evidence of Keller's predictions in three visions of the future of biology in the wake of the HGP.

The genomic era

The first vision has its roots firmly in genomics. It is the vision of Francis Collins and colleagues writing on behalf of the US National Human Genome Research Institute (NHGRI) (Collins *et al.* 2003). The NHGRI was established in 1997 as the successor to the National Center for Human Genome Research (NCHGR), but elevated to the status of an NIH Research Institute with grant-awarding authority. James Watson, the original director of the NCHGR, was replaced by Francis Collins in 1993. This vision, 'the outcome of two years of discussion with hundreds of scientists and members of the public' (Collins *et al.* 2003: 836), was printed in *Nature* in April 2003, the month in which the NHGRI celebrated the formal completion of the HGP.

The title, 'A vision for the future of genomics research: A blueprint for the genomic era', makes it immediately clear that, for these authors, genomics did not end with the completion of the HGP. Collins had previously expressed these feelings in February 2002 at a symposium at the US National Academy of Sciences, where he took issue with symposium title: 'Defining the mandate of proteomics in the post-genomics era'.

> [Collins] queried whether [the term post-genome era] means that from the beginning of the universe until 2001 we were in the pre-genome era, and then suddenly 'bang' we moved into the post-genome era (leading one to wonder what happened to the genome era). He suggested that it was presumptuous to say that the Human Genome Project is already behind us. He pointed out that proteomics is a subset of genomics, and genomics is more than sequencing genomes.
>
> (Kenyon *et al.* 2002: 763)

In keeping with this viewpoint, the vision makes no mention of 'post-genomics', or post-anything else for that matter. Neither is the term 'genomics' qualified, as in 'functional genomics' or 'structural genomics', with the exception of one reference to 'comparative genomics'. Here, biological and biomedical research in the wake of the HGP is simply 'genomics'.

In this vision, although all of the initial objectives[9] of the HGP have been achieved, the completion of the HGP is not the end, but the end of the beginning; 'what's past is prologue' (Collins *et al.* 2003: 846). Their blueprint offers 'a broader and still more ambitious vision, appropriate for the true dawning of the genomic era' (Collins *et al.* 2003: 836). The vision is illustrated using a picture of a building inspired by Frank Lloyd Wright, with which Collins had already been making the conference rounds (Burke 2003). The picture is entitled: 'The future of genomics rests on the foundation of the Human Genome Project', and the foundation of the building is accordingly labelled 'Human Genome Project'. The building's three floors represent the three 'major themes' of their vision, namely: 'genomics to biology' situated on the ground floor; 'genomics to health' on the first floor; and 'genomics to society' on the top. Of their vision the authors write: 'If we, like bold architects, can design and build this unprecedented and noble structure, resting on the firm bedrock foundation of the HGP, then the true promise of genomics research for benefiting humankind can be realized' (Collins *et al.* 2003: 847).

One gets the impression that the authors of this vision have been reading Kuhn (1970) for Collins *et al.* write of a 'revolution in biological research' (Collins *et al.* 2003: 835), where the HGP is both foundation and blueprint for the new discipline of genomics.

> The project's new research strategies and experimental technologies have generated a steady stream of ever-larger and more complex genomic data sets that have poured into public databases and have transformed the study of virtually all life processes. The genomic approach of technology development and large-scale generation of

community resource data sets has introduced an important new dimension into biological and biomedical research. ... Genome sequences, the bounded sets of information that guide biological development and function, lie at the heart of this revolution. In short, genomics has become a central and cohesive discipline of biomedical research.

(Collins *et al.* 2003: 835)

At the 'heart of this revolution' are genome sequences, and it is the reproduction of this type of resource – large-scale 'omic' data sets available to the community in public databases – that is transforming biological and biomedical research into a genomics discipline. Furthermore, as the building suggests, the genomic era is not just about the realignment of biology and medicine in accordance with the genomics discipline, but the genomic disciplining of society.

Drawing on lessons from the HGP, the authors identify six 'elements' that are necessary for the reproduction of the genomics discipline. These are depicted as six vertical columns that rise through all three floors, crosscutting the three themes. For each theme, the authors identify a number of 'grand challenges'. The entire framework for Collins *et al.*'s vision of the genomic era is summarised in Table 14.1, the organisation of which is based on the architecture of the building.

One way of conceptualising this vision is as one whose core goal is the 'generation of large publicly available comprehensive' sets of data. From this perspective, its themes and grand challenges can be read as ever more ambitious expressions of this core goal, and its cross-cutting elements can be seen as the conditions of possibility for its realisation.

Theme I, Genomics to Biology, starts with the compilation of the DNA-based genome 'parts list' (GC I.1),[10] and then progresses to the generation of experimentally derived data sets of genetic networks, proteins and protein pathways (GC I.2). Understanding the relationship between genotype and individual variation in biological function (GC I.3) requires projects such as the International HapMap Project to catalogue all common DNA variants in the human population.[11] For establishing and assessing the risks that particular gene variants contribute to health, disease susceptibility and drug response, large, longitudinal, population-based cohort studies such as the UK Biobank, Marshfield Clinic's Personalized Medicine Research Project and the Estonian Genome Project are recommended (GC II.1, II.2).[12] Such studies, which are part of Theme II, Genomics to Health, make demands that the data sets collected include not only genomic data but other types of data including clinical data, and information on disease risk. GC II.3 adds data on 'diet, exercise, lifestyle and pharmaceutical intervention' that 'could potentially be used in individualised preventive medicine', and GC II.6 adds socioeconomic status, culture, and environmental exposures to the list of data types to be included. Furthermore, the translation

Table 14.1 Framework of Francis Collins et al.'s vision for the future of genomics research

Major themes	Grand Challenges (GC)
Genomics to Society: Promoting the use of genomics to maximise benefits and minimise harms	GC III.4: Assess how to define ethical boundaries for the uses of genomics GC III.3: Understand the consequences of uncovering the genomic contributions to human traits and behaviours GC III.2: Understand the relationships between genomics, race and ethnicity, and the consequences of uncovering these relationships GC III.1: Develop policy options for the uses of genomics in medical and non-medical settings
Genomics to Health: Translating genome-based knowledge into health benefits	GC II.6: Develop genome-based tools that improve the health of all GC II.5: Investigate how genetic risk information is conveyed in clinical settings, how that information influences health strategies and behaviours, and how these affect health outcomes and costs GC II.4: Use new understanding of genes and pathways to develop powerful new therapeutic approaches to disease GC II.3: Develop genome-based approaches to prediction of disease susceptibility and drug response, early detection of illness, and molecular taxonomy of disease states GC II.2: Develop strategies to identify gene variants that contribute to good health and resistance to disease GC II.1: Develop robust strategies for identifying the genetic contributions to disease and drug response
Genomics to Biology: Elucidating the structure and function of genomes	GC I.5: Develop policy options that facilitate the widespread use of genome information in both research and clinical settings GC I.4: Understand evolutionary variation across species and the mechanisms underlying it GC I.3: Develop a detailed understanding of the heritable variation in the human genome GC I.2: Elucidate the organisation of genetic networks and protein pathways and establish how they contribute to cellular and organismal phenotypes CG I.1: Comprehensively identify the structural and functional components encoded in the human genome
Cross-cutting elements	Resources Technology development Computational biology Training ELSI Education

HUMAN GENOME PROJECT

Source: Derived from Collins et al. 2003

of genomics to medicine also requires that the cohorts in longitudinal population-based studies include 'full representation of minority populations' (GC II.1), and 'a large epidemiologically robust group of individuals with good health' (GC II.2).

The translation of the genomic era's large-scale heterogeneous data sets into useable knowledge requires a new method of knowledge production. This is because even 'well-known classes of functional elements, such as protein-coding sequences, still cannot be accurately predicted from sequence information alone' (Collins *et al.* 2003: 837). Something is needed (to borrow Keller's metaphor) to 'fill the gap' between large-scale datasets and biological knowledge and understanding and, in this vision, that something is computational biology. In the genomic era, new computational tools will identify hypothetical functional elements, model pathways, and make hypotheses about how they affect phenotypes. The validity of the hypotheses derived from these models will then be tested experimentally, and the resulting experimental data will be used to refine the computational models used for hypothesis generation (GC I.1, I.2). Computer programs will interface a variety of databases and obtain 'unbiased' determination of the risk associated with a particular gene variant (GC II.3), and a combination of computational and experimental methods will detect gene–gene and gene–environment interactions on health and disease (GC II.1).

In addition to large-scale data sets and advances in computational biology to house and analyse them, the safe and effective realisation of the benefits of genomics to medicine, health and society requires the translation of a large and heterogeneous cast of actors. To realise healthcare benefits, basic scientists must be trained so that they not only have a 'genomic attitude', they also adopt a 'therapeutic mindset' (GC II.4). They must also acquire computational skills because 'all future biomedical research will integrate computational and experimental components'. Healthcare professionals and the public must be educated about the interplay of genetic and environmental factors in health and disease, so that they are 'well-informed participants in a new form of preventive medicine' (GC II.3). Human illnesses are to be reclassified on the basis of detailed molecular characterisation, resulting in a new molecular taxonomy of illness (GC II.3). As genomics data moves beyond the clinic and into society, 'Both the potential users of non-medical applications of genomics and the public need education to understand better the nature and limits of genomic information and to grasp the ethical, legal and social implications of its uses outside health care' (GC III.1). Social scientists are also enrolled in the vision. Their task is to 'analyse the impact of genomics on concepts of race, ethnicity, kinship, individual and group identity, health, disease and "normality" for traits and behaviours'; and to 'define policy options, and their potential consequences, for the use of the genomic information and for the ethical boundaries around genomics research' (GC III). And individuals from different cultures and religious traditions, from minority and disadvantaged populations, from the disability community and NGOs, should be included as researchers and as participants in research and policy making (GC III.1, GC III.4).

The HGP is sometimes characterised as marking the advent of 'Big Biology'. However, looked at from the perspective of Collins *et al.'s* vision, it might be more accurately classified as 'Big Genetics', a mere prelude to the larger project of Big Biology, which is yet to be realised through Theme I, Genomics to Biology. The defining feature of the genomic era is the genomic way of doing things through the collection and analysis of large-scale 'omic' data sets. Theme II, Genomics to Health, requires cohort population studies whose organisation and use will translate healthcare research and delivery into 'Big Medicine'. Moreover the full realisation of the benefits of the genomic era requires the collection and analysis of 'sociomic' data sets on ethnicity, sexuality, behaviour, lifestyles, legislation, policies and attitudes. In other words, it is also a vision for 'Big Sociology' and the omicisation of the social.

Underpinning investment in the HGP was the expectation that knowing the sequence of genomic DNA would reveal what it is that makes an organism the way it is. Collins *et al.'s* grand vision, with its promises of future (but not-yet-realised) benefits could be construed as evidence that the HGP has fallen short of its target, or as a challenge to belief in DNA as the 'Book of Life', or both. Either interpretation could be construed as grounds for abandoning further large-scale research projects based on genomics. Collins *et al.*, however, do not acknowledge any failure on the part of the HGP, arguing that it fulfilled all of its original goals. Regarding improvements in human healthcare that the HGP was supposed to yield, they acknowledge that they were predicted in the original vision and subsequent reports on the HGP, but argue that they were not realised owing to the absence of a clear strategy. The failure was therefore one of strategic oversight, amenable to correction. Themes II and III of their vision aim to correct this strategic weakness. In this way, the failure of the HGP to deliver the anticipated improvements in healthcare is reconciled with, and becomes an argument in favour of, the proposal that society should invest in more research in its image.

Collins *et al.* also uphold the value of DNA as the 'Book of Life', arguing that 'embedded within this as-yet poorly understood code are the genetic instructions for the entire repertoire of cellular components, knowledge of which is needed to unravel the complexities of biological systems' (Collins *et al.* 2003: 837). Therefore, although Collins *et al.* write of a revolution in the biological and biomedical sciences, and refer to a new era, their futuristic vision looks like a conservative continuation of, rather than an incommensurable conceptual break with, the genetic determinism characteristic of twentieth-century molecular genetics. Theme I, for example, is based upon the premise that the genome sequences of the HGP contain 'the genetic instructions for the entire repertoire of cellular components' (Collins *et al.* 2003: 837). Identifying all the functional structures in genomic DNA is the objective of GC I.1, a cataloguing exercise whose goal is to 'functionate' (Collins *et al.* 2003: 842) the genome by providing a genetic 'parts list' of all the functional elements encoded within genomic DNA. Other grand challenges also seem to be based on genetic/ genomic determinism, such as those which seek the genetic basis of individual differences in biological function (GC I.3), speciation (GC I.4) and genetic contributions to health (GC II.2), disease and drug response (GC II.1, II.3), race and ethnicity (GC III.2) and human traits and behaviour (GC II.3).

Yet, although the genetic/genomic determinist rhetoric persists, as genomics moves from the foundational bedrock of the HGP, the concepts of both genomics and the genome undergo a subtle shift. Tucked away in GC I.2 it is acknowledged that:

> Genes and gene products do not function independently, but participate in complex, interconnected pathways, networks and molecular systems that, taken together, give rise to the workings of cells, tissues, organs and organisms. Defining these systems and determining their properties and interactions is crucial to understanding how biological systems function.

At the 'genetic level', GC I.2 requires the identification of the regulatory interactions in different cell types, and at the 'gene-product level', the *in vivo*, real-time measurement of protein expression, localisation, modification and activity/kinetics (GC I.2). In contrast to the conceptualisation of DNA as a self-sufficient 'Book of Life', GC I.2 recognises that there is more to understanding the functioning of biological systems than can be deciphered from static, linear DNA sequences. GC I.2's research agenda, which requires the large-scale study of proteins, could be viewed as a challenge to the genetic/genomic determinism that underpins both the molecular biology of the late twentieth century and the HGP itself. However, an overt challenge is averted through Collins *et al.*'s choice of language. Throughout their vision Collins *et al.* tend to refer to proteins as 'gene products'. In their terminology, GC I.2 employs a 'genomic approach' to catalogue all 'gene products'. Their 'gene talk' downplays the identity of proteins as a class of biomolecule with structures and functions that transcend what can be predicted from the DNA sequences that encode their primary amino acid sequences. This language use also expands the concept of genomics to include the collection of datasets on proteins – creating what could be called the 'new genomics'.

This linguistic 'geneticisation' of proteins is simultaneously a 'proteinisation' of the genome. Proteinisation of genetics and the genome is also apparent in Theme II, where the concept of 'genetic factors' (used in contradistinction to 'non-genetic factors') includes data on gene products as well as genes. Such language continues the use of 'gene talk' and 'genome talk', at the same time as allowing their molecular referents to shift.

This use of language in Collins *et al.* liberates the concept of the gene from the increasingly apparent limitations of DNA, and concomitantly increases its sphere of biological action. Uncoupled from its identity with DNA, the gene is free to associate adventitiously with whichever biomolecular entities appear to be the most functional. By breaking its fifty-year monogamy with DNA, the position of the gene as constituting the centre of biological action can be sustained. Through this manoeuvre, genetic/genomic determinist discourse is able to continue.

From genomics to proteomics

The vision in this section is from the perspective of writers located in proteomics. It is based on two publications co-authored by Matthias Mann: one with Walter Blackstock, Director of Proteomics at GlaxoSmithKline (Blackstock and Mann 2001); and the other with Mike Tyers of the Samuel Lunenfield Research Institute, University of Toronto (Tyers and Mann 2003). Mann was a doctoral student under John B. Fenn, joint winner of the Nobel Prize in Chemistry in 2002 for the development of the electrospray ionisation method used in protein mass spectrometry. Since then he has worked with Peter Roepstorff in Denmark, been Group Leader at the European Molecular Biology Laboratory in Heidelberg, Professor at the University of Southern Denmark, Odense, and is currently a Director at the Max-Planck Institute for Biochemistry. Like Aebersold and Roepstorff, Mann has been a member of HUPO's Council since its inauguration.

The Blackstock and Mann (B&M 2001) publication, entitled 'A boundless future for proteomics?' is the editors' introduction to a proteomics supplement in *TRENDS in Biotechnology*, published in October 2001, the year in which the first draft of the human genome was published. The Tyers and Mann (T&M 2003) article, entitled 'From genomics to proteomics', introduced a *Nature* Insight on proteomics, published in March 2003, the month before the formal completion of the HGP was announced.

Both articles acknowledge the importance of genome sequences and genomics for proteomics and for biology, writing that 'The human genome sequence is a wonderful resource that will underpin biology for decades to come ...' (B&M 2001); and that 'Proteomics would not be possible without the previous achievements of genomics, which provided the "blueprint" of possible gene products that are the focal point of proteomics studies' (T&M 2003: 193). With the completion of the HGP, they categorise the present as the 'post genomic era', an era interested in 'extracting the information embedded in the DNA sequence' (B&M 2001).

The authors note that the definition of the proteome has expanded since the term was first coined, from meaning the set of proteins encoded by the genome to encompass 'most of what was previously known as functional genomics' (B&M 2001). The result is that 'The study of the proteome, called proteomics, now evokes not only all the proteins in any given cell, but also the set of all protein isoforms and modifications, the interactions between them, the structural description of proteins and their higher-order complexes, and for that matter almost everything "post-genomic"' (T&M 2003: 193). They hope that proteomics will contribute to a description of cellular function, 'yield direct biological insights', 'have a profound impact on clinical diagnosis', and 'inevitably accelerate drug discovery' (T&M 2003: 193–5).

The authors categorise proteomics, along with transcriptomics and meta-bolomics, as 'functional genomics' approaches which have been created to explore gene function using the 'systems biology approach'. This involves using 'robots and automation to amass large datasets in an "unbiased" way', and then employs 'mathematical tools to extract information' (B&M 2001). The data sets produced by the various functional genomics approaches are expected to complement each other and to provide a type of methodolo-gical triangulation ('orthogonal omics') as a cross-validation that compen-sates for the difficulty of repeating high-throughput large-scale experiments (T&M 2003: 195). Their bioinformatic integration 'will yield a compre-hensive database of gene function', and provide a 'useful tool' for the individual researcher to generate and test non-obvious hypotheses that would otherwise not arise from any 'individual approach' (T&M 2003: 193, 195). Integrated omic databases will also be 'crucial' for systems biology, the goal of which is to comprehensively model cellular behaviour at the whole-system level (T&M 2003: 193).

The authors make a distinction between large-scale automated proteomics experiments and systems biology on the one hand, and 'hypothesis-driven' biology performed in individual biology laboratories on the other. Noting the importance of economics as a member of the 'omics' family, they pre-dict that most of biology in universities will remain essentially hypothesis-driven for the foreseeable future (B&M 2001). However, they call for a way to integrate large-scale experiments with the activities of individual laboratories so that they can functionally validate the output of proteomics experiments – the 'final key step in the discovery process that may always defy automation' (T&M 2003: 197).

In summary, Mann and colleagues write of a new era that is the product of the various genome projects. Called the post-genomic era, it is based upon genomics as a model and attributes great significance to genomics sequences as a resource. The focus of the post-genomic era is the explora-tion of gene function. Mann and colleagues identify two complementary models of biological research: hypothesis-driven research on the one hand, and automated large-scale omic research and systems biology on the other. Both are necessary if the anticipated benefits of proteomics and other functional genomics approaches of the post-genomic era are to be realised.

The new biology

The analysis in this section is of two articles co-authored by Ruedi H. Aebersold. One is a short commentary co-authored with Leroy E. Hood and Julian D. Watts, published in *Nature Biotechnology* in April 2000 (Aebersold *et al.* 2000). The other (Patterson and Aebersold 2003), published in a ten-year retrospective of *Nature Genetics* in March 2003, is co-authored with Scott D. Patterson, Senior Director for Proteomics at Celera Genomics, the main private-sector rival to the public HGP. It was at Hood's publicly-financed laboratory at the California Institute of Technology (Caltech) that the first automated DNA sequencing machine was developed, a machine subsequently developed and commercialised by Applied Biosystems Inc., whose ABI Prism™ 377 Sequencer became the staple machine for the HGP (Zweiger 2000). Aebersold worked in Hood's Caltech laboratory and later joined him at the University of Washington before they, together with Alan Aderem, co-founded the Institute for Systems Biology in 2000, where Watts is a Senior Research Scientist. In 2004 Aebersold moved to the Swiss Federal Institute of Technology (ETH) in Zurich. A member of the Human Proteome Organisation's (HUPO) Council since its inauguration in 2001, in 2005 Aebersold became Co-chair of HUPO's Proteomics Standards Initiative. Collectively, the authors of the vision analysed here are, or have been, deeply involved in genomics, proteomics and systems biology.

The 'new biology' that Aebersold and colleagues write about is the fusion of 'hypothesis-driven science', 'discovery science' and 'systems biology'. In 'hypothesis-driven science', which has historically been the mainstay of university-based research programmes, 'small, autonomous, highly specialized' research groups conduct experiments designed to test hypotheses (Aebersold *et al.* 2000). However in their view, the various genome mapping and sequencing projects have 'catalysed' the emergence of a new 'technology-driven' research method for biological and biomedical sciences called 'discovery science' (Aebersold *et al.* 2000). Included among the discovery sciences are proteomics (a 'suite of relatively mature tools that support protein cataloguing and quantitative proteome measurement ... at high throughput') (Patterson and Aebersold 2003: 317), genomic sequencing, microarray analysis and metabolite profiling. The defining feature of discovery sciences as a research method is that they investigate a 'biological system or process by enumerating the elements of a system irrespective of any hypothesis on how the system might function' (Patterson and Aebersold 2003: Box 1).

Discovery sciences produce large sets of 'systematically-collected data' (Patterson and Aebersold 2003: 319), the availability of which has 'catalysed the adoption of "reverse" research approaches' (Patterson and Aebersold 2003: 312). These are contrasted with 'forward' or 'reductionist' approaches which were the 'mainstay of biology in the 1980s' (Patterson and

Aebersold 2003: 312), characterised by research which moves from an observed function or phenotype to the identification of the relevant causal genes or gene products. In reverse approaches, by contrast, the direction of research is from the identification of uncharacterised biomolecules which appear to have significance as biomarkers to their functional validation.

According to Aebersold and colleagues, the 'exploding cascades of information' (Aebersold *et al.* 2000) generated by the discovery sciences are 'difficult to analyze using traditional knowledge-based interpretation' and require computational tools to 'extract biological insights or to formulate hypotheses' regarding the 'structure, function and control of biological systems' (Patterson and Aebersold 2003: 317). One of the ways in which hypotheses are generated is by comparative pattern analysis to detect differences, for example in proteome profiles between samples taken from different tissue types, or from disease tissue and healthy tissue.[13]

Aebersold and colleagues are optimistic regarding what can be achieved with the discovery sciences, believing that 'for any given species, the space of possible biomolecules and their organization into pathways and processes is large but finite. In theory, therefore, the biological systems operating in a species can be described comprehensively if a sufficient density of observations on all of the elements that constitute the system can be obtained' (Patterson and Aebersold 2003: 311). This belief underpins the emerging 'systems biology' approach that the discovery sciences have given rise to and are an important part of. Of the various discovery sciences, proteomics is regarded as a 'particularly rich source of information' for systems biology. In the future, Aebersold and colleagues predict the different data types produced by the various discovery sciences will be integrated and collectively interpreted to achieve a 'comprehensive understanding of the workings of biological systems' (Patterson and Aebersold 2003: 311). They anticipate that 'systems biology approaches will detect connections between broad cellular functions and pathways that were neither apparent nor predictable despite decades of biochemical and genetic analysis of the biological system in question' (Patterson and Aebersold 2003: 319). In their view, the ultimate goal of systems biology is to produce computational models and simulations of complex biological systems that are 'predictive of the behaviour of the system or its emergent properties in response to any given perturbation' (Aebersold *et al.* 2000).

Aebersold and colleagues foresee the convergence of discovery science and hypothesis-driven research, the beginnings of which is already apparent in the 'union of data' in information resources which combine 'systematically collected data' (i.e. data from discovery science) and the results of hypothesis-driven research published in the scientific literature (Patterson and Aebersold 2003: 319). Such information resources have been used in a new 'experimental strategy' involving the integration of different data types from discovery sciences into mathematical models consistent with information in the scientific literature. They give an example of how this approach has already predicted molecular pathways that were previously unknown, some of which were subsequently verified experimentally.

In their vision of the new biology, hypothesis-driven research programmes will provide the testing ground for systems biology-derived hypotheses, and hypothesis-driven research programmes will be fuelled by data generated by the discovery sciences. However, the facilities required for discovery science and systems biology are expensive and space-intensive, and their inter-disciplinarity does not fit into the traditional departmental structure of universities. Aebersold and colleagues therefore propose a reconfiguration of university administration, university research funding, and even the legal system so as to align with the demands of the new biology. Under the new regime, 'universities would create new administrative bodies outside and independent of their departmental structures' to run the systems biology programmes, whose set-up and operational cost would be supported by university–private-sector partnerships. In return for their financial support, the private sector would acquire trained personnel and new technologies. They also envisage a role for legislative bodies and legislation, perhaps mandating corporate investment in public research and protecting academic freedom from 'undue influence by the private sector' (Aebersold *et al.* 2000).

Aebersold and colleagues maintain that knowledge of the complete genomic sequence has resulted in a 'revolution in genetics' (Aebersold *et al.* 2000), has created a 'resource' that greatly accelerates protein identification (Patterson and Aebersold 2003: 313), and has led to the emergence of a new 'technology-driven approach to biological and biomedical sciences'. In the new biology of the future, the production of knowledge using traditional hypothesis-driven methods will be complementary to, rather than incommensurable with, the production of knowledge using the newer approaches of discovery science and systems biology. They identify barriers to the integration of the new approaches with the old in terms of incompatibilities between the resource, space and interdisciplinarity demands of discovery science and systems biology on the one hand, and the existing organisation and funding of university research on the other. To facilitate the development of the new biology, which combines hypothesis-driven science, discovery science and systems biology, Aebersold and colleagues propose reconfiguring the administration, financing and governance of university research and its relations with the private sector.

Discussion

The sections above analysed three visions of the future: one from the perspective of genomics, one from the perspective of proteomics and one from the combined perspectives of genomics, proteomics and systems biology. A comparison of the three visions suggests something of a turf war in the emergent fields of the twenty-first century biosciences, with a certain amount of boundary work being performed through the naming of things (Gieryn 1995). In Collins *et al.* the concept of genomics is being expanded beyond its original meaning, which was the sequencing and mapping of the DNA genome, to encompass the structure and function of proteins. Mann and colleagues highlight expansionism on the proteomics side so that proteomics is now more than just the study of the proteins encoded by the genome but encompasses almost all of what is described as functional genomics or post-genomics. Indeed, Paul Gilman, Director of Policy Planning at Celera Genomics Group, provocatively suggested that the US National Institutes of Health (NIH) should be 'folded into one big Institute for Proteomics' (Steinberg 2001). Aebersold and colleagues write of an all-encompassing 'new biology' that includes genomics, proteomics, transcriptomics, metabolomics and any other discovery science that can be used in systems biology, as well as traditional hypothesis-driven research.

Whilst the three visions differ in the naming of things in the new era of research that is emerging in the wake of the HGP, they each pay tribute to the legacy of and the model of research exemplified by the HGP and other genome initiatives. In each vision, genomic DNA data sets are foundational resources and genome research is the exemplar of a new era in biological and biomedical research. In each of the visions, research in the wake of the HGP is characterised by a new research approach or method referred to as the 'genomics discipline' (Collins *et al.*), 'functional genomics' (Mann), and 'discovery science' (Aebersold), which is distinguished from traditional 'hypothesis-driven' research. Although these visions do not use this terminology, the contrast made between the new and the old research approaches is reminiscent of the contrast made between 'deductive' and 'inductive' research in research methods books. In contrast to 'deductive', traditional hypothesis-driven research, knowledge production in the new 'inductive' era starts with the amassing of large datasets of various kinds in an 'unbiased way', irrespective of any hypothesis, the result of which is not knowledge but data which is a resource, or a tool. Hypotheses are generated subsequently, out of the data, through systematic 'unbiased' analysis, and then tested through traditional hypothesis-driven experiments. Therefore, although a new era of research is talked about, in these visions traditional hypothesis-driven research has not been dispensed with but plays a necessary and complementary role to discovery science approaches in the production of knowledge.

Another feature of knowledge production in the new era that the visions share is the dependency on computation. The volume of data produced by the discovery sciences is such that their storage, interchange, integration and analysis defy traditional approaches and require a computational infrastructure, bioinformatic standards and software tools. Furthermore, the 'omicisation' of the unit of data collection and analysis has transformed and requiring large budgets, automation, computerisation, shared and sometimes distributed facilities, collaborations between different disciplines (including the social sciences), and between the public and private sectors. In other words, the new era of research modelled on the HGP is 'Big Science' (Glasner 2002) requiring the reconfiguration of the administration, financing and governance of biological and biomedical research, and the training and education of researchers, physicians, and even the public.

The aim of this new era of research is the familiar one of increased biological understanding and healthcare benefits. Yet such an aim is tantamount to an admission that, despite knowing the DNA sequence of the human genome, science has still not revealed the promised secrets of biological understanding supposedly encoded in the 'Book of Life', nor delivered the promised improvements in healthcare and quality of life. How is it, then, that, despite failing to deliver what was initially promised, the HGP has become the stimulus for more research in its image? The characterisation of genomics as a 'discovery science' is one way in which the (disappointed) social expectations have been managed. They have been recast as unrealistic expectations for this type of scientific research, the product of which is resources or tools, which are not themselves the end products but means to an end. It is now argued that the promised biological insights and advances in medicine will ensue only once a sufficient density of data of complementary types has been amassed and subjected to sophisticated computerised analysis. Reconfigured so as to readjust social expectations, both retrospectively and prospectively (Brown and Michael 2003), this deferral of social benefits becomes not just a way of managing the HGP's past, but also a rationale for future research in its likeness.

Keller's book is an extended exposition of the 'gaps' (Keller 2000: 8, 138, 69, 70) that exist between genetic information on the one hand, and biological meaning on the other, through which she demonstrates the inconsistencies and inadequacies of various twentieth-century conceptualisations of the gene. Throughout her book, she uses the 'gap' metaphor to convey the shortcomings between the gene as a concept and the biological phenomenon it is expected to explain. Each chapter of her book replays the paradox that the more geneticists have tried to close the gap, the bigger it has become; not so much 'Black Box' as 'Black Hole'. Yet despite these repeated narratives, which culminate in the ultimate disappointment of the HGP, Keller seems to be of the opinion that, just because the approaches tried so far have not yet come up with an adequate account of life, it does not mean that life defies scientific explanation. Somehow the 'gap' will be closed, and biology will become a mature science, able to make predictions based on general principles just like physics and chemistry (Kanehisa and Bork 2003). It is just a question of finding the right new laws or concepts to fill the gap.

Given the shortcomings of the gene, Keller's suggestion is that geneticists might fill the gap by borrowing from the conceptual toolkit developed by engineers for the design of systems like aeroplanes or computers (2000: 147). However, genetics is not the only field of biological science to have looked so promising to so many before dramatically 'hitting the wall'. Yuri Lazebnik (2002), for example, writes of experiencing a similar trajectory whilst studying 'apoptosis'. He characterises this as a typical Kuhnian trajectory whereby at some point all fields of biological research reach a stage where it becomes apparent that 'models that seemed so complete fall apart, predictions that were considered so obvious are found to be wrong, and attempts to develop wonder drugs largely fail'; a stage that 'can be summarized by the paradox that the more facts we learn the less we understand the process we study' (2002: 179). His diagnosis is that this is due to an inherent weakness in reductionist, experimental approaches to biological research, and, like Keller, he too advocates the turn to engineering as a way through the apparent impasse of a paradigm in crisis. However, it is not clear what he and Keller foresee as the long-term outcome of this turn to engineering. Leaping forward a hundred years, one possible outcome is that the formal engineering approach is so radically different from the reductionist, experimental approaches it replaces or enhances, that it does not open up ever-widening gaps between 'facts' produced and phenomena they are supposed to explain. An entirely different paradigm, the engineering approach is immune to crisis. Alternatively, a hundred years from now the formal engineering approach may have reached the same paradoxical stage as reductionist, experimental approaches, a stage at which the gaps between facts and understanding call into question the entire research trajectory, resulting in the call for the transition to yet another new mode of knowledge production.

Regarding the more immediate future, the first of Keller's three predictions is that biologists will develop an expanded array of conceptual tools. Turning to the visions of the scientists analysed here, they seem to share Keller's optimism regarding the emergence of new tools. Specifically, they foresee new principles arising out of the combination of data sets and computational biology. They anticipate that computerised analysis of integrated omic data sets in an 'unbiased way' will result in the emergence of new biological hypotheses and principles that are neither apparent nor predictable from the body of knowledge from hypothesis-driven research. Science and technology studies, however, caution against the idea of the 'spontaneous generation' of biological principles, arising out of the omic data sets 'speaking for themselves', as it were. Whilst the emergent hypotheses and principles may be unpredicted, the choices of sample, experimental protocol and method of analysis are never neutral; each is laden with assumptions and expectations that shape the resultant data, hypotheses and principles. As Fujimura writes: 'Each model of biology incorporates what researchers hope to accomplish' (2005: 196, 221). The implication is that our knowledge of nature is never unmediated, but is always shaped by and a reflection of the methods, concepts and aspirations we use to produce it. Each of the visions reviewed here foresees a major role for computation in the future production of biological knowledge. In these visions it is computation that will integrate heterogeneous data sets, extract bio-logical insights, identify hypothetical functional elements, model pathways and make hypotheses. It is also computation that will realise the ultimate goal of systems biology, which is the production of models and simulations of complex biological systems that are predictive of the behaviour of the system. The product of the union of computation and biological data, these models and simulations constitute the 'siliconisation' of living matter and its functions – which brings us to the second of Keller's three predictions.

Keller's second prediction is that the new era will include numerous ele-ments that defy categorisation as animate or inanimate. Computer simulations of living systems could be examples of the ambiguous elements predicted by Keller. But is this anything new? Haraway (1985) argues that we have never had access to a Garden of Eden where we can know and recognise Nature in a pure and unadulterated way. The concept of an unambiguous and unadulterated Nature as an external standard that can be used for the measurement of naturalness and what separates life from non-life is itself an achievement, the result of a purification (Latour 1993; Grint and Woolgar 1997).

Keller's third prediction is that, although the gene as a core explanatory concept may have outlived its usefulness, 'gene talk' as an operational shorthand between scientists, or to acquire funding, promote research agen-das, and market new products, is not going to go away because of its unpre-cedented 'persuasive' 'rhetorical power' (2000: 143). However, Keller's distinction, between the survival of the gene on the one hand and the sur-vival of 'gene talk' on the other, is problematic. In her final footnote she explains that, when social scientists use the word 'rhetoric' in relation to the analysis of how language works in science, they do not imply that it is 'just' rhetoric, meaning something disreputable. Rather the word rhetoric is used to capture the 'complex and multipurpose ways in which scientific language not only does function, but also, and inescapably, has perforce to function in the real world of human actors and human interests' (see Keller 2000: 168). Yet in making a distinction between the fate of the gene and the fate of gene talk, Keller seems to be creating two types of rhetoric: the gene talk that is going to disappear with the gene; and the gene talk that is going to persist because it is useful. Unclear about how to make a distinction between the two types of gene talk apart from resorting to tautology, our preference is to regard all gene talk, wherever, whenever and for whatever reason it is used, as gene talk.[14] Moreover, regardless of the 'interpretative repertoire' (Gilbert and Mulkay 1984; Mulkay 1993) in which it is used, we regard the continued existence of any kind of gene talk as evidence that the concept of the gene has yet to outlive its usefulness. With this as our criterion, in our final assessment of the visions we have analysed, we find evidence both for and against the demise of the gene.

What the three visions analysed here have in common (and what distin-guishes them from the HGP and other genome initiatives), is that they each encompass a larger molecular repertoire than just DNA. Although they pay tribute to the importance of the availability of genomic DNA sets, none of these visions expects the Book of Life to be revealed through knowledge of DNA data sets alone. Instead they envisage that biological knowledge and medical benefits will only come from the integration of a broad range of heterogeneous data sets, even including social data. Whether or not this pluralisation of data types to be collected beyond just DNA indicates the demise of the gene is at present unclear.

On the one hand, the emergence of new research fields with 'degeneticised' names, such as 'systems biology' and 'proteomics', could be viewed as signs of the demise of the gene and the dawn of the post-genomic age. Interestingly, evidence of linguistic degeneticisation is also present in Collins *et al.*, where, rather than the geneticisation of illness (Hedgecoe 2002), they advocate its 'molecularisation'. The disappearance of the prefix 'geno' in the titles of major research programmes and fields such as these indicates the demise of the persuasive power of gene talk and of the gene.

On the other hand, there is Collins *et al.*'s vision in which proteins and other biomolecules are treated as gene products, and their study as part of genomics. This expansion of genomics effectively uncouples the concept of the gene from DNA. With the break-up of its special relationship with the gene, DNA may (once again) become just one biomolecule amongst equals, shrinking in importance in the Big Biological Picture, whilst the gene is free to associate promiscuously wherever the biological or medical action is. Through what could be called the 'extended genome' (Dawkins 1982) and the 'new genomics', the (redefined) gene may be able to retain its status as a core explanatory concept and not only survive into, but even thrive in, the twenty-first century.

Acknowledgements

The support of the Economic and Social Research Council (ESRC) is gratefully acknowledged. The work was part of the programme of research of the ESRC Centre for Economic and Social Aspects of Genomics (CESA-Gen), Flagship Project 'Transcending the Genome: The Paradigm Shift to Proteomics', Principal Investigator: Peter Glasner; Senior Research Associate: Ruth McNally. We would also like to thank those who kindly read this in draft and gave us the benefit of their suggestions and criticisms, whilst making it clear that we are entirely responsible for all shortcomings that remain.

Notes

1 Omicization has even transcended the boundaries of the biosciences, with the coining of 'sociomics' (McNally 2005). See also the section headed 'The genomic era'.
2 Two-dimensional gel electrophoresis (2DGE) uses mass and electrical charge to separate mixtures of proteins in a tissue sample on a polyacrylamide gel matrix.

When visualised, individual proteins appear as spots whose location and intensity are indicative of their identity and relative abundance, enabling qualitative and quantitative comparison of protein profiles between different samples. Looked at retrospectively, 2DGE satisfies one of the defining features of the omics, namely the aggregate analysis of all of the molecules of a certain kind in a given sample.

3 The basis of this genetic code is a sequence of three bases, called a 'codon'. The four nucleotide bases in DNA can occur in 64 possible 'codons'. Because there are only twenty amino acids the genetic code is 'redundant', meaning that several different codons code for the same amino acid.

4 'Big Science', as de Solla Price (1965) outlines, typically requires regulatory and policy changes, with costly facilities, large and highly differentiated teams of scientists and technicians, and publications listing often dozens of co-authors. The two published completed maps of the human genome, for example, named 520 scientists as authors, with those in the public consortium spread across forty-eight laboratories around the world (Glasner and Rothman 2004).

5 The Human Protein Index was envisaged as a 'reference database that every practicing physician, pathologist, clinical chemist, and biomedical researcher could access by satellite' (Zweiger 2000: 5). Data in the database would be managed and interpreted by computer. The Andersons estimated that the project would take five years to complete and cost $350 million. Leigh Anderson, John Taylor and their colleagues at the Argonne National Laboratory even undertook pioneering research to demonstrate the feasibility of their ideas. Using 2DGE, they profiled the expression of 285 proteins in five tumour cell lines, converted the profile data into electronic form, stored it in a database, and then analysed it by computer.

6 Brown identifies 'kairos', meaning 'the right time', as being a central component in the construction of a scientific event as a 'breakthrough' (Brown 2000).

7 Defined as 'open reading frames'.

8 A number of developments contributed to the favourable reception the Human Proteome Project received compared to the Human Protein Index proposal of the late 1970s. One was the development in the late 1980s of two methods for treating (ionising) proteins so that they could be analysed using mass spectrometry. Another was a direct consequence of the HGP and other genome initiatives. The availability of complete genome reference sequences allows proteins to be identified using bioinformatics search algorithms which correlate partial amino acid sequences from mass spectrometry with proteins predicted to be encoded by reference genome sequences (Twyman 2004).

9 These include the development of genome-analysis technologies, the physical and genetic mapping of genomes, the sequencing of model organism genomes and of the human genome, and the creation of a 'talented cohort of scholars' in ethical, legal and social implications (Collins *et al.* 2003: 835–6).

10 Grand challenges I.1 (genome functioning) and I.4 (definition of the genetic basis of speciation) require the collection and analysis of genomic DNA that is generic; in other words it represents the genome that is typical of a species. GCs I.3, II.1 and II.2 (individual variation in genotype and phenotypes), by contrast, require the collection of genomic data which include all common DNA variants within a species. The distinction between the two is that the former, as exemplified by the HGP, is the collection of a generic representative of all the genes typical of a species, whereas the latter, as exemplified by the International HapMap Project and longitudinal, population-based cohort studies, aims to collect a generic representative of all the genomes typical of a species. The difference could be expressed as the difference between genomics and metagenomics, except that the latter term has already been used to refer to the study of all the

genomes in a given ecosystem, such as the human body or the Sargasso Sea. Given that this is the case, if a genome is all the genes typical of a species, should all the genomes typical of a species be called a genome-ome?

11 http://www.genome.gov/Pages/Research/HapMap (accessed 14 June 2006).

12 http://www.ukbiobank.ac.uk; http://www.geenivaramu.ee (accessed 14 June 2006).

13 The principles of comparative pattern analysis were developed before genomics, through protein analysis projects such as the Human Protein Index (Patterson and Aebersold 2003: 314).

14 Wynne (2005) similarly argues for a symmetrical approach to the treatment of reductionist rhetoric in the context of systems biology.

References

Aebersold, R., Hood, L. E. and Watts, J. D. (2000) 'Equipping scientists for the new biology', *Nature Biotechnology*, 18 (April): 359.

Blackstock, W. and Mann, M. (2001) 'A boundless future for proteomics?', *Trends in Biotechnology*, 19/10 (Supplement) A TRENDS Guide to Proteomics: S1–S2.

Brown, N. and Michael, M. (2003) 'A sociology of expectations: retrospecting prospects and prospecting retrospect', *Technology Analysis and Strategic Management*, 15(1): 3–18.

Brown, N., Rappert, B. and Webster, A. (eds), (2000) *Contested Futures: a sociology of prospective techno-science*. Aldershot: Ashgate.

Burke, A. (2003) 'Know who coined the word genome? McKusick recalls HUGO's orgins at HGM', *Genome Web Daily News*, 30 April.

Chitty, M. (2003) *-Omes and -omics glossary: evolving terminology for evolving technologies*. Cambridge: Healthtech Institute.

Cohen, P. (2000) 'High in protein', *New Scientist*, 168 (4 November): 38–41.

Collins, F. S., Green, E. D., Guttmacher, A. E. and Guyer, M. S. (2003) 'A vision for the future of genomics research: a blueprint for the genomic era', *Nature*, 422 (24 April): 835–47.

Dawkins, R. (1982) *The Extended Phenotype: the long reach of the gene*. Oxford: Oxford University Press.

de Solla Price, D. J. (1965) *Little Science, Big Science*. New York and London: Columbia University Press.

Fujimura, J. H. (2005) 'Postgenomic futures: translations across the machine-nature border in systems biology', *New Genetics and Society*, 24(2): 195–225.

Gieryn, T. F. (1995) 'Boundaries of Science', in S. Jasanoff, G. E. Markle, J. C. Petersen and T. Pinch (eds), *Handbook of Science and Technology Studies*. Thousand Oaks, CA.: Sage.

Gilbert, N. and Mulkay, M. (1984) *Opening Pandora's Box: a sociological analysis of scientists' discourse*. Cambridge: Cambridge University Press.

Glasner, P. (2002) 'Beyond the genome: reconstituting the new genetics', *New Genetics and Society*, 21(3): 267–77.

Glasner, P. and Rothman, H. (2004) 'Splicing life?' *New Genetics & Society*. Basingstoke: Ashgate.

Grint, K and Woolgar, S. (1997) *The Machine at Work*. Cambridge: Polity Press.

Haraway, D. (1985) 'A manifesto for cyborgs', *Socialist Review*, 15(2): 65–107.

Hedgecoe, A. (2002) 'Reinventing diabetes: classification, division and the geneticization of disease', *New Genetics and Society*, 21(1): 7–27.

Kanehisa, M. and Bork, P. (2003) 'Bioinformatics in the post-sequence era', *Nature Genetics Supplement*, 33 (March): 305–10.

Keller, E. F. (2000) *The Century of the Gene*. Cambridge, MA.: Harvard University Press.

Kenyon, G. L., DeMarini, D. M., Fuchs, E., Galas, D. J., Kirsch, J. F. *et al.* (2002) 'Defining the mandate of proteomics in the post-genomics era: workshop report', *Molecular & Cellular Proteomics*,1(10): 763–80.

Krulwich, R. (2001) 'The next big thing: First came the genome … Now the pro-teome', vol. 2003, 26 April 2001 Edition: abcNews.

Kuhn, T. S. (1970) *The Structure of Scientific Revolutions*, 2nd Edition. Chicago, IL: University of Chicago Press.

Lander, E. S., Linton, L. M., Birren, B., Nusbaum, C., Zody, M. C., *et al.* (2001) 'Initial sequencing and analysis of the human genome', *Nature*, 409: 860–921.

Latour, B. (1993) *We Have Never Been Modern*. London: Harvester Wheatsheaf.

Lederberg, J. and McCray, A. T. (2001) 'Ome sweet 'omics: a genealogical treasury of words', *The Scientist*, 15(7): 2.

Lazebnik, Y. (2002) 'Can a biologist fix a radio? – Or, what I learned while study-ing apoptosis', *Cancer Cell*, 2: 179–82.

Magner, L. (1979) *A History of the Life Sciences*. New York and Basel: Marcel Dekker.

McKusick, V. A. (1989) 'The human genome organisation: history, purposes and membership', *Genomics*, 5: 385–7.

McNally, R. M. (2003) 'Beyond eugenics – post-eugenics and eubionics: discourse analysis of the handicap ground for abortion', *Faculty of Humanities, Languages and Social Sciences*. Bristol: University of the West of England.

—— (2005) 'Sociomics! Using the IssueCrawler to map, monitor & engage with the global proteomics research network', *Proteomics*, 5(12): 3010–16.

Mennella, T. (2003) 'The out-of-hand omnipresent ome', *The Scientist I*, 17 (12): 52.

Moss, L. (2004) *What Genes Can't Do*. Cambridge, MA and London: MIT Press.

Mulkay, M. (1993) 'Rhetorics of hope and fear in the Great Embryo Debate', *Social Studies of Science*, 23: 721–42.

Patterson, S. D. and Aebersold, R. H. (2003) 'Proteomics: the first decade and beyond', *Nature*, 33(March): 311–21.

Rabinow, P. (1996) *Essays on the Anthropology of Reason*. Princeton, NJ: Prince-ton University Press.

Righetti, P. G. (2004) 'Bioanalysis: its past, present & some future', *Electrophoresis*, 25: 2111–27.

Schroedinger, E. (1944) *What is Life?*. Cambridge: Cambridge University Press.

Steinberg, D. (2001) 'Is a human proteome project next?' *The Scientist*, 15 (7): 13.

Stent, G. S. (1971) *Molecular Genetics: an introductory narrative*. San Francisco, CA: W. H. Freeman & Company.

Twyman, R. M. (2004) *Principles of Proteomics*. Abingdon and New York: Garland Science/BIOS Scientific Publishers.

Tyers, M. and Mann, M. (2003) 'From genomics to proteomics', *Nature (Supplement: Insight Proteomics)* 424(13 March): 193–7.

Venter, C. J., Levy, S., Stockwell, T., Karin, R. and Halpern, A. (2003) 'Massive parallelism, randomness and genomic advances', *Nature Genetics Supplement* 33(March): 219–27.

Venter, J. C., Adams, M. D., Myers, E. W., Li, P. W., Mural, R. J., *et al.* (2001) 'The sequence of the human genome', *Science*, 291(16 February): 1304–51.

Watson, J. D. and Crick, F. H. (1953) 'Molecular structure of nucleic acids: a structure for deoxyribose nucleic acid', *Nature*, 171: 63–6.

Weinstein, J. (2001) 'Searching for pharmacogenomic markers: the synergy between omic and hypothesis-driven research', *Disease Markers*, 17(2): 77–88.

Wheale, P. R. and McNally, R. M. (1988) *Genetic Engineering: catastrophe or utopia?*. Brighton: Wheatsheaf Books.

Wilkins, M. R., Sanchez, J.-C., Gooley, A. A., Appel, R. D., Humphery-Smith, I., *et al.* (1995) 'Progress with proteome projects: why all proteins expressed by a genome should be identified and how to do it', *Biotechnology and Genetic Engineering Review*, 13: 19–50.

Wynne, B. (2005) 'Reflexing complexity: post-genomic knowledge and reductionist returns in public science', *Theory, Culture and Society*, 22(5): 67–94.

Zweiger, G. (2000) *Transducing the Genome: information, anarchy, & revolution in the biomedical sciences*. New York: McGraw-Hill.

Index